Fruit Metabolism and Metabolomics

Fruit Metabolism and Metabolomics

Editors

Annick Moing
Sonia Osorio
Pierre Pétriacq

MDPI • Basel • Beijing • Wuhan • Barcelona • Belgrade • Manchester • Tokyo • Cluj • Tianjin

Editors
Annick Moing
INRAE
France

Sonia Osorio
University of Málaga
Spain

Pierre Pétriacq
Univ. Bordeaux
France

Editorial Office
MDPI
St. Alban-Anlage 66
4052 Basel, Switzerland

This is a reprint of articles from the Special Issue published online in the open access journal *Metabolites* (ISSN 2218-1989) (available at: https://www.mdpi.com/journal/metabolites/special_issues/Fruit_Metabolism).

For citation purposes, cite each article independently as indicated on the article page online and as indicated below:

LastName, A.A.; LastName, B.B.; LastName, C.C. Article Title. *Journal Name* **Year**, *Article Number*, Page Range.

ISBN 978-3-03943-014-7 (Hbk)
ISBN 978-3-03943-015-4 (PDF)

Cover image courtesy of Annick Moing.

Contents

About the Editors

Annick Moing (Dr) gained an Engineer Diploma in Agronomy (ENSA Montpellier, France) and a Ph.D. in Plant Science (University of Bordeaux, France) and has subsequently completed more than 30 years research experience. After working on whole plant physiology, quantitative genetics, and fruit metabolism, her primary research activities have been centered on developing and using metabolomic strategies for the study of fruit and grain quality and metabolism, and the search for metabolite markers of plant response or performance. She has led the Metabolome Facility of Bordeaux Functional Genomics Centre for about 12 years, and is involved in MetaboHUB, the French Infrastructure in Metabolomics and Fluxomics. She has completed about 100 publications in peer reviewed journals and eight book chapters.

Sonia Osorio (Dr) graduated in Chemistry and attained a Ph.D. in Plant Science from the University of Malaga (Spain). She was a postdoc fellow at the Max Planck Institute of Molecular Plant Physiology. She is currently a Professor at the University of Malaga leading the laboratory of Biotechnology of Fruits (https://www.ihsm.uma-csic.es/investigadores/118). Her research interests center on uncovering the molecular mechanisms underlying the production of plant metabolites related to aroma, taste, and postharvest properties. Her activities are centered on understanding early events in the regulation of the ripening transition using strawberry and tomato as a model, and exploring the conservation of ripening mechanisms in those fruits. Her lab combines cutting-edge metabolomics in fruit development and postharvest. She has made a series of breakthrough achievements in fruit development and maturity molecule regulation networks, published in more than 80 papers in international mainstream journals such as *Nature, Nature Communications, The Plant Cell,* and *Plant Physiology* and more than 15 book chapters.

Pierre Pétriacq (Dr) obtained his Ph.D. in Plant Science from Paris-Sud University (France) where he was then appointed Assistant Professor in Plant Physiology. After a Leverhulme Trust postdoctoral fellowship at Sheffield University (UK), he became a Fellow of the Higher Education Academy then Group leader in Plant Metabolomics at P3 Institute, Sheffield (UK). Since 2017, he teaches as an Associate Professor at Bordeaux University and studies Plant Biochemistry at the Fruit Biology and Pathology Unit (Bordeaux, France). He is also head of the Bordeaux Metabolome facility, which provides tools and methods to investigate plant metabolism. His research focuses on redox biology and metabolomics applications for the study of plant performance. He is involved in the French Metabolomics and Fluxomics Infrastructure MetaboHUB, and has published valuable research in high-impact journals such as *Nature Chemical Biology, Plant Physiology, and The Plant Journal.*

metabolites

MDPI

Editorial

Special Issue on "Fruit Metabolism and Metabolomics"

Annick Moing [1,2,*], Pierre Pétriacq [1,2] and Sonia Osorio [3]

1 UMR Biologie du Fruit et Pathologie, Centre INRAE de Nouvelle Aquitaine-Bordeaux, University Bordeaux, INRAE, 71 av Edouard Bourlaux, 33140 Villenave d'Ornon, France; pierre.petriacq@inrae.fr
2 Bordeaux Metabolome, MetaboHUB, PHENOME-EMPHASIS, Centre INRAE de Nouvelle Aquitaine-Bordeaux, IBVM, 71 av Edouard Bourlaux, 33140 Villenave d'Ornon, France
3 Department of Molecular Biology and Biochemistry, Instituto de Hortofruticultura Subtropical y Mediterránea "La Mayora", University of Málaga—Consejo Superior de Investigaciones Científicas (IHSM-UMA-CSIC), 29071 Málaga, Spain; sosorio@uma.es
* Correspondence: annick.moing@inrae.fr

Received: 29 May 2020; Accepted: 1 June 2020; Published: 3 June 2020

Abstract: Over the past 10 years, knowledge about several aspects of fruit metabolism has been greatly improved. Notably, high-throughput metabolomic technologies have allowed quantifying metabolite levels across various biological processes, and identifying the genes that underly fruit development and ripening. This Special Issue is designed to exemplify the current use of metabolomics studies of temperate and tropical fruit for basic research as well as practical applications. It includes articles about different aspects of fruit biochemical phenotyping, fruit metabolism before and after harvest, including primary and specialized metabolisms, and bioactive compounds involved in growth and environmental responses. The effect of genotype, stages of development or fruit tissue on metabolomic profiles and corresponding metabolism regulations are addressed, as well as the combination of other omics with metabolomics for fruit metabolism studies.

Keywords: fruit metabolomics; developmental metabolomics; stress metabolomics; spatial metabolomics; central metabolism; specialized metabolism; mass spectrometry; nuclear magnetic resonance spectroscopy; omics; multi-omics integration

---

The growth and ripening of fruit are multifaceted and highly regulated developmental processes which yield colorful and flavorful tissues for organisms that consume and disperse the seeds therein [1]. Fruits are economically essential and vital for human nutrition and health owing to their content in sugars, organic acids, pigments, volatiles and other nutraceutical compounds [2,3], the metabolisms of which have been widely studied (e.g., [4–6]). The shift from single-metabolite analyses to analytical platforms that provide information on hundreds of metabolites has allowed researchers to better describe the links both within metabolites and between metabolism and important agronomic-associated traits. Metabolomics has permitted the identification of changes in the chemical composition of transgenic plants, mutants or populations and has allowed for identifying genomic regions associated with metabolite traits of agronomical value in model fruit species such as tomato and strawberry [7]. Tomato metabolomic studies have been numerous in the past decade [8,9] and have allowed the development of a comprehensive understanding of primary and specialized metabolism pathways and their interplay during fruit growth and development, and in diverse environmental conditions.

The number of teams involved in, and of articles published on, fruit metabolomics has been regularly and progressively increasing. Searching for articles in the Web of Science core collection for metabolomic(s) or metabolome, and fruit in the last decade (Figure 1) revealed an increment from about 20 articles per year in 2010 to about 200 in 2019. The application domains are also increasingly diverse, as illustrated by very recent works ranging, for instance, from the study of the regulation of steroidal glycoalkaloids biosynthesis in Solanaceae [10] to that of the effect of scion/rootstock interaction on the metabolic composition in citrus fruit [11].

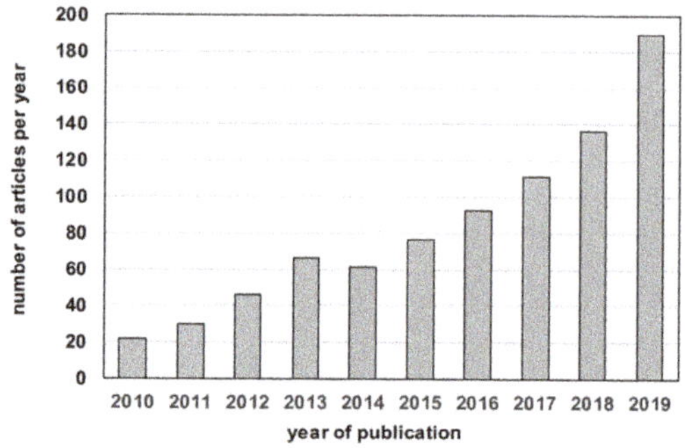

Figure 1. Number of articles published about fruit metabolomics over the past 10 years. Search in Web of Science core collection for TS = (metabolome OR metabolomics *) AND (TS = fruit) NOT (TS=mushroom * OR TS = fruit-fly).

This Special Issue focuses on applications of metabolomics within the field of plant sciences for the study of fruit. It contains 12 original research articles and one review article. The contributed research articles are exemplary studies covering primary research applied to fruit species from a basic understanding of metabolism regulation to applications for phenotyping for breeding or defence priming (Figure 2).

This Special Issue covers a range of fruit species, including model fruit (tomato [12–14]), temperate (kiwifruit [15], mulberry [16]) or tropical fruit crops (pineapple [17], cashew [18]), genetic resources (melon [19]), and indigenous fruit (Davidson's plum, finger lime and native pepper berry [20]). Different analytical approaches are covered also: near-infrared spectroscopy (NIRS) [18], gas chromatography coupled to mass spectrometry (GC-MS) [17], liquid chromatography coupled to mass spectrometry (LC-MS) [16,18,21,22], LC-MS/MS [20], and a combination of several analytical strategies including nuclear magnetic resonance spectroscopy (NMR), GC-MS and LC-MS [19] or several omics [15].

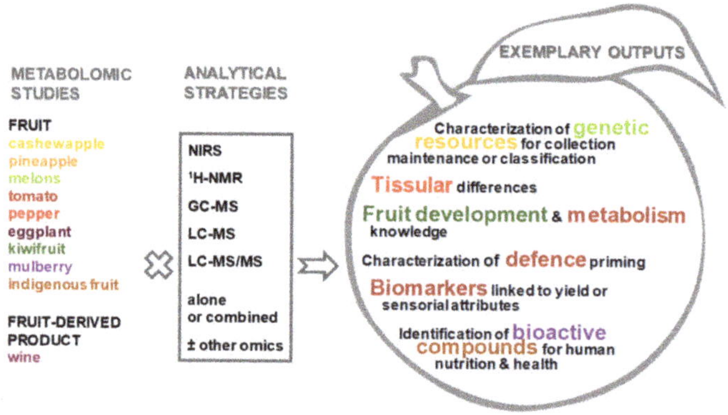

Figure 2. Summary of the works presented in the Special Issue on "Fruit Metabolism and Metabolomics". GC-MS, gas chromatography coupled to mass spectrometry; ^{1}H-NMR, proton nuclear magnetic resonance spectroscopy; LC-MS, liquid chromatography coupled to mass spectrometry; NIRS, near-infrared spectroscopy.

The original research articles can be classified according to scientific domains: biochemical phenotyping of genetic resources for collection maintenance or intraspecific classification, generating knowledge about fruit growth, ripening and post-harvest, understanding metabolism, characterizing defence priming, identifying bioactive compounds for human nutrition and health.

Biochemical phenotyping of cashew apple based on LC-MS and near-infrared spectroscopy [18] allowed the identification of a group of accessions, as a proof of concept for their use for the maintenance of a germplasm bank. For melon genetic resources [19], a combination of metabolome approaches enabled an evaluation of metabolomic relationship and its correlation with the genetic genome-by-sequencing distance of melon accessions, revealing that several melon groups, such as Inodorous, grouped in parallel with the genetic classifications, while other genome to metabolome and mineral element associations appeared less clear. A multi-approach study of color mutations introduced in a tomato Italian landrace [12] revealed unexpected biochemical changes. The double mutants expanded the effect on the metabolism of the single mutations by revealing additional additive or epistatic effects that could be of interest for the improvement and diversification of the landrace.

Fruit are complex organs constituted of different tissues. The distribution of metabolites throughout the different fruit tissues, pericarp, placenta, and seeds, was described in chilli pepper [22]. Pericarp showed a higher content of glycosides and terpenoids compared with other parts of the fruit. Placenta was the tissue having the highest content in alkaloids related to capsaicinoids and in tocopherols.

The mechanisms underlying fruit growth during kiwifruit development were investigated using phytohormone profiling with transcriptomics under two carbon supply levels [15]. The results suggest that cytokinins, known to be implicated in cell division, are also involved in fruit cell expansion and growth in kiwifruit. Pineapple ripening was characterized using GC-MS of crown, flesh, and peel samples [17]. Samples could be separated according to ripening phases from early-ripening to late-ripening phases for flesh and peel, which allowed highlighting metabolites correlated to ripening.

Concerning metabolism regulation, a cross-species study using LC-MS was performed to clarify the metabolic regulation of fruit phenolics among three Solanaceae crops, tomato, eggplant and pepper [21]. It allowed identifying the metabolic signatures of phenolics in each species from different fruit tissue-types and ripening stages and provided information for future functional genomics or breeding approaches. The metabolic changes during fruit postharvest storage were reviewed [23], emphasizing the roles that metabolomic platforms can play for better understanding the biochemical bases of postharvest physiology and identifying the numerous, often species-dependent, pathways affected during fruit senescence.

Metabolic correlation networks covering a variety of metabolic traits such as lipophilic and volatile compounds in tomato fruit have shown how the individual metabolic classes are related to yield-related phenotypic traits [13]. Moreover, metabolite-transcript correlation analysis exposed crucial putative genes involved in the biosynthesis of lipids.

Fruit metabolism is tuned in interaction with the abiotic and biotic environments. Immune priming of tomato fruit by the defence inducer ß-aminobutyric acid (BABA) was investigated against three different fruit pathogens using a combination of targeted and untargeted metabolomics [14]. While BABA resulted in a long-lasting induced resistance in tomato fruit against those pathogenic microbes, the primed responses were likely specific to the infection nature, rather than characterizing a common pattern of BABA-induced priming. Further, modelling approaches established that soluble sugars were essential to predict resistance to fruit pathogens.

In relation to human health, in white and black mulberry fruits, combining an LC-MS method with assays for (i) α-glucosidase inhibitory activity and (ii) antioxidant activity [16], allowed for identifying a list of key bioactive compounds with antioxidant and/or α-glucosidase inhibitory activities. Similarly, based on targeted and untargeted metabolite profiling and antioxidant assays on selected Australian native fruits, a number of compounds that provide the antioxidant activities were identified [20].

The latter results open new possibilities of using these indigenous fruits in the nutraceutical and food industries.

Wine is a famous product derived from fruit for which a range of metabolomic studies have been performed [24,25]. Here, combining non-targeted metabolite profiling by UPLC-Orbitrap-MS/MS profiling and sensory analyses revealed commonalities and differences of wines made with different grape cluster types obtained each from three *Vitis vinifera* cv. Pinot noir putative-clones grown in identical field environments [26]. In addition, a set of molecular-feature (metabolite) markers was selected as being associated with these sensorial attributes.

Overall, these 13 articles demonstrate that metabolomics is a powerful and insightful tool to address numerous biological issues that link to fruit biology before or after harvest. The future improvement of metabolite annotation workflows will certainly help to decipher more ambitious questions when metabolomics biomarker discovery broadens our understanding of fruit processes. A greater number of fruit metabolomics studies, especially for tropical species or species other than major crops, will allow for a better depiction of fruit mechanisms that govern development and responses to environmental fluctuations.

To conclude, we would like to thank each of the authors who contributed to this Special Issue on fruit metabolism and metabolomics, the peer reviewers who allowed for improving the quality of submitted manuscripts, and the staff members of the Metabolite Editorial Office for their support.

Funding: A.M. and P.P. were partly funded by MetaboHUB (ANR-11-INBS-0010) and PHENOME (ANR-11-INBS-0012) French Agence Nationale de la Recherche projects. S.O. was parcially supported by grants RTI2018-099797-B-100 (Ministerio de ciencia, Innovación y Universidades, Spain) and UMA18-DEDERJA-179 (Consejería de Economía, Conocimiento, Empresas y Universidades, Junta de Andalucía, Spain).

Conflicts of Interest: The authors declare no conflict of interest.

References

1. van der Knaap, E.; Østergaard, L. Shaping a fruit: Developmental pathways that impact growth patterns. *Semin. Cell Dev. Biol.* **2018**, *79*, 27–36. [CrossRef] [PubMed]
2. Pott, D.M.; Osorio, S.; Vallarino, J.G. From central to specialized metabolism: An overview of some secondary compounds derived from the primary metabolism for their role in conferring nutritional and organoleptic characteristics to fruit. *Front. Plant Sci.* **2019**, *10*. [CrossRef] [PubMed]
3. Rodriguez-Casado, A. The health potential of fruits and vegetables phytochemicals: Notable examples. *Crit. Rev. Food Sci. Nutr.* **2016**, *56*, 1097–1107. [CrossRef] [PubMed]
4. Sun, P.; Schuurink, R.C.; Caissard, J.-C.; Hugueney, P.; Baudino, S. My way: Noncanonical biosynthesis pathways for plant volatiles. *Trends Plant Sci.* **2016**, *21*, 884–894. [CrossRef]
5. Beauvoit, B.; Belouah, I.; Bertin, N.; Cakpo, C.B.; Colombié, S.; Dai, Z.; Gautier, H.; Génard, M.; Moing, A.; Roch, L.; et al. Putting primary metabolism into perspective to obtain better fruits. *Ann. Bot.* **2018**, *122*, 1–21. [CrossRef]
6. Ilahy, R.; Tlili, I.; Siddiqui, M.W.; Hdider, C.; Lenucci, M.S. Inside and beyond color: Comparative overview of functional quality of tomato and watermelon fruits. *Front. Plant Sci.* **2019**, *10*, 769. [CrossRef]
7. Hanhineva, K.; Aharoni, A. Metabolomics in Fruit Development. In *Molecular Techniques in Crop Improvement*, 2nd ed.; Jain, S.M., Brar, D.S., Eds.; Springer: Heidelberg, Germany, 2010; pp. 675–693.
8. de Vos, R.C.H.; Hall, R.; Moing, A. *Metabolomics of a model fruit: Tomato In Biology of Plant Metabolomics*; Hall, R., Ed.; Wiley-Blackwell Ltd.: Oxford, UK, 2011; Volume 43, pp. 109–155.
9. Tohge, T.; Fernie, A.R. Metabolomics-inspired insight into developmental, environmental and genetic aspects of tomato fruit chemical composition and quality. *Plant Cell Physiol.* **2015**, *56*, 1681–1696. [CrossRef]
10. Yu, G.; Li, C.; Zhang, L.; Zhu, G.; Munir, S.; Shi, C.; Zhang, H.; Ai, G.; Gao, S.; Zhang, Y.; et al. An allelic variant of GAME9 determines its binding capacity with the GAME17 promoter in the regulation of steroidal glycoalkaloid biosynthesis in tomato. *J. Exp. Bot.* **2020**, *71*, 2527–2536. [CrossRef]
11. Tietel, Z.; Srivastava, S.; Fait, A.; Tel-Zur, N.; Carmi, N.; Raveh, E. Impact of scion/rootstock reciprocal effects on metabolomics of fruit juice and phloem sap in grafted *Citrus reticulata*. *PLoS ONE* **2020**, *15*, e0227192. [CrossRef]

12. Dono, G.; Rambla, J.L.; Frusciante, S.; Granell, A.; Diretto, G.; Mazzucato, A. Color mutations alter the biochemical composition in the San Marzano tomato fruit. *Metabolites* **2020**, *10*, 110. [CrossRef]
13. Kuhalskaya, A.; Wijesingha Ahchige, M.; Perez de Souza, L.; Vallarino, J.; Brotman, Y.; Alseekh, S. Network analysis provides insight into tomato lipid metabolism. *Metabolites* **2020**, *10*, 152. [CrossRef] [PubMed]
14. Luna, E.; Flandin, A.; Cassan, C.; Prigent, S.; Chevanne, C.; Kadiri, C.F.; Gibon, Y.; Pétriacq, P. Metabolomics to exploit the primed immune system of tomato fruit. *Metabolites* **2020**, *10*, 96. [CrossRef] [PubMed]
15. Nardozza, S.; Cooney, J.; Boldingh, H.L.; Hewitt, K.G.; Trower, T.; Jones, D.; Thrimawithana, A.H.; Allan, A.C.; Richardson, A.C. Phytohormone and transcriptomic analysis reveals endogenous cytokinins affect kiwifruit growth under restricted carbon supply. *Metabolites* **2020**, *10*, 23. [CrossRef] [PubMed]
16. D'Urso, G.; Mes, J.J.; Montoro, P.; Hall, R.D.; de Vos, R.C.H. Identification of bioactive phytochemicals in mulberries. *Metabolites* **2020**, *10*, 7. [CrossRef]
17. Ikram, M.M.M.; Ridwani, S.; Putri, S.P.; Fukusaki, E. GC-MS based metabolite profiling to monitor ripening-specific metabolites in pineapple (*Ananas comosus*). *Metabolites* **2020**, *10*, 134. [CrossRef]
18. Alves Filho, E.; Silva, L.M.; Lima, Y.; Ribeiro, P.; Silva, E.; Zocolo, G.; Canuto, K.; Morais, S.; Castro, A.C.; de Brito, E. Metabolomic variability of different genotypes of cashew by LC-MS and correlation with near-infrared spectroscopy as a tool for fast phenotyping. *Metabolites* **2019**, *9*, 121. [CrossRef]
19. Moing, A.; Allwood, J.W.; Aharoni, A.; Baker, J.; Beale, M.H.; Ben-Dor, S.; Biais, B.; Brigante, F.; Burger, Y.; Deborde, C.; et al. Comparative metabolomics and molecular phylogenetics of melon (*Cucumis melo*, Cucurbitaceae) biodiversity. *Metabolites* **2020**, *10*, 121. [CrossRef]
20. Lim, V.; Gorji, S.G.; Daygon, V.D.; Fitzgerald, M. Untargeted and targeted metabolomic profiling of Australian indigenous fruits. *Metabolites* **2020**, *10*, 114. [CrossRef]
21. Calumpang, C.L.F.; Saigo, T.; Watanabe, M.; Tohge, T. Cross-species comparison of fruit-metabolomics to elucidate metabolic regulation of fruit polyphenolics among Solanaceous crops. *Metabolites* **2020**, *10*, 209. [CrossRef]
22. Cervantes-Hernández, F.; Alcalá-González, P.; Martínez, O.; Ordaz-Ortiz, J.J. Placenta, pericarp, and seeds of tabasco chili pepper fruits show a contrasting diversity of bioactive metabolites. *Metabolites* **2019**, *9*, 206. [CrossRef]
23. Pott, D.M.; Vallarino, J.G.; Osorio, S. Metabolite changes during postharvest storage: Effects on fruit quality traits. *Metabolites* **2020**, *10*, 187. [CrossRef] [PubMed]
24. Cozzolino, D. Metabolomics in grape and wine: Definition, current status and future prospects. *Food Anal. Methods* **2016**, *9*, 2986–2997. [CrossRef]
25. Pinu, F.R. Grape and wine metabolomics to develop new insights using untargeted and targeted approaches. *Fermentation* **2018**, *4*, 92. [CrossRef]
26. Cuadros-Inostroza, A.; Verdugo-Alegría, C.; Willmitzer, L.; Moreno, Y.; Vallarino, J.G. Non-targeted metabolite profiles and sensory properties elucidate commonalities and differences of wines made with the same variety but different cultivar clones. *Metabolites* **2020**, *10*, 220. [CrossRef] [PubMed]

metabolites

MDPI

Article

Metabolomic Variability of Different Genotypes of Cashew by LC-Ms and Correlation with Near-Infrared Spectroscopy as a Tool for Fast Phenotyping

Elenilson Alves Filho [1], Lorena Mara Silva [2], Ynayara Lima [3], Paulo Ribeiro [2], Ebenézer Silva [2], Guilherme Zocolo [2], Kirley Canuto [2], Selene Morais [3], Ana Cecília Castro [2] and Edy de Brito [2,*]

1 Department of Food Engineering, Universidade Federal do Ceará, Fortaleza-CE 60020-181, Brazil; elenilson.godoy@yahoo.com.br
2 Embrapa Agroindústria Tropical, Rua Dra Sara Mesquita, 2270, Pici, Fortaleza-CE 60511-110, Brazil; lorena.mara@embrapa.br (L.M.S.); paulo.riceli@embrapa.br (P.R.); ebenezer.silva@embrapa.br (E.S.); guilherme.zocolo@embrapa.br (G.Z.); kirley.canuto@embrapa.br (K.C.); cecilia.castro@embrapa.br (A.C.C.)
3 Centro de Ciências e Tecnologia, Universidade Estadual do Ceará, Fortaleza-CE 60020-181, Brazil; yna.colares@gmail.com (Y.L.); selenemaiademorais@gmail.com (S.M.)
* Correspondence: edy.brito@embrapa.br; Tel.: +55-082-33917393

Received: 6 May 2019; Accepted: 20 June 2019; Published: 25 June 2019

Abstract: The objective of the present work was to develop an advanced fast phenotyping tool to explore the cashew apple compositions from different genotypes, based on a portable near-infrared (MicroNIR) spectroscopy. This will be in addition to associating the variability of the respective cashew apple pulps with the genotypes by ultra-performance liquid chromatography (UPLC), coupled with high-resolution mass spectrometry (HRMS). The NIR analysis is a non-destructive, low-cost procedure that provides prompt results, while considering the morphology of different cashew apples (shape, size, and color). The UPLC-HRMS analysis is characterized by specific bioactive compounds, such as the derivatives of hydroxybutanoic acid, galloyl, and flavonoids. Furthermore, both techniques allowed the identification of a group of accessions, which presented similarities among the chemical profiling. However, to improve the understanding of cashew chemical and physical variability, further variables related to the cashew apple composition, such as edaphoclimatic conditions, should be considered for future studies. These approaches lead to the conclusion that these two tools are useful for the maintenance of BAG-Caju (Cashew Germplasm Bank) and for the cashew-breeding program.

Keywords: Anacardium occidentale; fast phenotyping; NIR; UPLC-HRMS; chemometrics

1. Introduction

Cashew (*Anacardium occidentale* L.) is a fruit tree originating from the north of South America, with greater dispersion in the coastal regions, from the state of Rio de Janeiro to the Amazon, where it is possible to find spontaneous populations with great genetic and, consequently, phenotypic variability. To conserve part of this variability, Embrapa maintains a germplasm bank (BAG-Caju) holding almost seven hundred accessions with important variability [1].

Nuts are true fruits with hard shells that are inedible due to the presence of cardol and cardanol (caustic substances). The cashew nut consists of nut kernels that are edible and commercially known as cashew nuts. This complete fruit forms on the distal end of the false fruit that is also edible (called cashew apple or peduncle) [2]. Cashew nut (kernel) is the main product of this crop marketed globally. However, Brazil, by tradition and culture, has the habit of consuming the cashew apple (false fruit) in different ways, e.g., as fresh fruit, juice, or sweets. Thus, BAG-Caju that was initially established based on the size and weight of the cashew nut began to include the variability related to the false fruit, such as vitamin C, sugars, acidity, astringency, color, and bioactive compounds [3].

Due to the elevated number of cashew apple accessions, we have investigated the applicable potential of various technologies to develop simple, precise, and low-cost methods for studying the physico-chemical and nutritional features of this fruit. Among the technologies, the near infrared (NIR) spectroscopy is considered to be an economical, fast, and efficient technique for the evaluation of foodstuffs [4–6]. The NIR analysis produces an electromagnetic spectrum (reflectance or transmittance) between 780 nm and 2,500 nm, where the wavelength depends on the scattering and absorption processes according to the chemical compositions: molecular functional groups, C-H, N-H, S-H, or O-H bonds. A MicroNIR spectrometer is a portable NIR device for real-time, in situ, and non-destructive chemical and physical analyses [7].

In particular, the cashew apple exhibits beneficial characteristics to human health, such as antitumor, antimicrobial, urease inhibitory, and lipoxygenase activities [3,8–10]. The genetic enhancement of the quality and attributes of the cashew apple is still under development in Brazil, although the knowledge of genetic diversity for the construction of germplasm banks exists. Most cashew breeding programs are based on traditional selection approaches, including the size and weight of nuts or yield of a cashew tree, and efforts have also been applied in the prospection of dwarf genotypes with enhanced fruit quality and resistance to pests and diseases [11]. Therefore, the identification of the characteristics with economic interest, related to conserved or cultivated accessions, may be appropriated by breeding programs aimed at launching high-performance cultivars to meet specific demands, such as cultivars with high levels of bioactive compounds. The analysis of unique chemical fingerprints, based on ultra-performance liquid chromatography (UPLC), coupled with high-resolution mass spectrometry (HRMS) has been successfully applied for the study of fruit pulps. In such foodstuff studies, a large number of samples have to be screened for identifying metabolites and their classification according to the geographic origin, climate conditions, genotype, and cultural practices using multivariate statistics [3,12–15]. Usually, multivariate analyses are applied, in order to explore complex matrices, such as foodstuff to determine the variations and relationships among the compositions of the samples [16,17]. This untargeted approach is advantageous when the compounds and degradation products are not known.

From the foregoing studies, the aim of the present study was to evaluate the potential of a MicroNIR spectrometer to explore the composition of different genotypes of intact cashew apple from the Embrapa germplasm bank, as well as to create multivariate regression models considering °Brix, total acidity, and concentrations of ascorbic acid (vitamin C). Furthermore, due to the need to investigate plants that produce fruits with numerous compounds beneficial to human health, the study of metabolite profiling (nutritional and functional compounds) to reveal the variability of pulps from different genotypes of cashew apple was also investigated.

2. Results

The results were divided according to the analytical technique applied to evaluate the cashew apple variability: Section 2.1 for MicroNIR and Section 2.2 for UPLC-HRMS.

2.1. Exploratory Multivariate Analysis of the MicroNIR Dataset

The NIR analysis contains undesirable sources of error (intrinsic imperfections) for multivariate studies, and therefore, pre-treatments can correct many systematic or random errors. The information from the cashew apple composition was partly obscured (overlapped), and the distribution of the variables was highly skewed by usual and expected spectral imperfections. Three pre-treatments of the MicroNIR dataset were tested to assess the features regarding the cashew apple composition, minimizing irrelevant variations within the spectra to obtain quality data: Multiplicative scatter correction (MSC), standard normal variate (SNV), and first derivatives, using Savitzky–Golay filter with a second order polynomial for five points (Figure S1 in Supporting Information). Therefore, the pre-processing by MSC was chosen as an input for the following chemometric analysis.

To comprehend and classify the variability of the organic compounds in intact cashew apples according to the genotype, via detecting the tendencies related to the composition, a method based on hierarchical clustering was developed to segregate the samples in groups according to the composition similarity. Figure 1 presents a 3D dendrogram in heat map form: Genotypes with the harvest year in columns; wavelength in rows, and the signals intensities illustrated in colors. The relatively deep red represents the high relative intensity (wavelength); the relatively deep blue, the low relative intensity; and white, the intermediate intensity. Important tendencies for four cluster formations were observed in the heat map based on genotype dissimilarity (Figure S3 in Supporting Information). Cluster 4 presented the most dissimilar genotypes (zero similarity). The results reflected the natural differences among the groups, which were formed by different spectroscopic information from cashew genotypes pulp composition.

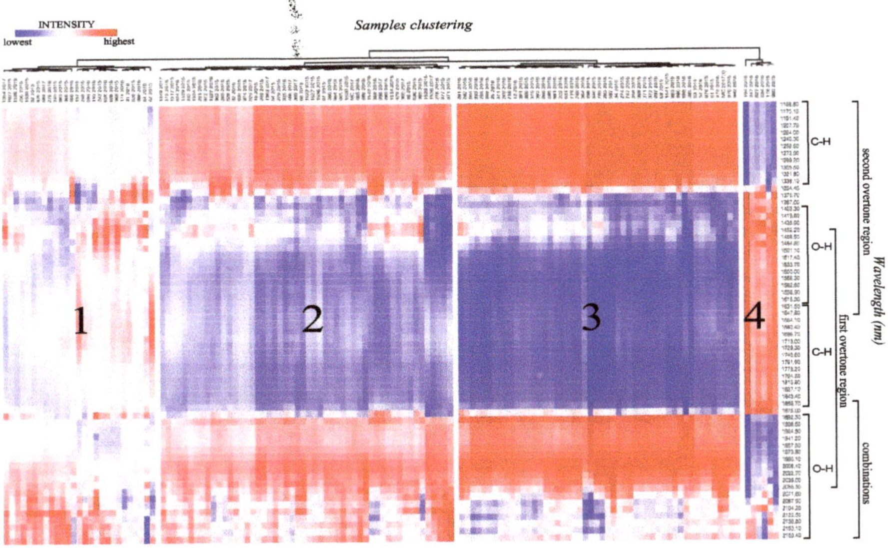

Figure 1. Three-dimensional (3D) dendrogram (sample × wavelength from MicroNIR × intensity) representing the chemical composition similarity relationships among the genotypes.

Multivariate Regression Analysis of the MicroNIR Dataset

Prior to performing the partial least squares (PLS) analysis, the °Brix value, total acidity, and concentrations of ascorbic acid (vitamin C) were determined in 31 cashew apples randomly chosen from the set of accessions. Three regression models were created for each dependent variable (°Brix, total acidity, and ascorbic acid) with spectroscopic data to evaluate the association between the chemical variability, genotype and NIR profile. Sequential to these supervised regression modeling, the PLS by intervals, known as iPLS (interval PLS), was developed to maximize the covariance between the independent variables on the MicroNIR dataset (X matrix) and each dependent variable. This method realizes individuals PLS models for each pre-defined spectra intervals, optimizing the predictive capacity of the model, while assisting the interpretation by reducing the number of variables, thereby providing superior prediction capacity using all the variables [18]. Figure 2 illustrates the relevant intervals (highlighted in green color) for regression modeling, based on RMSECV using the Brix values (a) and total acidities (b). The modeling using the concentrations of ascorbic acid was weakly adjusted based on statistical parameters (data not shown).

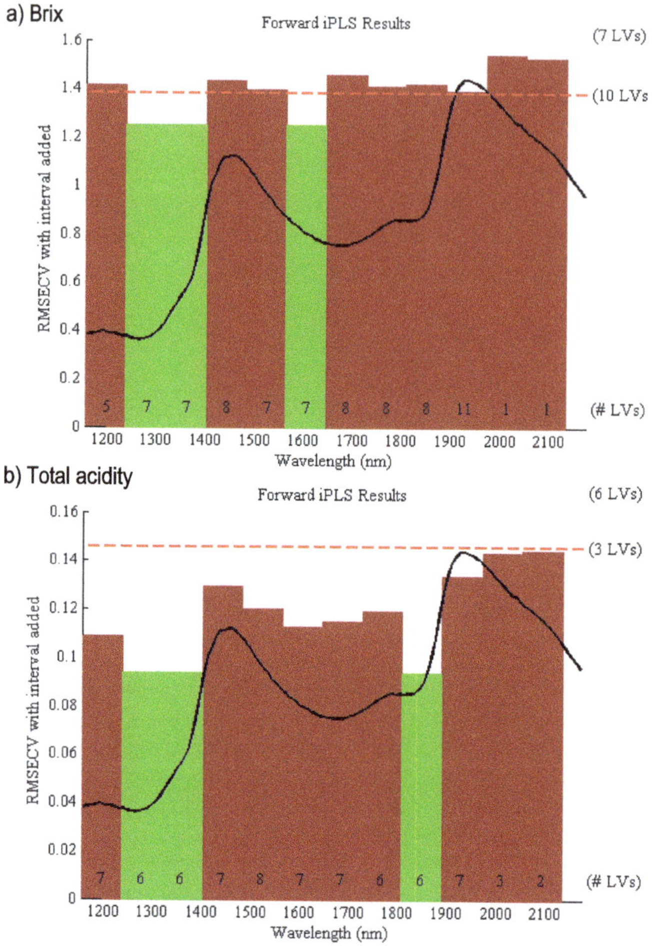

Figure 2. Spectral average (black line) of the MicroNIR spectra from different genotypes of cashew. The relevant absorbance selected by the iPLS model (green) for the genotype discrimination based on (**a**) °Brix and (**b**) total acidity.

The merit graph obtained by regression modeling, using °Brix and total acidities, were evaluated to assess the quality of the calibration models, which are illustrated in Supplementary Information (Figures S4 and S5). The Hotelling's $T^2 \times Q$ residuals graph indicates that the sample did not negatively influence the modeling [19]. The leverage plot revealed the influence of each genotype on models based on Hotelling's T^2, and the studentized residuals (mean zero and unit variance) indicated the lack of fit of some quantitative parameters [20]. However, despite the high leverage and high residuals, the respective genotypes expressed a very low studentized Y residual, and therefore, the regression model was still able to sufficiently predict the °Brix and total acidity based on selected MicroNIR spectra (green region in Figure 2). The robustness of both prediction models (°Brix and total acidity) was achieved by the proximity between both regression curves from calibration and cross-validation (red and green lines). Furthermore, the statistical parameters used to assess the modeling quality (Table 1) indicated a well-adjusted models for both °Brix and total acidity, according to the high total variance cumulated using three latent variables (LVs) (higher than 94%), low bias model, low calibration and cross-validation errors, and proximity between the calibration and cross-validation errors. Additionally, the root mean square error of calibration (RMSEC) and root mean square error of cross-validation (RMSECV) ratio, close to 0.75, is indicative of a well-adjusted model [21,22].

Table 1. Statistical parameters obtained by multivariate regression modeling of MicroNIR spectra with °Brix and total acidity using 3 LV.

Model	3 LV[1] (%)	r^2 cal[2]	RMSEC[3]	r^2 CV[4]	RMSECV[5]	RMSEC/RMSECV[6]	Bias[7]	CV Bias[8]
°Brix	96.5	0.74	0.11	0.46	0.16	0.69	3.3×10^{-15}	−0.004
Acidity	97.1	0.66	1.19	0.46	1.53	0.78	1.8×10^{-15}	0.064

[1] Percentage variance captured by the regression model; [2] Coefficient of correlation between the real and predicted values during the calibration; [3] Root mean square error of calibration; [4] Coefficient of correlation between the real and predicted values during the cross-validation; [5] Root mean square error of cross-validation; [6] Similarity criterion; [7] Average difference between the estimator and real values during the calibration; [8] Average difference between the estimator and real values during the cross-calibration.

2.2. UPLC-HRMS

Commonly, in UPLC-HRMS analysis, some organic compounds may preferentially ionize in the positive or negative ionization mode, as phenolic and carboxylic derivatives ionized well in negative ionization mode, while flavonoids and alkaloids ionize better in positive ionization mode. Therefore, the cashew apple pulps were analyzed under a negative ionization mode to screen the aforementioned compounds and differentiate each cashew apple pulp according to the genotypes. Due to the complexity of the chromatographic data, visual differentiation of the sample composition could not be achieved. Therefore, non-targeted multivariate analyses by hierarchical cluster analysis (HCA) and principal component analysis (PCA) were applied to comprehend the variability of the secondary metabolites in cashew apple pulps according to the genotype.

Initially, the unsupervised method HCA was applied: Segregating the pulps in groups according to similarity. Important tendencies for four cluster formations were observed in the dendrogram based on the genotype at a similarity index of 0.362 (Figure S3 in Supporting information). Cluster 4 presented the most distant samples included in the study, with zero similarity (as also observed in Figure 1). The results reflected the natural differences among the groups, which were formed as a function of pulp composition, dependent on the cashew genotype. In addition, the PCA method was applied to assist the modeling and interpretation of the multivariate data, with the scores graph presented in Figure 3a, and the respective loadings (plotted in lines) in Figure 3b. The tentatively identified biomarkers are presented in Table S1 (Supporting information). The main separation tendencies of the samples were observed with respect to PC1 and PC2 axes with a total explained the variance of 36.7%. The clusters observed in HCA were important for identifying the resultant grouping in the scores graph. The pulps were assigned symbols according to the clustering tendency: Blue triangles for negative scores of PC1 and PC2; red squares for positive scores of PC1 and negative scores of PC2; black stars for positive scores of PC1 and PC2; and green circles for positive scores of PC2. The samples with no relevant result were symbolized by gray circles considering the unrepresentative replicates according to the year.

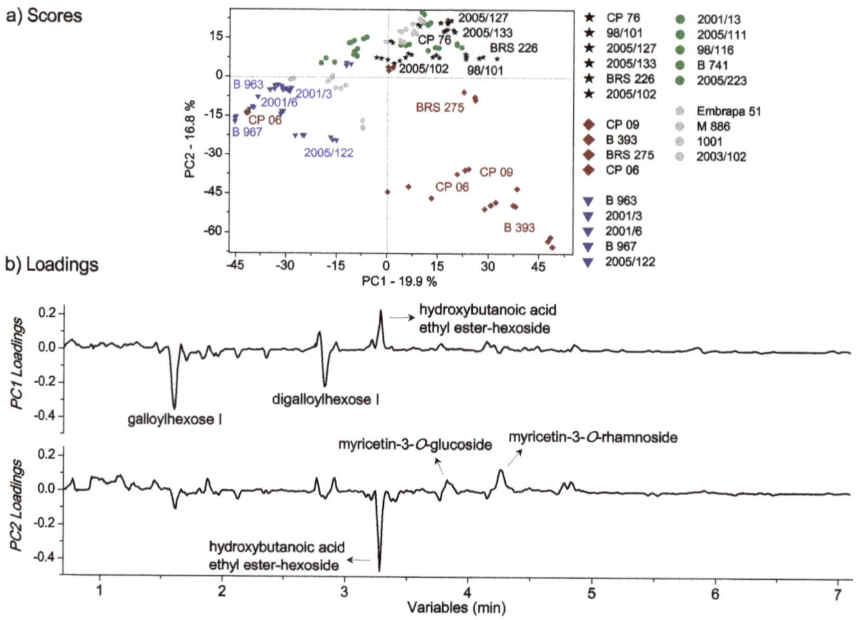

Figure 3. Principal component analysis (PCA) multivariate analysis of UPLC-HRMS data from cashew apple pulps of different genotypes: a) PC1 × PC2 scores coordinate system for the cashew apple pulps from different genotypes; b) respective loadings plotted in lines form. The samples were assigned symbols according to the clustering tendency: blue triangles for negative values of PC1 and PC2; red squares for positive values of PC1 and negative values of PC2; black stars for positive values of PC1 and PC2; green circles for positive values of PC2. The samples that did not present relevant results were symbolized by gray circles.

Compounds with significant changes, based on genotypes according to the chemometric evaluation, and not exhibiting overlapping signals, were integrated (details in Experimental Section 4.2.2), and their variations were expressed as the relative contribution. The relative peaks areas were calculated for quantitative expression of the chemical properties, with the differences evaluated by ANOVA single factor. Figure 4 illustrates the relative contributions of the total ion abundance of the peaks in the pulps, since the relative amplitude of the peaks provided the relative population of the isotopic forms in the chromatograms. The results from the signal area of the base peak intensity (BPI) corroborated the PCA results, considering the deviation of the method from the three replicates of sampling during two years, which totaled 6 pulps for each genotype, with the hydroxybutanoic acid ethyl ester-hexoside (at 3.28 min) being responsible for the pulps clustering at the red group, and galloylhexose I (at 1.60 min) and digalloylhexoside I (at 2.82 min) for the pulps clustering in the blue group.

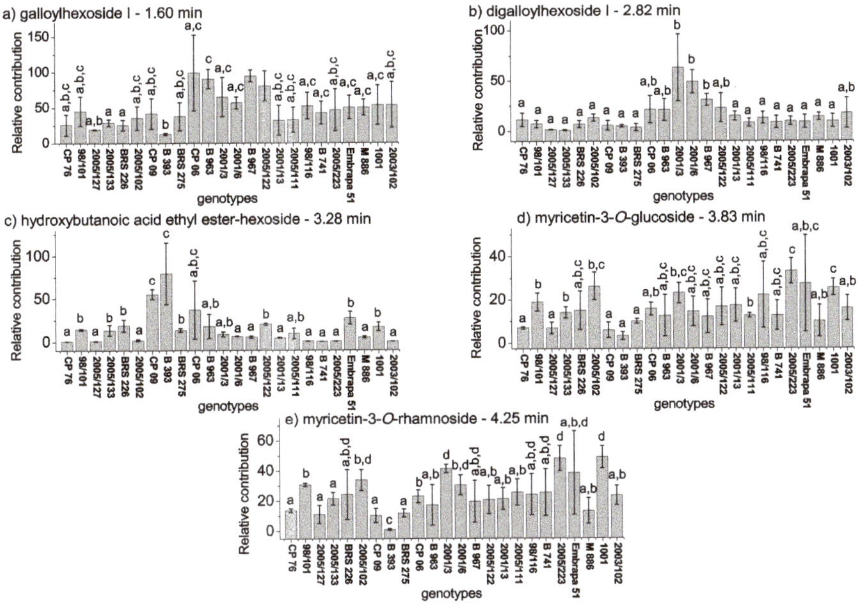

Figure 4. Relative contributions of the isotopic forms in chromatograms (UPLC-HRMS) for the compounds at retention times of 1.60, 2.82, 3.23, 3.82, and 4.25 min.

2.2.1. Multivariate Classification Analysis of the UPLC-HRMS Dataset

Based on the non-targeted chemometrics results and the relative quantification, the classification modeling by partial least squares-discriminant analysis (PLS-DA) was employed to improve the association of the chemical variability of the pulps according to the cashew genotype, which is illustrated in Figure 5. The model presented a classification ability of 88.05% using 3 LVs, considering the year replicate. The statistical parameters used to assess the quality of the modeling (Table 2) indicated a well-adjusted classification, with an RMSEC/RMSECV ratio close to 0.75 (similar values) [22]. The low calibration and cross-validation errors expressed a suitable predictive performance of the model estimated as a function of the global error, samples leverage, and the sample residual X-variance. The low error values indicated the similarity between the pulps used for prediction and those used to make the calibration model.

Table 2. Parameters from partial least squares-discriminant analysis (PLS-DA) classification model of UPLC-HRMS data from cashew apple pulps of different genotypes.

Model	LV1+LV2+LV3 [1]	r^2 cal [2]	RMSEC [3]	r^2 val [4]	RMSECV [5]	RMSEC/RMSECV [6]
PLS-DA	88.05%	0.88	0.298	0.85	0.341	0.874

[1] Total variance percent in X matrix refer to 3 LVs; [2] Coefficient of correlation between the real and predicted groups during the calibration; [3] Root mean square error of calibration; [4] Coefficient of correlation between the real and predicted groups during the cross-validation; [5] Root mean square error of cross-validation; [6] Similarity criterion.

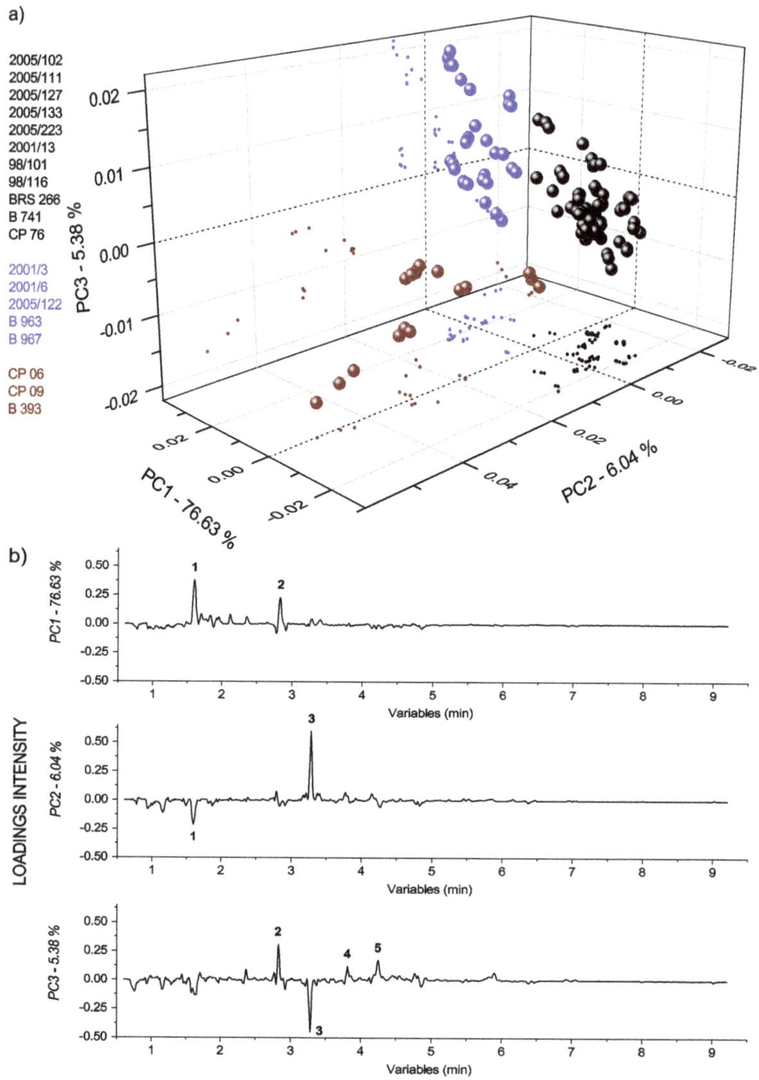

Figure 5. PLS-DA classification model of UPLC-HRMS data from cashew apple pulps of different genotypes: (**a**) LV1 × LV2 × LV3 scores 3D plot; (**b**) respective loadings potted in lines form (b) from PLS-DA for classification of the cashew apple pulps. Legend: 1-galloylhexose I (1.60 min); 2-digalloylhexoside I (2.82 min); 3-hydroxybutanoic acid ethyl ester-hexoside (3.28 min); 4-myricetin-3-*O*-glucoside (3.83 min); and 5-myricetin-3-*O*-rhamnoside (4.25 min).

3. Discussion

It is known that NIR spectroscopy is a non-destructive, low-cost, and non-invasive procedure providing prompt results for a sample composition: Any molecule containing C-H, N-H, S-H, or O-H bonds. However, due to the intrinsic overlapping of the signals, pre-processing is mandatory. The MSC and SNV algorithms presented similar effects, which were considered as exchangeable (Figure S1 in Supporting Information). After the application of the MSC algorithm, the spectra were adjusted to present the same scatter level estimated by a mean spectrum. Conversely, the SNV algorithm treated each spectrum separately by absorbance autoscaling [23]. The derivative algorithm emphasized steep peaks and enhanced the overlapping peaks, as well as reduced the measurement variations, thereby improving the differentiation of the bands. The first and/or second derivatives are more commonly applied, and the second derivative NIR spectrum may result in sharp signals [24,25]. Therefore, the MSC algorithm was chosen for the development of the experiments.

According to the HCA-heat map (Figure 1), important tendencies for four main cluster formations were observed based on the genotype, even with differences in fruit morphology, such as shape, size, and peel color. Another disadvantage of NIR analysis was the penetration of radiation into the tissues of fruits, which skin may reduce the light penetration that decreases with the depth [26]. The results reflected the natural differences among the cashew apple groups formed as a function of the composition intrinsically related to the genotype. Cluster 4 presented the most dissimilar samples compared to those of the other groups. The main absorbance peaks, related to the clustering, were located between 1150 and 1340 nm, 1370 and 1850 nm, and 1900 and 2020 nm. The absorbance from 2040 to 2170 nm was characterized as indicative of non-relevant functional groups. The absorbance from 1400 to 1620 nm and 1900 to 2020 nm were characterized as the second and first overtone of the OH stretch, respectively. The absorbance from 1150 to 1340 nm and 1650 to 1850 nm were due to the C-H stretches related to the second, and first overtone, respectively, which may be from carbohydrates and other organic compounds present in the cashew apple skin and/or pulp. The band at 1340 nm is attributed to the CH group from cellulose [27]. Additionally, it was highlighted that the O-H group from monomeric organic acids presented the first overtone at 1445 nm; and a characteristic overtone around 1890 nm indicative of the O-H stretching combined with C-O stretching from organic acids [26,28]. However, it is known that all these absorption bands are close to the stronger water absorption regions, hindering the signals observation [29].

The iPLS evaluation highlighted the region between 1250 and 1400 nm as the most important for fruit discrimination, based on the calibration modeling by °Brix values and total acidity (statistical parameters, described in Table 1. The modeling for ascorbic acid was weakly adjusted based on statistical parameters. In particular, the absorbance range between 1570 and 1650 nm was important for the °Brix model, while the absorbance between 1800 and 1900 nm was important for the total acidity. The absorbance between 1650 and 1850 nm may be related to the C-H stretches from the second, and first overtone, respectively (Figure 1), which may be from carbohydrates and other organic species in cashew apple skin and/or pulp; and between 1800–1900 nm may be related to carbonyl and carboxyl groups from carboxylic acids [27]. The regression modeling °Brix values (a) and total acidities (b) were well adjusted based on statistical parameters, and the modeling for ascorbic acid was weakly adjusted. This may be because the MicroNIR was operated between 1150–2170 nm, and some organic acids found in fruits typically show bands from O-H group related to the second and third overtones approximately at 1000, and 800 nm, respectively, as well as starch and sugars to the second (920 nm) and the third (720 nm) overtones of O-H stretching, and the third (910 nm) and the fourth (750 nm) overtones of C-H stretching [28].

The relatively low correlation coefficients (r^2) for both models (°Brix values and total acidity) express the rather weak calibration performances estimated as a function of the global model error, samples leverage, and the sample residual X-variance. Therefore, the cross-validation results indicated that further parameters related to the fruits morphology (shape, size, and color), non-homogeneous distribution of particles into the fruits (density variations), and environmental factors that may affect the instrument performance as the illumination (since all the analysis was developed during the year), must be taken into account for improving the comprehension of the chemical and physical variability of the cashew apples from the germplasm bank of Embrapa. For instance, small physical variations from sample to sample may lead to light scattering that influences the MicroNIR measurement, resulting in baseline shifts and scaling variations (intensity variations), and consequently, disturbing the future predictions evaluation.

Prior to the chemometric analysis of the HPLC-HRMS analysis, a valuable feature of chromatograms acquisition was taken into account, since the retention times of the chromatographic peaks are sensitive to minor fluctuations in temperature, pH, flow, pump operation, etc. To solve the problem of small peaks shifts related to the same compound inter-chromatograms, some different peak alignment methods were tested. The alignment practice can be performed manually, using COW (correlation optimized warping) [30], or using the bucketing method, which reduces the chromatogram dimensionally by slicing it into equal sized regions [31], making it easier to analyze the respective loadings. Therefore, all the chromatogram peaks were previously aligned using the COW method, which is illustrated in Figure S2 in Supporting Information.

Significant composition variability was detected in PCA regarding cashew genotypes despite the low cumulated variance, which indicated the existence of further factors beyond the scope of this study, such as seasonality. An examination of the loadings provided evidence of the variables (compounds) responsible for the separations or clustering observed in scores. The signals from galloylhexose I (at 1.60 min, 331.0650 *m/z*), digalloylhexoside I (at 2.82 min, 483.0741 *m/z*), hydroxybutanoic acid ethyl ester-hexoside (at 3.28 min, 293.1242 *m/z*), myricetin-3-*O*-glucoside (at 3.83 min, 479.0826 *m/z*), myricetin-3-*O*-rhamnoside (at 4.25 min, 463.0875 *m/z*), and a mixture of unknown compounds (at 4.82 min) were responsible for the pulps placement based on genotypes. Minor compounds were irrelevant due to the pretreatments of mean centering applied over the samples [16]. According to Figure 3, the negative loadings of PC1 are the pulps symbolized by blue triangles (2001/3, 2001/6, 2005/122, B 963, and B967) with relatively high amounts of galloylhexose I and digalloylhexoside I. This latter compound is a trihydroxybenzoic acid derivative, which provides astringent flavor and can contribute to the characteristic bitter taste of immature cashew apple [32]. The pulps symbolized by red squares (CP 06, CP 09, B 393, and BRS 275) and black stars (2005/127, 2005/133, 2005/102, BRS 226, CP 76, 98/101) exhibited tendencies of having relatively high amounts of hydroxybutanoic acid ethyl ester-hexoside at a retention time of 3.28 min according to positive loadings of PC1. Hydroxybutanoic acid ethyl ester-hexoside has been previously described in melon fruit and it is considered a precursor of volatile compounds [33]. The presence of this compound has been associated with some amino acids, such as alanine, glutamine, isoleucine, phenylalanine, tryptophan, and tyrosine. Alanine has been reported as one of the key amino acids of the characteristic profile of cashew apple. Phenylalanine and tyrosine have also been detected but in small amounts [34]. The positive loadings of PC2 exhibited the tendency of the cashew apple pulps symbolized by green circle (2005/111, 2005/223, 2001/13, 98/116, and B 741) and black stars (2005/127, 2005/133, 2005/102, BRS 226, CP 76, 98/101) to have relatively high amounts of the flavonoids, myricetin-3-*O*-glucoside, myricetin-3-*O*-rhamnoside, and an unknown mixture of compounds between the retention times of 4.79 and 4.85 min. The presence of myricetin-derived and other flavonoids compounds in pulps may offer biological benefits, including the reduction of cardiovascular disease and risks of cancer [3], as well as antihyperglycemic property [35]. Furthermore, these compounds have been reported in methanol-water extracts of cashew apple [3]. The corroborative results between quantitative and chemometric analyses confirm the advantages of an untargeted multivariate analysis, since the compounds and degradation products are not always known, and it is sometimes difficult to find certified standards.

According to Figure 5, the LV1 axis was most relevant in clustering the cashew apple pulps in black color (2005/102, 2005/111, 2005/127, 2005/133, 2005/223, 2001/13, 98/101, 98/116, BRS 266, B 741, CP 76), and in separating them from those in red (CP 06, CP 09, B 393) and blue (2001/3, 2001/6, 2005/122, B 963, B 967). In addition, the LV2 axis was important in clustering the red samples at the positive scores, and LV3 was relevant in the separation of the red and blue cashew apple pulps. The interpretation of the loadings revealed that the pulps in black had higher amounts of galloylhexose I and digalloylhexoside I than those of in red and blue. The results corroborated that the pulps in red color have a higher amount of hydroxybutanoic acid ethyl ester-hexoside than those in the black and blue pulps; in addition, opposite behavior between the hydroxybutanoic acid ethyl ester-hexoside and galloylhexose I was presented. Finally, the LV3 axis confirmed the relatively high amount of the flavonoids, myricetin-3-O-glucoside and myricetin-3-O-rhamnoside, in the pulps in blue and presented the opposite behavior between these flavonoids and hydroxybutanoic acid ethyl ester-hexoside. Therefore, the statistics data indicated that the model could be acceptable to classify new or unknown cashew apple pulps based on the main secondary metabolites.

4. Materials and Methods

4.1. Sampling and Experimental Planning

Based on experimental viability, 764 cashew apples (*Anacardium occidentale*, L.) from 24 different accessions were randomly harvested at the Embrapa experimental station (Pacajus, Ceará, Brazil)—coordinates 4°11′07″S; 38°30′07″W; altitude: 70 m. The region has a tropical climate, average temperatures of 26 to 28 °C, and 1020 mm average annual rainfall. The soil is classified as Ultisol and has a sandy/medium texture with low organic matter content. The cashew samplings were collected between August and December in two years. Table 3 presents the different genotypes with their respective accession numbers and morphoagronomic characteristics.

Table 3. Illustration of the cashew genotypes associated with the accession number and morphoagronomic characteristics: plant size, tree appearance, fruits color and shape, and origin (county-state).

Accession Number	Plant Size	Tree Appearance	Fruit Color	Fruit Shape	Sampling Origin	Illustration
CP 76	tall	Open erect	orange	pyriform	crop*/Maranguape-CE	
Clone 98/101	semi tall	compact erect	orange	pyriform	breeding program*/Pacajus-CE	
Progeny 2005/127	semi tall	compact erect	dark red	pyriform	breeding program/Beberibe-CE	
Progeny 2005/133	semi tall	open erect	orange	spherical	breeding program/Cruz-CE	

Table 3. *Cont.*

Accession Number	Plant Size	Tree Appearance	Fruit Color	Fruit Shape	Sampling Origin	Illustration
BRS 226	dwarf	compact erect	orange	pyriform	crop/Pio IX-PI	
Clone 2005/102	tall	compact erect	orange	pyriform	breeding program/Beberibe-CE	
CP 09	semi tall	compact erect	orange	pyriform	crop/Maranguape-CE	
B 393	tall	compact erect	light red	spherical	germplasm*/Aracati- CE	
BRS 275	semi tall	open erect	orange	pyriform	crop/Pacajus-CE and Maranguape-CE	
B 963	tall	open erect	yellow orange	pyriform	germplasm/Maranguape-CE	
Hybrid 2001/3	semi tall	compact erect	orange	pyriform	breeding program/Maranguape-CE and Pio IX–PI	

Table 3. *Cont.*

Accession Number	Plant Size	Tree Appearance	Fruit Color	Fruit Shape	Sampling Origin	Illustration
Hybrid 2001/6	semi tall	compact erect	yellow orange	pyriform	breeding program/Maranguape-CE and Pio IX-PI	
B 967	tall	open erect	orange	cylindrical	germplasm/Pacajus-CE	
CP 06	tall	open erect	yellow	conical obovate	crop/Pacajus-CE	
Progeny 2005/122	semi tall	open erect	yellow	spherical	breeding program/Beberibe-CE	
Hybrid 2001/13	semi tall	open erect	orange	pyriform	breeding program/Pacajus-CE	
Clone 2005/111	semi tall	open erect	orange	pyriform	breeding program/Serra do Mel-RN	
Clone 98/116	semi tall	open erect	orange	pyriform	breeding program/São Luiz do Curu-CE	

Table 3. *Cont.*

Accession Number	Plant Size	Tree Appearance	Fruit Color	Fruit Shape	Sampling Origin	Illustration
B 741	semi tall	compact erect	orange	pyriform	breeding program (CP 76 x *A. microcarpum*)/Maranguape-CE	
Progeny 2005/223	semi tall	open erect	orange	pyriform	breeding program/Beberibe-CE	
Embrapa 51	semi tall	open erect	orange	pyriform	crop/Pacajus-CE	
M 886	tall	open erect	yellow	spherical	breeding program/Beberibe-CE	
1001	tall	open erect	orange	pyriform	crop/Pacajus-CE	
Clone 2003/102	semi tall	compact erect	orange	pyriform	breeding program/Pio IX -PI	

* Legends: crop means registered product on market; breeding program means plant before crossbreed; germplasm means plant collected and conserved in the germplasm bank.

4.2. Portable NIR Spectrometer Analysis

The portable NIR (MicroNIR) analysis of the cashew apple composition was divided into two stages. First, all the 764 intact fruits were analyzed by MicroNIR. After that, 31 fruits were randomly selected for the determination of quantitative parameters, such as °Brix, total acidity, and concentration of ascorbic acid (vitamin C) to develop multivariate regression models.

The NIR experiment was acquired using a portable NIR spectrometer (MicroNIR 1700, Viavi, Milpitas, CA, USA), which operated in a range between 1150 and 2170 nm (spectral resolution of 10 nm,), with dimensions of 45 mm diameter × 42 mm high, two tungsten sources for reflectance measurements, and a continuous monochromator based on a linear variable filter. The parameters for spectral data acquisition were set at 50 ms integration time and an average of 100 scans. The reference spectrum for the absorbance calculation was obtained from a piece of Spectralon™, and the dark signal was obtained by pointing the measurement window of the instrument to the ambient environment.

4.2.1. Determination of °Brix, Total Acidity, and Concentration of Ascorbic Acid

The °Brix, total acidity, and concentration of ascorbic acid (vitamin C) were determined in 31 cashew apples randomly chosen from the set of accessions. These experiments were carried out to use the quantitative results as categorical variables (Y column) to develop multivariate regression models by partial least square (PLS) analyses by maximizing the covariance between X matrix (NIR spectral data) and Y responses (°Brix, total acidity, and ascorbic acid as dependent variables).

The °Brix (concentration of sucrose w/w) was determined following the AOAC method (2005) [36] (soluble solids content), which was obtained by refractometry using a digital refractometer (ATAGO™ N1, Kirkland-WA-USA) with automatic temperature compensation.

The total acidity was determined as follows: Total titratable acidity (TTA) determined by titration with NaOH solution (0.1 N) in 1 g of the pulp diluted to approximately 50 mL of distilled water, containing 3 drops of 1% phenolphthalein until pink coloration, was observed. The results were expressed as the percentage of malic acid according to IAL (1985) [36].

The concentrations of ascorbic acid were determined by titration with 0.02% 2,6-dichloro-indophenol (DFI) as reported by Strohecker and Henning (1967) [37]. One gram of pulp was diluted to 100 mL with 0.5% oxalic acid and homogenized. Subsequently, 5 mL of this solution was diluted to 50 mL with distilled water and titrated. The results were expressed as mg·100 g^{-1} FW (fresh weight).

4.2.2. Chemometric Analysis of the MicroNIR Dataset

Different multivariate approaches were performed on the numerical matrix from 764 cashew apple fruits (Section 2.1). The averaging method was applied on those fruits from the same accession (from four to six replicates), resulting in 135 mean spectra. The spectral region between 1150 and 2170 nm was used for the modeling, and a matrix with the dimensionality of 16,875 data points (135 spectra × 125 variables into each spectrum) was generated. The samples were named according to the accession number and year of harvest.

For numerical matrix construction, each spectrum was converted to an American Standard Code for Information Interchange (ASCII) file and imported by the Origin™ program (version 9.4). To reduce the dimensionality of the original data and to assist the interpretation of the multivariate dataset, the matrix was averaged along variables by a factor of 2 using the PLS-Toolbox™ program (version 8.6.2, Eigenvector Research Incorporated, Manson, WA USA), and imported by GENE-E program (https://software.broadinstitute.org/GENE-E/index.html) for pattern recognition through the hierarchical clustering algorithm by heat map. The Euclidean distance was used to measure the proximity between the samples (columns), and the average linkage method (sum-of-squares approach in calculating intercluster distances) was applied. The results were presented as heat maps (three-dimensional (3D) dendrogram = sample × wavelength × intensity) [22].

In addition to the unsupervised analysis, supervised methods by PLS were developed using the °Brix values, total acidity, and concentrations of ascorbic acid previously calculated in Section 2.2.1 to improve the identification of chemical changes according to genotypes by MicroNIR. The simplified PLS (SIMPLS) algorithm was applied to build the models and the LVs were selected in accordance with the statistical parameters based on the full cross-validation method: RMSEC, RMSEV, and calibration and cross-validation coefficients (r^2) [16].

4.3. *UPLC-HRMS Analysis*

Due to the health benefits of the cashew apple, additional experiments were developed in parallel by non-targeted UPLC-HRMS analysis. A total of 24 different genotypes of cashew apple (as described in Section 2.1) were evaluated. The cashew apples were manually pressed to obtain the resultant pulp, followed by centrifuged for 5 min at 804.6 g (IEC clinical centrifuge, Damon/IEC Division, Needham, MA, USA). The samples were preserved at −80 °C until the analysis.

Prior to the UPLC-HRMS analysis, the samples were filtered using PTFE membranes of 0.22 μm. The analysis was performed on an Acquity system (Waters) coupled with quadrupole/TOF (Waters) equipped with an ESI source operated in the positive ion mode. The chromatographic separation was performed using Waters Acquity UPLC BEH (150.0 × 2.1 mm, 1.7 μm) column with the temperature set at 40 °C. Water and acetonitrile were used for the mobile phase, both with 0.1% formic acid. The gradient ranged from 2% to 95% of water in 15 min in a flow of 0.4 mL·min^{-1} and injection volume of 5.0 μL per sample. The desolvation gas was N_2. The desolvation temperature was set at 350 °C at a flow rate of 350 L·h^{-1} and a source temperature of 120 °C. The capillary voltage was set to 3200 V. The collision energies/cone voltages were set at 6 eV/15 V (low) and 30–50 eV/30 V (high) to achieve sufficient fragmentation. Data were collected using the negative ionization mode between 100 Da and 1180 Da, and the mode tandem was MS^E.

4.3.1. Chemometric Analysis of the HPLC-HRMS Dataset

Chemometric analysis was performed on the numerical matrix from 24 cashew apples harvested in duplicates this year, and analytical triplicate, resulting in 144 chromatograms. The chromatographic region between 0.65 and 7.12 min was selected. The samples were named according to the accession numbers (Table 3).

The same procedure applied for numerical matrix construction from MicroNIR (Section 4.2.2) dataset was also applied for the UPLC-HRMS dataset. Therefore the chromatograms were converted to an American Standard Code for Information Interchange (ASCII) file and import by the Origin™ program (version 9.4) in order to build the matrix. The final matrix was exported for chemometric analyses by HCA, PCA, and PLS-DA using PLS Toolbox™ program (version 8.6.2, Eigenvector Research Incorporated, Manson, WA, USA).

The normalized scaling parameter and baseline correction, using linear fit algorithms, were applied over the variables, and mean-centered processing was applied over the samples, which reduced the noise and minor analytical errors [38,39]. For HCA, the matrix was mean-centered, and the incremental linkage method (sum-of-squares approach in calculating the inter-cluster distances) was applied. The Euclidian distance was used for distance metric. The PCA was performed using singular value decomposition (SVD) algorithm. To improve the identification of the chemical constituents associated with cashew genotype, a supervised method by PLS-DA was employed using the SIMPLS algorithm. The number of LVs were selected in accordance with the statistical parameters: RMSEC; RMSECV; calibration and cross-validation coefficients (r^2); and similarity criterion RMSEC/RMSECV ratio above 0.75 [16,21].

4.3.2. Relative Contribution

The peaks detected as exactly as possible, in both m/z and retention time, were used for determining the peak area for achieving the relative contribution of the compounds with less overlapped signals in the chromatograms. The relative contribution of the areas was calculated based on the total ion abundance from the peaks in the samples, since the relative amplitude of the peaks provides the relative abundance of the isotopic forms in the chromatograms. Therefore, the normalized means in the base peak intensity (BPI) at the retention times of 1.60, 2.82, 3.28, 3.83, and 4.25 min were determined.

The results were evaluated using the analysis of variance ANOVA single factor (significance level of 0.05; means comparison using Tukey test; Levene's test for the homogeneity of the variance) to statistically certify the differences among the relative contributions. The deviation of the method was estimated based on the null hypothesis (p-value) from the three replicates of sampling for two years, totaling 6 samples for each cashew genotype.

5. Conclusions

It was demonstrated that MicroNIR spectroscopic analyses of the cashew apple composition provided a non-destructive and low-cost method for obtaining prompt results. In addition, important composition tendencies were observed with four fruits clusterings according to their composition similarity, and genotype, even considering the morphologic differences, including shape, size, and color. The multivariate regression results, obtained using °Brix and total acidity, showed that it is possible to satisfactorily predict °Brix and total acidity within the cashew genotypes. However, the parameters related to the fruit composition, and the environmental factors that affect the instrument performance must be taken into account to improve the comprehension of the chemical and physical variability of the cashew apples from the germplasm bank of Embrapa.

Additionally, the chemometrics evaluation of the UPLC-HRMS dataset was suitable to follow changes in the composition of cashew apple pulps according to genotype. The current study resulted in the identification of relatively high amounts of different bioactive compounds, including galloylhexose I, digalloylhexoside I, hydroxybutanoic acid ethyl ester-hexoside, and the flavonoids, myricetin-3-*O*-glucoside and myricetin-3-*O*-rhamnoside, in different genotypes. This information is useful for breeding programs to establish accessions with higher concentrations of important compounds for human health.

Supplementary Materials: The following are available online at http://www.mdpi.com/2218-1989/9/7/121/s1. Figure S1: (a) Raw absorbance spectra from 135 cashew apples obtained using the portable NIR spectrometer, and the same spectra after the following treatments, (b) MSC, (c) SNV, and (d) first derivative using the Savitzky–Golay filter with a second order polynomial for five points. Figure S2: The total ion chromatograms from 24 different genotypes of cashew apple pulps acquired under negative ionization mode. Figure S3: Dendrogram representing the chemical composition similarity relationships among the cashew apple pulps. Figure S4: Regression modeling using the MicroNIR dataset from different genotypes of cashew apples based on °Brix values: a) influence plot of Hotelling's $T^2 \times Q$ residuals, (b) leverage × studentized residuals, c) Y calibration × cross-validated Y with 95% confidence limits, and d) scores on LV1 × LV2. Figure S5: Regression modeling using the MicroNIR dataset from different genotypes of cashew apples based on total acidity: a) influence plot of Hotelling's $T^2 \times Q$ residuals, (b) leverage × studentized residuals, c) Y calibration × cross-validated Y with 95% confidence limits, and d) scores on LV1 × LV2.

Author Contributions: All authors provided critical feedback and helped shape the research, analysis, and manuscript writing. Particularly, the researchers E.d.B. (Chemist) and E.S. (Agronomist) conceived the presented idea, theory development, and investigated the study viability. The researchers S.M. (Chemist), G.Z. (Chemist) and K.C. (Chemist) helped to plan and supervise the project. The researchers A.C.C. (Biologist) and E.S. (Agronomist) planned and designed the agronomical experiments. The authors L.M.S. (Chemist) and P.R. (Chemist) verified and developed the analytical methods. The author Y.L. (graduate) helped in chemical experiments and contributed to sample preparation. The author E.A.F. (Chemist) developed the chemometric experiments and wrote the manuscript together with L.M.S. under the E.d.B supervising.

Funding: This research was funded by CNP*q*, grant number 314737/2018-9; and FUNCAP, grant number DCR-0024-01686.01.00/15.

Conflicts of Interest: The authors declare no conflict of interest.

References

1. Caramês, E.T.; Alamar, P.D.; Poppi, R.J.; Pallone, J.A.L. Quality control of cashew apple and guava nectar by near infrared spectroscopy. *J. Food. Compos. Anal.* **2017**, *56*, 41–46. [CrossRef]
2. Ribeiro, L.P.D.; da Silva, A.P.M.; de Lima, A.A.; de Oliveira Silva, E.; Rinnan, Å.; Pasquini, C. Non-destructive determination of quality traits of cashew apples (*Anacardium occidentale* L.) using a portable near infrared spectrophotometer. *J. Near. Infrared. Spectrosc.* **2016**, *24*, 77–82. [CrossRef]
3. De Brito, E.S.; de Araújo, M.C.P.; Lin, L.Z.; Harnly, J. Determination of the flavonoid components of cashew apple (*Anacardium occidentale*) by LC-DAD-ESI/MS. *Food Chem.* **2007**, *105*, 1112–1118. [CrossRef]
4. Mathlouthi, M. Water content, water activity, water structure and the stability of foodstuffs. *Food Control* **2001**, *12*, 409–417. [CrossRef]

5. Nicolai, B.M.; Beullens, K.; Bobelyn, E.; Peirs, A.; Saeys, W.; Theron, K.I.; Lammertyn, J. Nondestructive measurement of fruit and vegetable quality by means of NIR spectroscopy: A review. *Postharvest Biol. Technol.* **2007**, *46*, 99–118. [CrossRef]
6. Connolly, C. NIR spectroscopy for foodstuff monitoring. *Sens Rev.* **2005**, *25*, 192–194. [CrossRef]
7. Dos Santos, C.A.T.; Lopo, M.; Páscoa, R.N.; Lopes, J.A. A review on the applications of portable near-infrared spectrometers in the agro-food industry. *Appl. Spectrosc.* **2013**, *67*, 1215–1233. [CrossRef]
8. Cavalcante, A.A.M.; Rübensam, G.; Erdtmann, B.; Brendel, M.; Henriques, J.A. Cashew (*Anacardium occidentale*) apple juice lowers mutagenicity of aflatoxin B1 in *S. typhimurium* TA102. *Genet. Mol. Biol.* **2005**, *28*, 328–333. [CrossRef]
9. Kubo, I.; Nihei, K.I.; Tsujimoto, K. Antibacterial action of anacardic acids against methicillin resistant *Staphylococcus aureus* (MRSA). *J. Agric. Food Chem.* **2003**, *51*, 7624–7628. [CrossRef]
10. Ha, T.J.; Kubo, I. Lipoxygenase inhibitory activity of anacardic acids. *J. Agric. Food Chem.* **2005**, *53*, 4350–4354. [CrossRef]
11. Chaves, N.; Castro, A.; Vidal, R.; Albuquerque, A.; Moura, C.; Barros, L. Characterization of cashew accessions for ornamental purposes. In Proceedings of the X International Symposium on Postharvest Quality of Ornamental Plants 1060, Liege, Belgium, 19–24 May 2019; pp. 315–320.
12. Scherer, R.; Rybka, A.C.P.; Ballus, C.A.; Meinhart, A.D.; Filho, J.T.; Godoy, H.T. Validation of a HPLC method for simultaneous determination of main organic acids in fruits and juices. *Food Chem.* **2012**, *135*, 150–154. [CrossRef]
13. Zepka, L.Q.; Mercadante, A.Z. Degradation compounds of carotenoids formed during heating of a simulated cashew apple juice. *Food Chem.* **2009**, *117*, 28–34. [CrossRef]
14. Pereira, A.L.F.; Almeida, F.D.L.; de Jesus, A.L.T.; da Costa, J.M.C.; Rodrigues, S. Storage stability and acceptance of probiotic beverage from cashew apple juice. *Food Bioprocess Tech.* **2013**, *6*, 3155–3165. [CrossRef]
15. Assunção, R.B.; Mercadante, A.Z. Carotenoids and ascorbic acid composition from commercial products of cashew apple (*Anacardium occidentale* L.). *J. Food Compos. Anal.* **2003**, *16*, 647–657. [CrossRef]
16. Beebe, K.R.; Pell, R.J.; Seasholtz, M.B. *Chemometrics, a Practical Guide*; Wiley: New York, NY, USA, 1998; Volume 4.
17. Hotelling, H. Analysis of a complex of statistical variables into principal components. *J. Educ. Psychol.* **1933**, *24*, 417–441. [CrossRef]
18. Andersen, C.M.; Bro, R. Variable selection in regression—A tutorial. *J. Chemom.* **2010**, *24*, 728–737. [CrossRef]
19. Ballabio, D. A MATLAB toolbox for Principal Component Analysis and unsupervised exploration of data structure. *Chemom. Intell. Lab. Syst.* **2015**, *149*, 1–9. [CrossRef]
20. Wise, B.M.; Gallagher, N.; Bro, R.; Shaver, J.; Windig, W.; Koch, R.S. *PLS Toolbox 4.0*; Eigenvector Research Incorporated: Wenatchee, WA, USA, 2007.
21. Wold, S.; Sjöström, M.; Eriksson, L. PLS-regression, a basic tool of chemometrics. *Chemom. Intell. Lab. Syst.* **2001**, *58*, 109–130. [CrossRef]
22. Freitas, J.V.B.; Alves Filho, E.G.; Silva, L.M.A.; Zocolo, G.J.; de Brito, E.S.; Gramosa, N.V. Chemometric analysis of NMR and GC datasets for chemotype characterization of essential oils from different species of Ocimum. *Talanta* **2018**, *180*, 329–336. [CrossRef]
23. Varmuza, K.; Filzmoser, P. *Introduction to Multivariate Statistical Analysis in Chemometrics*; CRC Press: Boka Raton, FL, USA, 2016.
24. Press, W.H.; Teukolsky, S.A.; Vetterling, W.T.; Flannery, B.P. *Numerical recipes. The Art of Scientific Computing*, 3rd ed.; Cambridge University Press: Cambridge, UK, 2007; Volume 3.
25. Gorry, P.A. General least—Squares smoothing and differentiation by the convolution (Savitzky–Golay) method. *Anal. Chem.* **1990**, *62*, 570–573. [CrossRef]
26. Bureau, S.; Ruiz, D.; Reich, M.; Gouble, B.; Bertrand, D.; Audergon, J.M.; Renard, C.M.G.C. Rapid and non-destructive analysis of apricot fruit quality using FT-near-infrared spectroscopy. *Food Chem.* **2009**, *113*, 1323–1328. [CrossRef]
27. Shenk, J.S.; Workman, J.J.; Westerhaus, M.O. Application of NIR spectroscopy to agricultural products. *Pract. Spectroscopy Ser.* **2001**, *27*, 419–474.
28. Malegori, C.; Nascimento Marques, E.J.; de Freitas, S.T.; Pimentel, M.F.; Pasquini, C.; Casiraghi, E. Comparing the analytical performances of Micro-NIR and FT-NIR spectrometers in the evaluation of acerola fruit quality, using PLS and SVM regression algorithms. *Talanta* **2017**, *165*, 112–116. [CrossRef]

29. Magwaza, L.S.; Opara, U.L.; Nieuwoudt, H.; Cronje, P.J.; Saeys, W.; Nicolaï, B. NIR spectroscopy applications for internal and external quality analysis of citrus fruit—A review. *Food Bioprocess Tech.* **2012**, *5*, 425–444. [CrossRef]
30. Nielsen, N.-P.V.; Carstensen, J.M.; Smedsgaard, J. Aligning of single and multiple wavelength chromatographic profiles for chemometric data analysis using correlation optimised warping. *J. Chromatogr. A* **1998**, *805*, 17–35. [CrossRef]
31. Alves Filho, E.G.; Almeida, F.D.; Cavalcante, R.S.; de Brito, E.S.; Cullen, P.J.; Frias, J.M.; Bourke, P.; Fernandes, F.A.; Rodrigues, S. 1H NMR spectroscopy and chemometrics evaluation of non-thermal processing of orange juice. *Food Chem.* **2016**, *204*, 102–107. [CrossRef]
32. Cunha, A.G.; Brito, E.S.; Moura, C.F.; Ribeiro, P.R.; Miranda, M.R.A. UPLC–qTOF-MS/MS-based phenolic profile and their biosynthetic enzyme activity used to discriminate between cashew apple (*Anacardium occidentale* L.) maturation stages. *J. Chromatogr. B* **2017**, *1051*, 24–32. [CrossRef]
33. Bernillon, S.; Biais, B.; Deborde, C.; Maucourt, M.; Cabasson, C.; Gibon, Y.; Hansen, T.H.; Husted, S.; de Vos, R.C.; Mumm, R. Metabolomic and elemental profiling of melon fruit quality as affected by genotype and environment. *Metabolomics* **2013**, *9*, 57–77. [CrossRef]
34. Okpanachi, U.; Attah, S.; Shaahu, D. A Comparative study between vitamins and amino acid profile of sun-dried red and yellow cashew pulp. *Int. J. An. Biol* **2015**, *1*, 23.
35. Arumugam, B.; Palanisamy, U.D.; Chua, K.H.; Kuppusamy, U.R. Potential antihyperglycaemic effect of myricetin derivatives from *Syzygium malaccense*. *J. Funct. Foods* **2016**, *22*, 325–336. [CrossRef]
36. Lutz, I.A. *Normas Analíticas, Métodos Químicos e Físicos para Análise de Alimentos*; Instituto Adolfo Lutz: Sao Paulo, Brazil, 1985; Volume 1.
37. Strohecker, R.; Henning, H.M. *Análisis de Vitaminas Métodos Comprobados*; Madrid Paz Montalvo: Madrid, Spain, 1967.
38. Silva, L.M.A.; Alves Filho, E.G.; Choze, R.; Lião, L.M.; Alcantara, G.B. 1H HRMAS NMR spectroscopy and chemometrics for evaluation of metabolic changes in citrus sinensis Caused by *Xanthomonas axonopodis* pv. citri. *J. Braz. Chem. Soc.* **2012**, *23*, 1054–1061. [CrossRef]
39. Costa, D.P.; Alves Filho, E.G.; Silva, L.M.A.; Santos, S.C.; Passos, X.S.; Silva, M.d.R.R.; Seraphin, J.C.; Ferri, P.H. Influence of fruit biotypes on the chemical composition and antifungal activity of the essential oils of *Eugenia uniflora* leaves. *J. Braz. Chem. Soc.* **2010**, *21*, 851–858. [CrossRef]

metabolites MDPI

Article

Comparative Metabolomics and Molecular Phylogenetics of Melon (*Cucumis melo*, Cucurbitaceae) Biodiversity

Annick Moing [1], J. William Allwood [2], Asaph Aharoni [3], John Baker [4], Michael H. Beale [4], Shifra Ben-Dor [3], Benoît Biais [1], Federico Brigante [5,6,7], Yosef Burger [8], Catherine Deborde [1], Alexander Erban [5], Adi Faigenboim [8], Amit Gur [9], Royston Goodacre [10], Thomas H. Hansen [11], Daniel Jacob [1], Nurit Katzir [9], Joachim Kopka [5], Efraim Lewinsohn [9], Mickael Maucourt [1], Sagit Meir [3], Sonia Miller [4], Roland Mumm [12], Elad Oren [9], Harry S. Paris [9], Ilana Rogachev [3], Dominique Rolin [1], Uzi Saar [9], Jan K. Schjoerring [11], Yaakov Tadmor [9], Galil Tzuri [9], Ric C.H. de Vos [12], Jane L. Ward [4], Elena Yeselson [8], Robert D. Hall [12,13,†] and Arthur A. Schaffer [8,*,†]

1 INRAE, Univ. Bordeaux, UMR1332 Fruit Biology and Pathology, Bordeaux Metabolome Facility MetaboHUB, Centre INRAE de Nouvelle Aquitaine - Bordeaux, 33140 Villenave d'Ornon, France; annick.moing@inrae.fr (A.M.); bbiais@free.fr (B.B.); catherine.deborde@inrae.fr (C.D.); daniel.jacob@inrae.fr (D.J.); mickael.maucourt@inrae.fr (M.M.); dominique.rolin@inrae.fr (D.R.)
2 The James Hutton Institute, Environmental & Biochemical Sciences, Invergowrie, Dundee, DD2 5DA Scotland, UK; william.allwood@hutton.ac.uk
3 Department of Plant and Environmental Sciences, Weizmann Institute of Science, Rehovot 7610001, Israel; asaph.aharoni@weizmann.ac.il (A.A.); sagit.meir@weizmann.ac.il (S.M.); Ilana.rogachov@weizmann.ac.il (I.R.); Shifra.ben-dor@weizmann.ac.il (S.B.-D.)
4 Rothamsted Research, Harpenden, Herts AL5 2JQ, UK; john.baker@pfizer.com (J.B.); mike.beale@rothamsted.ac.uk (M.H.B.); sonia.miller83@yahoo.co.uk (S.M.); jane.ward@rothamsted.ac.uk (J.L.W.)
5 Max Planck Institute of Molecular Plant Physiology, Potsdam-Golm 14476, Germany; federicobrigante@outlook.com (F.B.); Erban@mpimp-golm.mpg.de (A.E.); Kopka@mpimp-golm.mpg.de (J.K.)
6 Universidad Nacional de Córdoba, Facultad de Ciencias Químicas, Dto. Química Orgánica, Córdoba 5000, Argentina
7 CONICET, ICYTAC (Instituto de Ciencia y Tecnologia de Alimentos Córdoba), Córdoba 5000, Argentina
8 Institute of Plant Science, Agricultural Research Organization—Volcani Center, Rishon LeZiyyon 7515101, Israel; burgery@agri.gov.il (Y.B.); adif@agri.gov.il (A.F.); elenae@agri.gov.il (E.Y.)
9 Newe Ya'ar Research Center, Agricultural Research Organization, P. O. Box 1021, Ramat Yishay 3009500, Israel; amitgur@volcani.agri.gov.il (A.G.); katzirn@agri.gov.il (N.K.); twefraim@agri.gov.il (E.L.); elad.oren@mail.huji.ac.il (E.O.); hsparis@agri.gov.il (H.S.P.); yarden@agri.gov.il (U.S.); tadmory@agri.gov.il (Y.T.); galilt@agri.gov.il (G.T.)
10 Department of Biochemistry, Institute of Integrative Biology, University of Liverpool, Liverpool L69 7ZB, UK; roy.goodacre@liverpool.ac.uk
11 Department of Plant and Environmental Sciences & Copenhagen Plant Science Center, Faculty of Science, University of Copenhagen, DK-1871 Frederiksberg C, Denmark; thh@plen.ku.dk (T.H.H.); jks@plen.ku.dk (J.K.S.)
12 Business Unit Bioscience, Wageningen University & Research, Post Box 16, 6700AA, Wageningen, Netherlands; roland.mumm@wur.nl (R.M.); ric.devos@wur.nl (R.C.H.d.V.); robert.hall@wur.nl (R.D.H.)
13 Department of Plant Physiology, Wageningen University & Research, Laboratory of Plant Physiology, Post Box 16, 6700AA, Wageningen, Netherlands
* Correspondence: vcaris@volcani.agri.gov.il; Tel.: + 972(3)9683646
† These authors have contributed equally to this work.

Received: 27 February 2020; Accepted: 20 March 2020; Published: 24 March 2020

Abstract: The broad variability of *Cucumis melo* (melon, Cucurbitaceae) presents a challenge to conventional classification and organization within the species. To shed further light on the infraspecific relationships within C. *melo*, we compared genotypic and metabolomic similarities among 44 accessions representative of most of the cultivar-groups. Genotyping-by-sequencing (GBS) provided over 20,000 single-nucleotide polymorphisms (SNPs). Metabolomics data of the mature fruit flesh and rind provided over 80,000 metabolomic and elemental features via an orchestra of six complementary metabolomic platforms. These technologies probed polar, semi-polar, and non-polar metabolite fractions as well as a set of mineral elements and included both flavor- and taste-relevant volatile and non-volatile metabolites. Together these results enabled an estimate of "metabolomic/elemental distance" and its correlation with the genetic GBS distance of melon accessions. This study indicates that extensive and non-targeted metabolomics/elemental characterization produced classifications that strongly, but not completely, reflect the current and extensive genetic classification. Certain melon Groups, such as Inodorous, clustered in parallel with the genetic classifications while other genome to metabolome/element associations proved less clear. We suggest that the combined genomic, metabolic, and element data reflect the extensive sexual compatibility among melon accessions and the breeding history that has, for example, targeted metabolic quality traits, such as taste and flavor.

Keywords: genetic resources; melon; genotype by sequencing; elemental analysis; metabolome; *Cucumis melo*

---

1. Introduction

Cucumis melo L., melon, is a phenotypically highly variable species with respect to fruit characteristics [1]. Melon fruits vary not only in size and shape but also in the accumulation of various metabolites, the most obvious of which include the horticulturally important metabolites of external and internal pigmentation, volatiles responsible for fruit aroma, and carbohydrates and organic acids accounting for sweetness and acidity [2]. Melon fruit range in sizes up to 20 kg; in shape from spherical to very long; in taste from insipid to sweet, acidic, or bitter; have external colors of green, yellow, orange, and red, with internal flesh colors of white, green, orange, or cream; and encompass a broad range, from highly aromatic to almost non-aromatic types. The species is unique in that it has representatives of both climacteric and non-climacteric ripening physiology [3], which further impacts on the metabolite components of the ripe fruit.

The extreme fruit variation within *Cucumis melo* does not easily lend itself to conventional infraspecific classification. Kirkbride [4] suggested that the species should be considered to consist of two subspecies, based on the pubescence of ovaries and young fruits. Accordingly, ovaries and young fruits having appressed short hairs were assigned to C. *melo* subspecies *agrestis* and those having pilose or lanate, spreading, long hairs were assigned to C. *melo* subsp. *melo*. However, it is often not easy to reconcile young fruit pubescence with the melon fruits seen in marketplaces in various parts of the world. Although markets in some regions feature young cucumber-like melons, most markets feature mature ripe sweet dessert melons, others ripe and highly aromatic but insipid "duda'im" melons, and yet others fully grown but unripe "snap" melons [1,5].

Attempts to classify melons according to fruit characteristics date to the first half of the 19th century, and over the past two centuries, a great number of proposed infraspecific classifications for melons have accumulated (reviewed in [6]). Among these proposed classifications are "lumpers", which have not indicated enough divisions within the species to adequately reflect its wide variation, and "splitters", which have fragmented the species into an unwieldy number of units. To be feasible, any useful infraspecific classification must at once be reflective of genetic relationships and also be easily observed by those for whom the classification is intended, which should be a wide international audience that includes scientists and non-scientists alike. For melons, this means that the characteristics of fruits should serve as the basis for any such classification. Given all of the above considerations, our view is that the infraspecific breakdown to 16 varietas first proposed by Pitrat et al. [6] or cultivar-groups

("Groups") (adopted by Burger et al., [2]) is currently the most useful. Groups are lower ranking than the subspecies, but each Group can consist of more than one market type.

The economically most important Groups are the Reticulatus (climacteric, netted), Cantalupensis (climacteric, non-netted), Inodorus (non-climacteric), and Ameri (climacteric, dryland cultivation, Asian), all of which have sweet flesh when ripe and belong to subsp. *melo*. Other Groups include the Flexuosus (snake melon), Duda'im (pocket melon), Momordica (snap melon), and Khandalak, and the East Asian Conomon (Oriental pickling), Makuwa, and Chinensis Groups. However, even the placement of the various Groups into the two subspecies is not fully agreed upon, as may be expected from a somewhat subjective classification based on the single trait of ovary pubescence. The recent attempts at making order within the melon species, at the subspecies level, literally focused on hair-splitting. For example, while the *agrestis* subspecies proposed by Pitrat et al. [6] encompasses five Groups (Conomon, Makuwa, Chinensis, Momordica, and Acidulus), the *agrestis* as described by Decker-Walters et al. [7] combines Momordica, Conomon, Duda'im, and Chito. The latter two Groups, Duda'im and Chito, were placed in subsp. *melo* by Pitrat et al. [6], and Zhao et al. [8] recently traced the Momordica Group to subsp. *melo*. Furthermore, the distinction between Reticulatus (climacteric, netted) and Cantalupensis (climacteric, non-netted) based on netting is particularly unclear since the netting phenotype is expressed as a continuum of what is basically a quantitative trait. Thus, semi-netted Charentais-type melons have vacillated between the two infraspecific Groups [6,9,10].

The difficulty of clear infraspecific classification of the species is further confounded by the ease of crossbreeding within the species, as well as the undocumented history, both early and more recent, of most accessions, and the widespread occurrence of feral melons. Even the natural range of melon species has been the subject of ongoing studies. Traditionally, melon has been grouped with other *Cucumis* spp. associated with African origin [4]. However, recent studies indicate a closer relationship to an Australian/Asian clade, suggesting that *C. melo* has origins in Asia [11,12]. Inodorous Group melons have a Central Asian origin, traceable to the mid-9th century and had arrived in Spain by the 11th century [13]; Duda'im Group melons have been traced to mid-9th century Persia [14]; and Flexuosus Group melons have been traced back 4000 years to ancient Egypt [15,16].

Molecular-based phylogenetics contributed to the infraspecific classification of *C. melo*, consistently supporting a subspecies separation, with small differences in the placement of some of the Groups [17,18]. Recently, Endl et al. [19] pointed to the complexity within the *agrestis* subspecies and showed that there is likely a complex African/Asian/Australian origin of the *agrestis* types of *C. melo*, based on polymorphisms in seven DNA regions. The most recent and encompassing phylogenetic classification of the species was reported by Zhao et al. [8], which corroborated the subspecies classification but placed the Indian Momordica Group within the *melo* subspecies. Interestingly, Leida et al. [20], based on ca. 200 single nucleotide polymorphisms (SNPs), identified what they refer to as Spanish and European Inodorus Group landraces within the *agrestis* subspecies, distinct from the well-known cultivated Inodorus Group, suggesting that the non-climacteric genetic trait may have evolved independently multiple times.

Metabolomic-based phylogeny, or chemosystematics, has regularly been attempted from the early part of the last century [21], and was especially popular in the 1970s. However, these studies focused on intra-family classification at the species level and were generally based on the measurement of individual components of single biochemical families, frequently alkaloids, as allowed by the technologies of the period. For example, the chemical systematics of the family Rutaceae and the order Rutales attracted much research attention. This chemosystematic classification has been compared to the phylogeny determined by molecular polymorphism strategies and found to be generally confirmatory [22]. However, chemical systematics has not been attempted at the infraspecific level for the obvious reason that a limited number of chemicals are unlikely to be adequate for describing the extent of varietal differences. Furthermore, relatively minor differences due to few, or even single, genes may still be causal of large differences in metabolic components. For example, the genetic polymorphism controlling ripening physiology in melons, whether climacteric or non-climacteric, although essentially a simply inherited trait [3], will likely be accompanied by major changes in the primary and specialized metabolomes.

While small subsets of metabolites may not be useful for infraspecific systematics and phylogenetics, the number of metabolites available that can be analyzed by large-scale metabolomic strategies has potential discriminatory power. The advances in metabolomics technologies now allow for these techniques to be used and to attempt to analyze infraspecific relationships [23]. Recently, we described a metabolomic comparison of two melon cultivars [24], and the results were encouraging regarding the relevance of such analyses to a larger infraspecific classification. Furthermore, we performed a targeted metabolomic characterization of 77 metabolites, primarily of quality-determining volatiles, sugars, and carotenoids, of a novel segregating genetic population derived from random introgressions between two distantly related melon accessions [25]. The results indicated the potential of metabolomic analyses of genetic variability for the discovery of associations between metabolites and metabolic pathways. Esteras et al. [26] also recently surveyed the volatile and carotenoid profiles of broad genetic variability of melon and emphasized the value of unidentified and untapped diversity for melon fruit quality improvement.

Considering the presence of hundreds of thousands of primary and specialized metabolites in the plant kingdom, no single metabolomics platform is capable of describing the plant metabolome; however, the combination of multiple platforms can approach that goal. In this study, the metabolite profile of 51 melon accessions, representing the broad genetic variability in the species, was characterized in depth using an orchestra of metabolomic strategies. The metabolomic technologies used comprised flow injection electrospray mass spectrometry (FIE-MS) fingerprinting of semi-polar extracts, untargeted proton nuclear magnetic resonance (^{1}H-NMR) profiling of polar or semi-polar extracts, liquid chromatography coupled to QTOF (quadrupole time-of-flight) mass spectrometry (LC-QTOF-MS) of semi-polar compounds, gas chromatography coupled to mass spectrometry (GC-MS) of polar extracts, solid-phase microextraction (SPME) GC-MS of volatiles, and inductively coupled plasma mass spectrometry (ICP-MS) of mineral elements [27]. This broad array of technologies collected over 80,000 metabolite signatures, i.e., molecular features.

The analyses detailed in this report are based on the comparison of the infraspecific classification as determined by metabolomic profiling of an extensive collection of accessions representing most of the cultivar-groups of *Cucumis melo*, with the classification derived from extensive comparative genotyping. The unique combination of >80,000 specific molecular features and >20,000 genetic polymorphisms obtained in this study allowed us to compare the inferred infraspecific phylogenies derived from each of the strategies and to test the relationships between metabolomic and genetic distance in this complex widely cultivated species.

2. Results

2.1. Infraspecific Structure of Cucumis melo

Fifty-one accessions of *Cucumis melo*, representing the cultivar-groups Reticulatus, Cantalupensis, Inodorus, Ameri, Flexuosus, Duda'im, Momordica, Khandalak, Conomon, Makuwa, and Chinensis, were grown in Israel during a spring-summer season and flesh and rind samples of mature fruit were harvested and distributed to the collaborating laboratories for the respective metabolomics analyses (Figure 1, Table 1; Table S1). In our classification into subspecies and cultivar-groups, we followed the proposed classifications of Pitrat et al. [6] and Burger et al. [2], with the exception of the placement of the single Momordica accession PI414723, which we included among the subspecies *melo*.

In parallel, 44 accessions could be genotyped by sequencing (GBS), which, in brief, is based on high-throughput sequencing of restriction enzyme fragments of sample DNA. For the present study, GBS provided 23,931 informative SNPs, which were selected for genome-wide analyses and phylogenetic classification. The accessions not included in the GBS analysis were provisionally genotyped by direct sequencing of over 11,000 bps, derived from 20 PCR reactions, yielding 116 informative SNPs (Figure S1).

Figure 1. Fruits of the accessions used for this study, representing most of the cultivar-groups of *Cucumis melo* (listed in Table 1 and Table S1): Cantalupensis, Reticulatus, Inodorus, Ameri, Flexuosus, Dudaim, Momordica, Khandalak, Conomon, Chinensis, and Makuwa.

Table 1. List of melon accessions used in this study. All accessions were analyzed for metabolites. The seven accessions not included in the GBS analysis are noted as NI. Further details regarding the accessions, including classification of climacteric behavior, are presented in Table S1.

Accession Name	Subspecies	Cultivar Group	Accession Abbreviation	Included in GBS	Clade (as in Figure 2)
Batley's Hermaphrodite	*agrestis*	Chinensis	BAH	x	I
PI161375	*agrestis*	Chinensis	PI161375	x	I
Black Skin	*agrestis*	Conomon	BSK	NI	
Freeman's Cucumber	*agrestis*	Conomon	FRC	x	I
Tokyo Giant	*agrestis*	Conomon	TOG	x	I
Early Silver Line	*agrestis*	Makuwa	ESL	x	I
Ohgon	*agrestis*	Makuwa	OHG	x	I
PI157071	*agrestis*	Makuwa	PI157071	x	I
PI157080	*agrestis*	Makuwa	PI157080	x	I
Sakata's Sweet	*agrestis*	Makuwa	SAS	x	I
Ananas Yoqne'am	*melo*	Ameri	AYO	x	IIa2
Ananas 'En Dor	*melo*	Ameri	EDO	x	IIa2
Bellegarde	*melo*	Cantalupensis	BEL	x	IIc2
Charentais Fom 2 Res.	*melo*	Cantalupensis	CHF	x	IIc2
Charentais	*melo*	Cantalupensis	CHT	x	IIc2
Doublon	*melo*	Cantalupensis	DOU	x	IIc2
Noy Yizre'el	*melo*	Cantalupensis	NOY	x	IIb1
Ogen	*melo*	Cantalupensis	OGE	x	IIb1
PH 406	*melo*	Cantalupensis	PH406	x	IIb1
Védrantais	*melo*	Cantalupensis	VEP	x	IIc2

Table 1. *Cont.*

Accession Name	Subspecies	Cultivar Group	Accession Abbreviation	Included in GBS	Clade (as in Figure 2)
Indian Best	*melo*	Khandalak	INB	x	II
Duda'im	*melo*	Duda'im	DUD	x	II
PI435288	*melo*	Flexuosus	PI435288	NI	
Armenian Yard Long	*melo*	Flexuosus	AYL	x	IIa3
Doya	*melo*	Flexuosus	DOY	x	IIa3
Faqqous	*melo*	Flexuosus	FAQ	x	IIa3
Striped Armenian	*melo*	Flexuosus	STA	NI	
Amarillo Pipa Blanca	*melo*	Inodorus	AMP	x	IIa1
Amarillo Oro	*melo*	Inodorus	ARO	x	IIa1
Crenshaw	*melo*	Inodorus	CRE	NI	
Golden Beauty	*melo*	Inodorus	GOB	x	IIa2
Gold King	*melo*	Inodorus	GOK	x	IIa1
Honey Dew Green	*melo*	Inodorus	HDG	x	IIa1
Noy Zahov	*melo*	Inodorus	NZA	x	IIa1
Piel de Sapo	*melo*	Inodorus	PDS	x	IIa1
PI 334107, Kirkagac	*melo*	Inodorus	PI334107	x	IIa2
PI 201581b, Kirkagac	*melo*	Inodorus	PI201581	NI	
Piel de Sapo Redon	*melo*	Inodorus	PSR	x	IIa1
Rochet	*melo*	Inodorus	ROC	x	IIa1
Tendral Verde Tardio	*melo*	Inodorus	TVT	x	IIa1
Yellow Canary	*melo*	Inodorus	YCA	x	IIa1
PI414723	*melo*	Momordica	PI414723	x	II
Bender's Surprise	*melo*	Reticulatus	BES	x	IIc1
Dulce	*melo*	Reticulatus	DUL	x	IIc1
Fordhook Gem	*melo*	Reticulatus	FOG	x	IIc1
Hale's Best Jumbo	*melo*	Reticulatus	HBJ	x	IIc1
Krymka	*melo*	Reticulatus	KRY	x	IIb2
Magyar Kincs	*melo*	Reticulatus	MAK	x	IIb2
PMR45	*melo*	Reticulatus	PMR	NI	
Top Mark	*melo*	Reticulatus	TPM	x	IIc1
PI149169	undecided	undecided	PI149169	NI	
Qishu Meshullash	(outlier)	(feral)	QME	x	(outlier)

Based on 23,931 genetic polymorphisms, the 44 accessions could be classified into two well-defined clusters (referred to as I and II) clearly distinguishing between the subspecies *agrestis* and *melo* (Figure 2). A more distant accession designated Qishu Meshullash (QME), either *Cucumis trigonus* or *C. colossus*, both of which have been included in *agrestis* by Endl et al. [19], was included in the GBS analysis as an outlier. The smaller cluster I consists entirely of the subsp. *agrestis* accessions of the Chinensis, Makuwa, and Conomon Groups. Within the Conomon Group, accessions FRC and TOG are weakly separated from the accessions of the Chinensis and Makuwa Groups, which are interspersed among themselves. The Momordica and Duda'im accessions are distinct from the *agrestis* subspecies but are also separated from the rest of cluster II, which contains accessions of subsp. *melo* and consists of a number of sub-clusters (Figure 2). Ten accessions of the Inodorus Group form a sub-cluster IIa1, which is distinct from, but closely related to, sub-cluster IIa2 comprising the Kirkagac-type Inodorus accessions and the two Ananas-type Ameri Group accessions. The Flexuosus Group accessions (IIa3) also Group together and are most closely related to the Inodorus and Ameri accessions. An additional cluster (IIb) includes both the green-fleshed Ha'Ogen-type Cantalupensis accessions (IIb1) and the closely related, yet distinct, green-fleshed Galia-type Reticulatus (IIb2). A third cluster IIc consists of two sub-clusters of orange-fleshed cultivars, one of which consists of the five orange-fleshed American accessions of the Reticulatus Group (IIc1) and, the other of four European accessions of the orange-fleshed Charentais-type Cantalupensis Group. The most unexpected placement was that of BEL, which we assumed to be a Charentais-type Cantalupensis melon. However, it was genetically distinct from the other subspecies melo types in clades IIa,b,c.

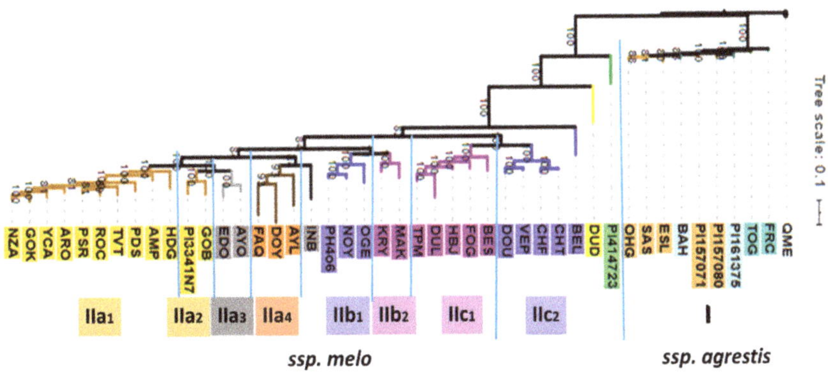

Figure 2. Phylogram representing the phylogenetic relationships of the 44 accessions of *C. melo* used for GBS analysis in this study. The feral QME serves as outlier. Bootstrap values are based on 100 iterations.

The seven accessions, which were solely characterized metabolomically, are PI2015801, STA, PI435288, BSK, PMR, PI149169, and CRE, and these were classified based on 116 informative SNPs as shown in Figure S1 and Table S2. PI2015801 is a Kirkagac type of Inodorus and the PCR-based sequencing placed it near the Kirkagac-type PI334107. CRE is a unique orange-fleshed Inodorus accession and was genotypically closer to the two Kirkagac Inodorus accessions than to the more typical Inodorus accessions, HDG and PSR. BSK was classed together with the two other Conomon accessions, FRC and OHG. PI435288 has the appearance of a characteristic long-fruited Flexuosus type (Figure 1) and was genotypically classed as being closely related to the FAQ variety. PMR was most similar to the American netted Reticulatus accessions, both in appearance and based on the genetic polymorphisms. STA and PI149169 classed closely together, and related to PI435288 and FAQ.

2.2. Combined Analysis of All Metabolomic and Elemental Data

We analyzed fruit flesh and rind samples of the 51 melon accessions (Table 1). An independent duplicate of one accession, namely subsp. *melo* Duda'im (DUD), was included for control purposes, i.e., for internal validation of profile similarity.

A total of >80,000 molecular features were collected from the combination of nine analytical profiling strategies applied to the samples of melon flesh and rind, yielding more than 36,000 and 46,000 features for the flesh and rind, respectively (Table 2). The vast majority of features were collected by the two non-targeted MS-TOF techniques, UPLC profiling of semi-polar metabolites, and GC profiling of a polar fraction enriched for primary metabolites, including taste-relevant sugars and organic acids. For the metabolic classification purpose of this study, these molecular features remained non-annotated. However, for the selection of molecular features that were relevant to build a classification model of melon accessions (refer to Section 2.4.), we used a subset of annotated features [27]. In addition, three global variables of fruit quality were included, percent dry matter (%DM), total soluble solids

(TSS), and pH, with mean values of 11.5 %DM, 9.7 °Brix, and pH 5.9, These showed considerable variability among the accessions as illustrated by their coefficients of variation of 10.4%, 26.2%, and 11.0%, respectively (n = 52, Table S1, Figure S2).

Table 2. Summary of the metabolome and elemental data measured using MS or NMR analytical strategies in extracts of fruit flesh or rind samples from of 52 melon accessions.

Analytical Strategies and Corresponding Examples of Covered Compounds or Compound Families	Number of Molecular Features	
	Flesh	Rind
GC-MS of polar extracts: soluble sugars, sugar-alcohols, organic acids, amino acids, polyamines	12 397	13 200
^{1}H-NMR fingerprints of polar extracts: major soluble sugars, organic acids, amino acids and other amino compounds	40	28
^{1}H-NMR quantitative profiles of polar extracts: major soluble sugars, organic acids, amino acids and other amino compounds	108	108
^{1}H-NMR fingerprints of semi-polar extracts: major soluble sugars, organic acids, amino acids and major semi-polar specialized metabolites	839	819
DI-ESI-MS of semi-polar extracts: positive ionization mode negative ionization mode semi-polar major and specialized metabolites	931931	931931
LC-QTOF-MS of semi-polar extracts: negative ionization mode non-volatile specialized metabolites and their conjugates including the flavonoid- and hydroxycinnamate-families	20 785	30 695
SPME GC-MS of volatiles: volatile specialized metabolites including alcohols, aldehydes, terpenoids	282	-
ICP-MS of mineral elements: mineral elements including P, K, Fe, Ni, and low-abundant trace elements	20	-

We combined the data sets of the profiled molecular features (Table 2) after 0.1–0.9 quantile range normalization of each of the sub-datasets, separately. Unique features of melon accessions were given high weight in our analyses through missing value substitution by zero. We combined all normalized sub-datasets into a single matrix of melon accessions characterized by all available molecular features of both flesh and rind samples.

Preliminary analyses by principal components analyses (PCA) and independent components analyses (ICA) verified the complex nature of the combined dataset and revealed that only 23.3% of the total variance was represented by the first three principal components. ICA of the first three principal components (Figure 3A,B) relaxed the orthogonality criterion of PCA and instead optimized independent component kurtosis as a measure of bi- or multi-modality of the scores' values of the resulting ICA axes [28]. The ICA scores plot array indicated separate clusters of subsp. *agrestis* and subsp. *melo*, with full or partial separation of respective accession Groups.

To transform the multi-dimensional ICA clusters of melon accessions into a mono-dimensional tree structure comparable to conventional genetic distance representations, e.g., by phylogenic or genetic distance trees, we created a covariance matrix of the complete accession profiles. For this purpose, we did not preselect features and did not apply dimension reduction methods to consider the complete molecular variance reflected by the data set. The covariance matrix was subsequently subjected to hierarchical cluster analysis (HCA) and bootstrapped to generate an HCA support tree (Figure 3C), with the node confidence validated by the bootstrap values. These values ranged from 2 to as high as 92 on a scale of 0–100. High bootstrap support was given to the terminal nodes of similar melon accessions with two of the Duda´im profiles giving an estimate of a bootstrap threshold of 82 that was indicative of the level of metabolic similarity of genetically identical melon accessions. The basal nodes of the HCA tree, however, received low bootstrap support. These results indicated that alternative basal classifications may be present in the data set.

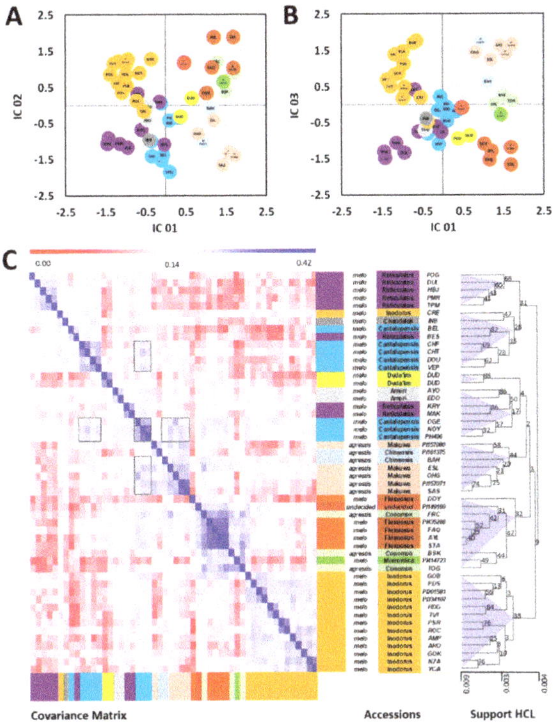

Figure 3. Metabolic classification of *Cucumis melo* accessions using the combined dataset of melon flesh and rind analyzed by multiplexed metabolome and elemental profiling. (**A**,**B**) ICA of the first three PCs obtained from the complete data set of >80,000 molecular features. The first three PCs comprised 23.3% of the total variance of the data set. (**C**) Covariance matrix of the complete molecular profiles of *C. melo* accessions, with HCA of the matrix using covariance distance metrics and complete linkage for clustering. HCA included support analysis by bootstrapping with 1000 iterations. Bootstrap support values of all HCA nodes and a scale of node height are included. Boxed regions of the covariance matrix indicate examples of additional alternative metabolic associations described in detail in the text. Cluster cones were arbitrarily set at the 75% tree height.

This interpretation was supported by alternative highly correlated groups of melon accessions that became apparent in the covariance matrix (Figure 3C). Multiple instances of alternative associations were apparent. One example was provided by the accessions of the Cantalupensis Group. This Group split between two HCA sub-branches, but alternative associations of the two Cantalupensis splits were also detected. In one HCA branch, the Cantalupensis accessions OGE, NOY, and PH406 clustered with the Reticulatus accessions KRY and MAK and the Ameri accessions AYO and EDO. In the second HCA branch, the remaining Cantalupensis accessions clustered with Inodorus accession CRE, Khandalak INB, and Reticulatus BES.

We concluded that the reduction of all paired metabolic similarities into a tree structure perhaps only reflects in part the complex metabolic similarities that were generated by the melon breeding process. This breeding process may have included crossbreeding beyond the Group or subspecies boundaries within *Cucumis melo*. Such breeding events may cause genetic introgressions with transfers of large gene sets. In addition, breeding for metabolic quality traits may have caused partial convergence of metabolic profiles from genetically distant accessions. We found a matrix representation of accession similarity more adequate in representing a complex intra-species breeding history. Consequently, we took this approach to also compare and correlate metabolic to genetic distance matrices instead of analyzing tree similarities only (Figures S3 and S4).

The highly supported nodes of our current HCA tree (Figure 3C) were useful to evaluate the closest metabolic neighbors among the GBS-characterized and the seven PCR-characterized accessions (Table 1). The Conomon cultivar BSK (Black Skin) of subsp. *agrestis* grouped at a medium bootstrap support of 44 with TOG, a Conomon Group accession of subsp. *agrestis*. A higher-level metabolic cluster with a bootstrap support of 32 contained all five Flexuosus accessions of subsp. *melo*, including non-characterized accession STA, PI435288, and accession PI149169 of undecided Group affiliation. A well-supported cluster consistently contained 13 Inodorus accessions. This cluster included both Kirkagac-type accessions, PI201581 and PI334107, and their closest neighbor, GOB. As reported above, non-characterized CRE, provisionally classified as Inodorus, grouped to a high-level cluster of diverse subsp. *melo* accessions and was metabolically most similar to INB of the Khandalak Group. Finally, PMR membership to the Reticulatus Group was confirmed with nearest neighbor TPM (Figure 3C). PMR and TPM had high similarity to a highly supported (bootstrap value 68) cluster containing three other Reticulatus accessions.

2.3. *Platform-Specific Metabolomic Analyses of Melon Flesh or Rind Tissue*

For each analytical strategy, an HCA of accession samples was separately performed, based on the range-normalized levels of all features detected per strategy (Table 2). This HCA was performed for both the flesh (eight technological platforms, Figure S3) and rind (six technological platforms, Figure S4) of the fruits, and the similarity of the grouping of the 52 accession samples based on the observed variation in their metabolome or element composition was compared to the results based on the variation in their genome (Figure 2, Table S3).

For flesh samples (Figure S3), the HCA showed a partial or complete clustering, depending on the analytical strategy, of accessions from the Inodorus Group. Likewise, the accessions of the Flexuosus Group were closely clustered, except for data from the NMR profiling and volatiles SPME GC-MS platforms. Grouping of Cantaloupe accessions was also clear, with the four analytical strategies detecting specialized metabolites (semi-polar extracts and volatiles) as well as for ^{1}H-NMR fingerprints of polar extracts and the mineral element platform. Overall, grouping was less clear for the three platforms, detecting mainly primary metabolites (polar extracts). The grouping of accessions based on metabolites detected by SPME-GCMS and especially LC-QTOF-MS mostly showed a relatively high correlation with the genetic distances between accessions, while grouping based on primary metabolites targeted by NMR profiling data showed a relatively lower correlation. For rind samples (Figure S4), the HCA showed a partial or complete clustering of accessions from the Inodorous Group for all six platforms, except for LC-QTOF MS of semi-polar extracts for which distances within and between clusters were similar. Grouping of Cantaloupe accessions was also clear with GC-MS of polar extracts, except for one accession.

For the 44 accessions having both GBS and metabolome or elemental data, the association between the genetic distance matrix and each metabolomic or elemental distance matrix was measured using a Mantel test, separately for flesh and rind data (Table 3). Among the 14 associations between genetic and metabolomic/elemental distances that could be tested, namely 8 for flesh, and 6 for rind (volatiles and microelements were not measured in rind), 13 including the flesh microelement-based matrix, showed a significant association with a p-value <0.0001. Only the metabolomic distance matrix based on LC-QTOF-MS of semi-polar extracts in rind was not related to genetic distances. For flesh, the highest two correlations between the genetic and a metabolic/elemental distance matrix were observed for specialized metabolites, i.e., for LC-QTOF-MS of semi-polar extracts ($r = 0.56$) and for SPME GC-MS of volatiles ($r = 0.39$). For rind, the highest two correlations between the genetic and metabolic/elemental distance matrix were observed for the ^{1}H-NMR fingerprints of semi-polar extracts ($r = 0.56$) and for FIE-MS of semi-polar extracts ($r = 0.47$).

Table 3. Link between genetic and compositional distances measured from melon fruit flesh or rind. Mantel test between the molecular distance matrix and each metabolomic or elemental distance matrix for the 44 accessions genotyped and phenotyped. '-' indicates that these analyses were not performed on the rind samples.

Analytical Strategy	GC-MS of Polar Extracts	^{1}H-NMR Fingerprints of Polar Extracts (0.04 ppm VS bucketting)	^{1}H-NMR Quantitative Profiles of Polar Extracts	^{1}H-NMR Fingerprints of Semi-Polar Extracts (0.01 ppm bucketting)	FIE-MS of Semi-Polar Extracts	LC-QTOF-MS of Semi-Polar Extracts	SPME GC-MS of Volatiles	ICP-MS of Mineral Elements
Flesh								
Pearson correlation (r) between the molecular distance matrix and the metabolomic or elemental distance matrix	0.207	0.117	0.224	0.202	0.149	0.560	0.387	0.177
Correlation p-value [a]	< 0.0001	< 0.0001	< 0.0001	< 0.0001	< 0.0001	< 0.0001	< 0.0001	< 0.0001
Rind								
Pearson correlation (r) between the molecular distance matrix and the metabolomic or elemental distance matrix [b]	0.315	0.273	0.267	0.561	0.468	−0.041	-	-
Correlation p-value	< 0.0001	< 0.0001	< 0.0001	< 0.0001	< 0.0001	0.189	-	-

[a] p-value calculated with 10000 Monte Carlo simulations. [b] Euclidian distance on range-normalized data for metabolomic or elemental distance matrices between accessions.

2.4. Feature Selection by Random Forest Technology

To select molecular features that were relevant for the metabolic classification of melon accessions, we applied random forest (RF) machine learning technology. RF technology tuned towards metabolic feature selection enables the selection of small sets of molecular features that, if manually supervised, can be relevant for sample classification [29]. For this purpose, we used the set of 605 provisionally annotated molecular features from the combined >80,000 data set (Table S4). To create approximately balanced classes, we split the available melon accessions into six subsets, namely 1) all accessions of subsp. *agrestis*, 2) the Cantalupensis accessions, 3) the Flexuosus accessions, 4) the Inodorus accessions, 5) the Reticulatus accessions of subsp. *melo*, respectively, and 6) a class that contained all other melon Groups that were represented by only one or two accessions (Figure 4A). Because of the diverse nature of the sixth class, we did not expect good classification results, but we added this class to contrast with the remaining more populated classes that had 5–14 members (Figure 4).

A

Actual class	Class size	Predicted class: sp. *agrestis*	sp. *melo* Cantalupensis	sp. *melo* Reticulatus	sp. *melo* Inodorus	sp. *melo* Flexuosus	sp. *melo* miscellaneous	Classification error
sp. *agrestis*	10	8.50	0.00	0.00	0.00	0.10	0.10	0.02
sp. *melo* Cantalupensis	8	0.30	5.80	0.60	0.20	0.00	0.10	0.17
sp. *melo* Reticulatus	7	0.00	0.50	5.80	0.00	0.00	0.90	0.19
sp. *melo* Inodorus	14	0.00	0.00	0.00	11.90	0.00	0.60	0.05
sp. *melo* Flexuosus	5	0.30	0.00	0.00	1.00	0.90	2.20	0.80
sp. *melo* miscellaneous	8	0.20	0.40	2.00	0.00	1.50	2.10	0.66

B

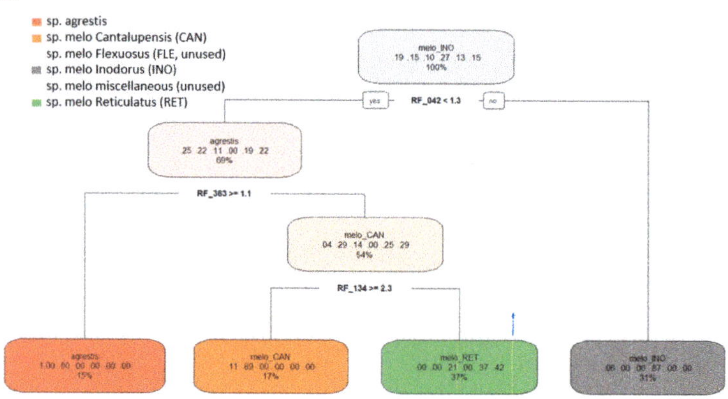

Figure 4. Random forest (RF) analysis (**A**) and decision tree classification (**B**) of six *C. melo* accession classes using a subset of 605 provisionally annotated molecular features. Classification of six pre-defined melon accession classes was performed. The classification table (A) lists classes, class size, the actual and predicted class membership, and the classification error (means of 10 iterations using hyperparameter-tuned RF settings). The decision tree uses the top 20 most informative molecular features ranked by the mean decrease in accuracy. The node information of the decision tree reports the used molecular feature code (Table S4) and threshold value. The branch information (colored ovals) lists the main class, the fraction of classified samples, left to right, subsp. *agrestis*, Cantalupensis, Flexuosus, Inodorus, miscellaneous, and Reticulatus accessions of subsp. *melo*. The percentage value indicates the fraction of the 52 accession samples that fall into each of the diagnostic categories.

Hyperparameter tuning of the RF procedure was applied to the six melon classes and based on the annotated data subset, provided the parameter settings: Sampled fraction of 0.899 for RF-classification, minimal node size of 2, and mtry value of 59. The estimate of the overall error rate across 10 repeated RF classifications was 23.9% ± 2.9 %, mean ± standard deviation (SD). The classification error differed, however, between tested classes. Subsp. *agrestis* classified best, with an average classification error of 0.02, followed by subsp. *melo* Inodorus with an error of 0.05. Subsp. *melo* Cantalupensis and subsp. *melo* Reticulatus classified with errors of 0.17 and 0.19, respectively. Unexpectedly, Flexuosus accessions were difficult to classify using the annotated metabolites. Molecular features relevant for the classification of the Flexuosus Group may be present among the non-annotated features (Figure 3B,C). Using annotated molecular features, the Flexuosus classification failed and had a high classification error, similar to the class of miscellaneous accessions (Figure 4A).

The annotated mass features were ranked by RF analysis according to the variable selection parameter mean decrease in accuracy (Table S4). A decision tree based on the top 20 ranks of annotated mass features used only three features (Figure 4B). This decision tree had four classes but failed to define a diagnostic rule for the Flexuosus and miscellaneous classes. One decision rule defined a class that only contained subsp. *agrestis* accessions to 100%. A second class defined Cantalupensis accessions to 89% with a minor contribution of subsp. *agrestis* accessions (11%). A third class contained 87% Inodorus accessions with 6% each of subsp. *agrestis* and Flexuosus. The fourth class was, with 42%, enriched for Reticulatus accessions but contained, in addition, 21% Flexuosus and 37% miscellaneous accessions. The top 20 informative molecular features according to RF analysis were glucose, formic acid, glycerol, arabitol, galactinol, and raffinose in rind; xylose and 1-kestose in flesh; 10 volatile organic compounds detected in melon flesh; and Cr and As elements detected in melon flesh (Table S4). These features provide additional supporting information for melon classification.

3. Discussion

3.1. Phylogenomic and Phytochemical Relationships Partly Coincide in C. melo

Besides contributing to the genetic classification of the species, the goal of this research was to determine whether phylogenomic and phytochemical relationships in melon coincide and are mutually supportive. Studies of this nature have been carried out for over 50 years but have been limited in two significant respects. Firstly, due to the limiting nature of the targeted chemical analyses performed in earlier studies, chemosystematics was limited to interspecific classification where biochemical differences were large. Thus, interspecific chemosystematics, in conjunction with genomic classification were successfully implemented to distinguish, for example, among *Capsicum* (pepper) species [30], *Brassica* species [31], and *Citrus* species [32]. However, large-scale infraspecific classification based on chemosystematics has not been successful to date. Secondly, the comprehensiveness of metabolomic analyses, as is presented here, was made possible only recently, and earlier studies attempting chemosystematics focused on only single metabolite families, such as volatiles and carotenoids [26]. This inevitably not only limited the breadth of the characterization but also could cause strong bias in the results, especially if the metabolites most impacting the clustering have also been of selective value, either through natural selection or human selection via breeding. The variability of select metabolite families is generally controlled by few genes, which are strongly affected by selection pressure. Hence, for crop varieties, targeted analyses of metabolite variation may not reflect the genetic distance. For example, the impact of a trait, such as non-climacteric ripening, although determined by few genes, is expected to have wide-ranging pleiotropic effects on numerous secondary metabolites, including volatiles [33].

Our study, which is based on the combination of large-scale unbiased genetic and similarly broad characterization of metabolomic phenotypes, allowed us to determine the degree of coincidence between the phylogenetic and chemosystematic classification of the *C. melo* species. We expected to have improved representation of phylogenetic/genomic relationships in our data set by including non-targeted metabolomic technologies, as compared to a multi-targeted approach that may cover predominantly those metabolic traits that were directly subject or linked to the selective breeding process. Human breeding in melon largely aims for metabolic traits and therefore, breeding can lead to convergence of such traits in cultivars of different genotypic backgrounds and evolutionary histories. We hypothesized that non-targeted metabolites, such as those detectable by the UPLC-TOF-MS and GC-TOF-MS technologies, were presumably of less selective value during domestication and breeding than the targeted metabolites of volatiles, sugar, and organic acid levels, which together comprise the major determinants of quality and hence selection. We therefore predicted that our broad non-targeted profiling would cover the potentially convergent breeding traits, but these would be diluted by metabolic and elemental traits that were not targeted by breeding and hence hypothesized to have higher discriminatory power. Thus, chemosystematics based on non-targeted metabolites would be more likely to reflect true genetic relationships.

Systemization of melon variation goes back at least as far as the 1832 monograph of Jacquin [34]. These early classifications and nearly all subsequent ones were based primarily on fruit traits. The introduction of molecular genetic technologies has helped greatly in evaluating prospective infraspecific relationships. For example, the infraspecific classification proposed for the extremely variable *Cucurbita pepo* L. (Cucurbitaceae) that was based mainly on fruit shape [35] was supported and clarified by polymorphisms in 134 simple sequence repeat loci [36]. Rapid development of new technologies has allowed for successively and increasingly more comprehensive and precise identification and analyses of genetic variability. Most recently, high throughput sequencing and massive GBS strategies have been applied to melon variability in representative GWAS populations [8,37], yielding 17,000 and 22,000 genetic polymorphisms, respectively. Such large-scale genotyping allows for a more precise assessment of genetic relationships within species. The genotype classification of 44 of the 51 accessions used in this study was derived from the GWAS GBS previously reported by Gur et al. [37]. The remaining accessions not included in that GWAS study but screened metabolomically in this study could be reliably placed within Groups based on an additional smaller set of polymorphisms identified by direct PCR product sequencing.

Our results present a mixed picture; on the one hand the metabolomics-based infraspecific classification indeed largely reflects the phylogenetic classification. The correlation between genetic and metabolomics distance increases with the addition of metabolite signals, as expected, and the >80,000 combined metabolite signals best mirror the genetic classification (Figure 3). However, on the other hand, there are exceptions to a straightforward correlation, as will be discussed here further, and the metabolomics-based classification at times may better reflect the market-type classification rather than the genotype classification. This may indicate that selection and breeding for particular fruit characteristics may have had a disproportional impact on metabolites for which breeding did not directly select.

The present phylogenetic results substantiate the dichotomy within *Cucumis melo* to subsp. *agrestis* and subsp. *melo* (Figure 2). Furthermore, the results suggest that only the East Asian cultivar-groupGroups, Chinensis, Makuwa, and Conomon, belong to subsp. *agrestis*. The Duda'im and Momordica accessions that were included in our study are allied with subsp. *melo* but form an outlying cluster within that subspecies, as recently suggested [8,17,19]. However, with respect to the comparison of phylogenetic and the combined metabolomics-based classification (Figures 2 and 3), the *agrestis* subspecies does not behave as a single Group. The two Chinensis and five Makuwa types of the *agrestis* subspecies are similar to each other based on their metabolite composition and clearly clade together. However, they are distinct metabolomically from the other *agrestis* subspecies Group Conomon. Instead, the three Conomon accessions, BSK, FRC, and TOG, clade with the Flexuosus Group, along with the single Momordica accession PI414723 and the undecided accession PI149169. Thus, the Flexuosus and Conomon accessions are metabolically more similar to each other than would be expected based on their genetic relationships. Fruits of both Groups are used primarily when young, either fresh or pickled, similar to cucumbers. The mature fruit of both Groups are characterized by an acidic pH and low sugar content, in contrast to the dessert melons with low acidity and high sugar content in the edible mature fruit [38]. In light of the fact that these two traits are governed by only a few genes [38–40], we expected that the large-scale metabolomics studies would overcome the bias of these few targeted metabolites and would more closely reflect the unbiased genetic relationships combining the three *agrestis* Groups. Our results with respect to *agrestis* suggest otherwise.

3.2. Cantalupensis and Reticulatus Accessions are Separated from Each Other Both Genotypically and Metabolomically

With regard to the Cantalupensis and Reticulatus accessions, the combined metabolomics-based classification does strongly parallel the phylogenetic classification. The separation of Reticulatus/Cantalupensis into two individual clades, IIb and IIc (Figure 2), that we observed in the genetic cladogram is similarly evident in the metabolomics classification (Figure 3). The Reticulatus Galia types (MAK, KRY) and the Cantalupensis Ha'Ogen types (OGE, NY, PH406) have a distinct metabolite profile from the Reticulatus/Cantalupensis Groups of American cantaloupes and Charantais. Furthermore, within each of the two Reticulatus/Cantalupensis Groups, the Cantalupensis and Reticulatus accessions are separated from each other, both genotypically and metabolomically. This is evident in the separation between the Ha'Ogen and Galia types, as well as the American cantaloupe and Charentais types from each other. The Ananas-type Ameri accessions, EDO and AYO, group with the Galia and Ha'Ogen accessions, as they did in the phylogenetic classification. Thus, the results relating to the climacteric, netted, and semi-netted *melo* subspecies (Reticulatus/Cantalupensis/Ameri) do support the similarity between genetic and metabolomics-based classification.

The Cantalupensis and Reticulatus Groups are horticulturally distinguished largely according to the extent of rind netting or reticulation, which is actually a continuous trait rather than two distinct phenotypes of netted versus smooth. The trait of rind reticulation is presumed to be relatively simply inherited [39,41,42], but there remains the possibility that the netting trait, present in the different clades of Reticulatus/Cantalupensis, may be under separate genetic control. The most visual distinction between the different clades of Reticulatus/Cantalupensis is the flesh color. While the Charentais-type Cantalupensis are orange-fleshed, the Ha'Ogen-type Cantalupensis are green-fleshed. Similarly, the American cantaloupe Reticulatus varieties are orange-fleshed while the Galia Reticulatus are green-fleshed. Flesh color is also determined by relatively few genes [40,43–45], but the associated pleiotropic metabolomic effects due to chloroplast to chromoplast transformations are expected to be large.

3.3. The Metabolomic-Based Classification May Indicate That Two Independent Evolutionary Events Led to Non-Climacteric Ripening

The results of the GC-MS analysis of melon volatiles (Figure S3) indeed showed that volatile metabolites alone distinguish between the non-climacteric and climacteric groups. However, our comprehensive genomic and metabolomic results also indicate that the Inodorus Group is genetically distant from the other cultivar-groups and this large genetic distance cannot be attributed only to the limited climacteric-related genetic differences but also primarily to other unrelated genetic evolution.

Concerning the Inodorus Group, most accessions are clearly sorted on the basis of their combined metabolites, similar to their phylogenetic classification. The inclusion of the Kirkagac Inodorus accessions clearly within the Inodorus metabolite group even though they were genetically distinct from it further indicates that the relationship between genetic and metabolomics-based classification is not simple, even when the latter is based on a large number of metabolic signatures as in our study. The genetic distinction between the characteristic Inodorus Group melons and non-characteristic Inodorus types referred to as European types was also noted by Leida et al. [20].

These results regarding the genetic relationships between most of the Inodorus accessions of clade IIa (Figure 2) and those of the Kirkagac accessions in clade IIb are significant as they indicate that there may have been two independent evolutionary events leading to non-climacteric ripening, one represented in the large Inodorus Group and the other more closely related to the clade IIb climacteric types. The genetic control of climacteric ripening in *C. melo* has been studied and found to be due to mutations in just a few genes [46,47]. However, these genetic studies were performed using typical Inodorus accessions as a genetic source for the non-climacteric trait and typical climacteric Cantalupensis (Charentais) accessions. The possibility remains that the genetic cause of non-climactericism in the IIa and IIb clades may not be identical and may even be due to mutations in different genes. In support of this possibility, Eduardo et al. [48] and Vegas et al. [49] reported that a supposed non-climacteric Conomon accession, crossed with the non-climacteric Inodorus PDS, yielded climacteric phenotypes, indicating independent genetic mutations. In fact, Saladie et al. [50] pointed out that classification of melon fruit ripening behavior into just two distinct types is an over-simplification, and that, in reality, there is a continuous spectrum of fruit ripening behavior. If so, further study of the relationship between the non-climacteric traits of Kirkagac and other Inodorus genotypes could broaden the genetic variability available for this important characteristic for genetically improving fruit quality by extending the harvest period as well as the shelf life of the horticultural product.

While practically all the Inodorus accessions metabolomically clade together, the CRE variety is a striking exception (Figure 3). CRE is most similar with respect to its metabolite signature to the single Khandalak accession INB, and together are most closely related to the Charentais Cantalupensis accessions. In fact, CRE is strikingly distinct phenotypically from the other Inodorus accessions, as can be seen in Figure 1. Whereas all the Inodorus lines studied are green-fleshed, as are non-climacteric Inodorus lines in general, the CRE variety is orange-fleshed and likely also climacteric. In fact, the GC-MS volatile results of flesh (Figure S4e) also distinguish between CRE and the other Inodorus lines and place CRE together with the BEL Ha'Ogen-type Cantalupensis. The CRE accession is a Crenshaw market type, which is considered to be a hybrid derivative of the non-climacteric, green-fleshed, smooth-skinned Casaba market-type melon and the climacteric cream/orange-fleshed Persian Cantalupensis melons and thus this may explain that metabolomically, it is similar to the latter.

The different analytical strategies and over all the different types of compounds targeted (primary or specialized metabolites, mineral elements) contributed differently to the classification of melon accessions. This may be related to the fact that some metabolite classes were under breeding selection and may represent convergence of metabolic traits rather than genetic distance. The latter seems probable for major primary metabolites, measured using ^{1}H-NMR or GC-MS of polar extracts in flesh and implicated in sweetness or acidity. Convergence may also be the case for different families of specialized metabolites measured using LC-MS of semi-polar extracts, such as phenolics or alkaloids with an astringent or bitter taste, or GC-MS of volatiles, such as esters and alcohols. Even mineral elements in flesh contributed to accession classification, possibly through a link with flesh acidity, for instance, for K, or metabolic links between metabolites and mineral elements as shown previously [27]. Selection may have been less stringent for rind than for flesh composition, as for five out of the six analytical strategies used for both flesh and rind analyses, the correlations between compositional distances and genetic distances were higher for rind than for flesh (Table 3).

The correlation between genetic and metabolomics distance increased with the addition of the >80,000 combined molecular features and best mirrored the genetic classification. Nevertheless, information reduction by RF analysis allowed pointing to 20 informative molecular features that differentiate several accession groups. In flesh, 10 volatiles and xylose are possibly linked with climactericity as discussed above. Of these top 10 volatiles, 6 could be reliably annotated and are well known components of melon fruit, including (Z,E)-3-hexen-ol and 1-octen-3-ol, which result from fatty acid degradation, and also two typical acetate esters (2-methyl butyl acetate and pentyl acetate) as well as alpha-ionene [27,51]. Xylose is a major non-cellulosic neutral sugar of cell walls in cucurbit fruit [52]. Fruit softening during maturation involves cell wall degradation. For cell wall-related genes, specific genes of each gene family can be categorized as totally ethylene dependent, totally ethylene independent, or partially ethylene dependent [3]. Surprisingly, two mineral elements were in the top 20 molecular features highlighted by the RF analysis. Among the six compounds highlighted in peel, glycerol may be related with the fruit surface, such as cuticular waxes [53] or suberization [54]. Galactinol and raffinose linked classification to the raffinose family and to oligosaccharides metabolism in the peel (Table S4). Galactinol and raffinose linked with the raffinose family of oligosaccharide metabolism after phloem unloading into the melon fruit may be related to fruit sweetness [55].

4. Conclusions

The main objective of this research was to determine whether classification by large-scale, non-targeted metabolomic and element profiling technologies recapitulates or extends the phylogenomic relationships within *Cucumis melo*. Our results indicate that, in general, metabolomic/elemental means of classification can indeed significantly reflect the genetic relationships. Nevertheless, there are deviations of metabolomics and elemental groupings from genetic classifications. These differences indicate that the selection for major phenotypic quality traits by the breeding process has been influential. Some of the quality traits that can be controlled by single or few enzymatic or regulatory genes, include changes of metabolite levels that define color, taste, and flavor. In addition, some traits change fruit development and thereby have pleiotropic effects on fruit metabolomics.

Large-scale information on genomic sequence and metabolomic/elemental variation of a broad range of genetic diversity can serve for dissecting the genetic basis of metabolic diversity. This process has been recently referred to as mGWAS (metabolite genome wide association study) [56]. mGWAS presents a powerful tool for attributing metabolite variation to particular genetic regions. In melon, GWAS based on GBS was even capable of mapping traits to the single candidate gene level [37] and it is expected that the comprehensive metabolomics data presented here will allow for a large-scale mapping of metabolic traits.

Furthermore, this work sets the stage for an unlimited number of metabolic QTL (mQTL) analyses based on recombinant populations generated from selected metabolically characterized parental lines. The identification of valuable genetic and metabolic variability forms the basis for directed crop diversification and genetic improvement by breeding. The future combination of the results of this study with gene expression data of the developing melon fruit rind and flesh provides an 'omics' blend of genomics, metabolomics, and transcriptomics that will be especially useful for the identification of the genetic basis of metabolic diversity [44].

5. Materials and Methods

5.1. Plant Material Description, Cultivation, and Sampling

An initial core collection comprised of 51 *Cucumis melo* accessions representing a broad spectrum of melon genetic variation was grown as a spring-summer crop in an open field at the Newe Ya'ar Research Center in the Yizre'el Valley, northern Israel. The accessions prospectively represented both subspecies and 11 of the cultivar-groups (Table 1; Figure 1; Table S1). These 51 accessions were a representative subset of a larger collection of 177 accessions that were used for detailed characterization of fruit trait variation and evaluation of the potential of genome-wide association (GWA) for trait mapping in melon [37].

Of the 51 accessions used for metabolomics analysis, 7 were not included in the GBS study but were genotypically classified in a less encompassing sequencing project based on direct sequencing of 20 PCR products of known genes. This direct sequencing comprised a total of nearly 12,000 bp and provided 116 polymorphisms for which haplotypes were produced and used for the phylogenetic analysis (Table S2). This phylogenetic analysis is not as comprehensive as the GBS but was useful in placing those accessions for which GBS data was missing.

Among these 51 accessions were included two closely related Duda'im melons (DUD2, DUD3) for a total of 52 sampled accessions. Each accession was represented in three replicated plots of five plants each. A minimum of 10 ripe fruits were harvested from each accession. Fruit were photographed prior to sampling. For sample preparation, a midsection of each fruit was excised, and its seed cavity removed. For the rind sample, the rind was removed using a vegetable peeler to a depth of approximately 2–3 mm and the rinds from all fruit of each accession were combined into a single sample. Similarly, the flesh samples of each accession were combined. Combined bulk samples were ground to a powder in liquid nitrogen. Falcon tubes were filled with approximately 50 mL of powdered sample and a total of 104 tubes (flesh and rind samples of 52 accessions) were stored at −80 °C until shipment under dry ice to each of the participating laboratories where two replicate samples were analyzed per accession/per platform.

5.2. Genotype and Phylogenetic Analysis

5.2.1. DNA Isolation for Genotype by Sequencing (GBS)

For the GBS of the melon collection, leaf tissue was taken from 44 of the 51 accessions, listed in Table 1. Total genomic DNA isolation was performed as described by Gur et al. [37] using the GenEluteTM Plant Genomic DNA miniprep kit (Sigma, St. Louis, MO, USA). The quality of the DNA was analyzed with an ND-1000 spectrophotometer (Nanodrop Technologies, Wilmington, DE) and by electrophoresis on agarose gel. The concentration of DNA was estimated using a Qubit® 2.0 Fluorometer (Life Technologies, City, Singapore) and a Qubit® dsDNA BR Assay Kit (Life Technologies, Eugene, OR, USA).

5.2.2. GBS Analysis

DNA was analyzed at the Institute for Genomic Diversity facility at Cornell University for GBS. GBS 96-plex libraries were prepared using the restriction enzyme ApeKI, following an established protocol [57]. Fragments were sequenced on an Illumina HiSeq 2500 as 100 bp, single-end reads and aligned to the reference genome of *C. melo* [58] available at https://melonomics.net/files/Genome/Melon_genome_v3.5.1/. TASSEL pipeline v3.0.173 was used for sequence alignment and single nucleotide polymorphism (SNP) calling [59]. Further filtration was performed using TASSEL v5.2.33 [60]; the SNP list was filtered to MAF > 0.05 and a maximum of 6% missing data per site.

5.2.3. Phylogenetic Analysis

TASSEL software (v5.2.33) was used to estimate the distance matrix for all pairwise combinations. The phylogenetic tree based on the GBS results of the 44 accessions was assembled using the neighbor-joining function. In addition, we constructed a more limited phylogenetic tree based on 116 polymorphic sites derived from direct sequencing of 11,950 bp derived from 23 genomic sequences (Table S2). The sites were concatenated to create a pseudo-sequence haplotype, which was then aligned (Figure S1a). The aligned sequences were used to build a neighbor-joining tree, and a Phylip distance matrix (Figure S1b). The final alignment, tree, and distance matrix were performed using Clustal Omega version 3.0, www.ebi.ac.uk/Tools/msa/clustalo.

5.3. Global Measurements of Fruit Quality

Fruit total soluble solids (TSS, as degrees Brix) and pH were measured on a set of the frozen powdered samples. Portions were allowed to defrost, and juice was measured using a hand-held refractometer (Atago A-10) and a pH meter for each of the 52 accessions used for metabolomic and elemental measurements. Percent of dry matter (%DM) was measured by drying a representative sample in a 60 °C oven and weighing before and after. Means per accession were calculated.

5.4. Metabolomics and Elementals Analysis

5.4.1. NMR-Based Metabolomic Analyses

For targeted ^{1}HNMR, polar metabolites were extracted from 50 mg of lyophilized powder using a hot ethanol/water series, and then analyzed, identified, and quantified by ^{1}H-NMR profiling as previously described [27,61]. For the preparation of extracts and NMR acquisition parameters, special care was taken to allow absolute quantification of individual metabolites. Quantitative ^{1}H-NMR spectra were recorded at 500.162 MHz and 300 K on a Bruker Avance spectrometer (Wissembourg, France) with a 5-mm inverse probe using a 90° pulse angle and an electronic reference for quantification with calibration. Two replicates were extracted and analyzed for each accession. Unknown metabolites were named using the mid-value of the chemical shift and the multiplicity of the corresponding resonance group and quantified in arbitrary units. The same spectra issued from polar extracts were also processed as fingerprints: The spectra from δ9.40–0.40 ppm (without the residual water resonance) were binned to chemical shift regions of 0.04 ppm and data were scaled to the total signal intensity.

For untargeted NMR fingerprinting of semi-polar extracts, metabolite extraction was carried out according to the methods described by Ward et al. [62]. Briefly, triplicate aliquots (15 mg) of freeze-dried powder were extracted with an 80:20 D_2O:CD_3OD mixture (1 mL) containing *d4*-TSP as the internal standard (0.01% *w/v*). Samples were extracted for 10 min at 50 °C. Supernatants were transferred to a clean tube and heated to 90°C for 2 min. Samples were cooled and centrifuged. For ^{1}H-NMR, 200 μL of the supernatant was evaporated to dryness and reconstituted in 650 μL of deuterated sodium phosphate buffer solution (pH 6.0, 200 mM). Samples were mixed and allowed to stand at room temperature for 1 h after which 600 μL were then transferred to a 5-mm NMR tube for ^{1}H-NMR analysis. ^{1}H-NMR spectra were collected at 600 MHz on a Bruker Avance spectrometer equipped with a 5-mm selective inverse probe. Parameters for data acquisition are as described in Ward et al. [62]. ^{1}H-NMR data (δ9.395–0.505 ppm) were binned to chemical shift regions of 0.01 ppm and data were scaled to the total signal intensity. Regions corresponding to residual water and methanol were removed (H_2O 4.775–4.865; MeOH 3.285–3.335 ppm set to zero).

5.4.2. GC-MS-Based Metabolomic Analysis of Polar Compounds

GC-MS profiling analysis of a polar metabolite fraction enriched for primary metabolites, including sugars, amino acids, and organic acids, was performed as described earlier [63] with modifications reported by Moing and co-authors [27]. Samples were extracted, partitioned into a polar liquid phase, and dried for chemical derivatization, as described. Methoxymation was performed with 40 μL of pyridine containing 1 mg/mL methoxyaminohydrochloride and 80 μL silylation-mixture, containing 7:1 (*v/v*) N,O-bis(trimethylsily)-trifluoroacetamide (BSTFA, Macherey-Nagel, Düren, Germany) and a mixture of alkanes in pyridine. GC-MS was by splitless mode after injection of 1 μL of chemically derivatized sample. Evaluated mass features had intensities of ≥ 50 arbitrary intensity units. Peak heights of mass features defined by nominal mass to charge ratios (*m/z*) and *n*-alkane based retention indices were normalized to the sample fresh weight and the internal standard, succinic-d_4 acid (Sigma-Aldrich, Deisenhofen, Germany). The internal standard was added to the extraction solution. Normalized mass spectral features were aligned, correlated across all recorded samples, and placed into clusters and time groups using TagFinder [64]. Metabolite annotations of mass spectral features were manually supervised using TagFinder visualizations for mass spectral matching [64]. Metabolite annotation required a match of at least three co-eluting and correlated mass features and a retention index deviation < 5% [65] compared to the reference data of the Golm Metabolome Database, http://gmd.mpimp-golm.mpg.de/ [66].

5.4.3. GC-MS-Based Metabolomic Analysis of Volatile Compounds

For GC-MS of volatile organic compounds, headspace-solid phase micro extraction (SPME) with a 65-mm polydimethylsiloxane-divinylbenzene fiber (Supelco, Bellefonte, USA) was used as described previously [67]. In short, 200 mg of frozen powder was mixed with 4 mL of 4.6 M $CaCl_2$ containing 5 mM EDTA in a 10-mL vial. Volatiles in the sample headspace were trapped for 20 min at 50 °C with agitation (CombiPAL autosampler, CTC Analytics, Switzerland) and thermally desorbed in the GC injection port for 1 min at 250 °C. A HP-5 column, 30 m x 0.25 mm ID, 1.05 μm – film thickness (Hewlett Packard, Palo Alto, CA, USA) was used to separate the volatile compounds, applying a temperature gradient from 44 to 250°C at a speed of 5 °C min^{-1}. All masses from m/z 35 to m/z 400 resulting from 70 eV electron impact ionization were recorded (MD800 electron impact MS, Fisons Instruments, Milan, Italy). Metalign software [68] was used to extract and align all mass features in an untargeted manner and masses originating from the same molecule were then clustered to reconstruct the relative intensity and mass spectrum of each detected compound, using MSClust software [69]. Volatiles were putatively annotated by matching the reconstructed mass spectra of detected compounds to the electron impact mass spectral libraries of NIST (National Institute of Standards and Technology, Gaithersburg, MD, USA) and Wiley. Subsequently, for the random forest (RF) machine learning analysis on the putatively annotated data (Supplementary Table S4), the top 10 volatiles arising as potential markers were then further manually checked for matching of the retention index and mass spectra. Annotations with an RI >20 and/or a match factor <750 were reclassified as 'unknowns'.

5.4.4. FIE- or LC-MS-Based Metabolomics Analysis

For FIE-MS, 50 µL of the semi-polar extract supernatant prepared for NMR fingerprinting was diluted with 80:20 H_2O:CH_3OH (950 µL). Samples were analyzed using an Esquire 3000 spectrometer (Bruker Daltonics, Coventry, UK) in both positive and negative ionization modes as described previously [24]. Spectral data were exported as ASCII files containing mass-intensity pairs and automatically reduced using AMIX software version 3.9.11 (Bruker Biospin, Coventry, UK), to a single CSV file for each ionization mode, containing integrated regions of equal width (*m/z* width = 1). Individual signal intensities were scaled to the total intensity and m/z regions relating to *d4*-TSP and its isotope peaks were removed from the data prior to statistical analysis.

For LC-QTOF-MS, ground frozen rind and flesh tissue (0.5 g) were extracted in 1.5 mL of methanol, containing 0.1% formic acid. The samples were vortexed vigorously, sonicated for 20 min, and centrifuged for 15 min at 4400 *g*. For rind samples, the supernatant was filtered through a 0.22-µm PVDF filter directly to HPLC vials. Flesh samples were further concentrated as follows: 1 mL of supernatant was freeze-dried and resuspended in 150 µL of 75% aqueous methanol, containing 0.1% formic acid, sonicated for 20 min, and filtered through a 0.22-µm PVDF filter directly to the insert of the HPLC vial. Melon samples were injected into a UPLC-QTOF-MS (HDMS-Synapt, Waters, Manchester, UK) in negative ionization mode as in [70] with some modifications: Short 9.5-min gradient was used for metabolite separation. The linear gradient program was as follows: 100% to 90% phase A over 1 min, 90% to 75% phase A over 3 min, 75% to 55% phase A over 2 min, and 55% to 0% phase A over 0.5 min; held at 100% phase B for 1 min; and then returned to the initial conditions (100% phase A) in 1 min and conditioning at 100% phase A for 1.0 min. A divert valve (Rheodine) excluded the first 1.3 min and last 1.8 min of injection. A mixture of 15 standard compounds, injected after each 10 samples, was used for quality control. XCMS software [71] was used for peak picking and peak alignment. Intensity values were $\log^e$-transformed.

5.4.5. Elemental Analysis

For microelements, frozen milled melon samples were freeze-dried and 200–250 mg dry material was wet digested (5 mL 65 % HNO_3 and 5 mL 15 % H_2O_2) at 210 °C in 100-mL closed vessels using a microwave oven (Multiwave 3000 software version 1.24, Anton Paar, Graz, Austria). Before analysis, the digests were diluted to a final concentration of 3.5% HNO_3. Multi-element analysis was performed using an ICP-MS (Agilent 7500ce, Agilent Technologies, Wokingham, UK) as described in Bernillon et al. [24]. The impact of spectral interferences was reduced using an octopole ion guide, pressurized with He or H_2 certified reference material (NIST 1515, apple leaves, particle size <75 µm; National Institute of Standards and Technology, Gaithersburg, MD, USA) were included. Only elements deviating less than ±10 % from the certified reference values are reported.

5.4.6. Data Handling and Mining

Metabolome and elemental data, i.e., the means of at least two technical replicates per accession sample and of each fruit tissue, were gathered from the consortium of laboratories. Initial response/abundance values below zero were judged to be missing data and removed. Data of each molecular feature were log_{10}-transformed response ratios relative to the median across all samples. Subsets of each analytical technology, including both concatenated flesh and rind profiles, were 0.1 - 0.9 quantile range-normalized to harmonize the diverse linear response ranges of each profiling technology. Missing data were replaced by zero to give enhanced weight to subspecies- or group-specific molecular features rather than inferring values by more elaborate means, e.g., [72,73]. The resulting data set of > 80,000 molecular features was analyzed either in combined mode or divided into subsets.

Independent component analysis (ICA) of 52 samples included 51 accessions (Table 1) and a second independent sample of the Duda´im accession. ICA was applied to the first three components of a preceding principal component analysis and was performed using the MetaGeneAlyse web-application, https://metagenealyse.mpimp-golm.mpg.de/ [28]. ICA scores values were plotted. The Multi Experiment Viewer software v 4.9 [74] generated the covariance matrix of the combined data set and a support hierarchical cluster analysis (support HCA) using covariance distance, complete linkage, and bootstrapping by 1000 iterations. Bootstrap values range from 1–100 and represent the frequency of tree-node occurrence across the iterations.

In order to compare accession clustering based on genetics with the ones based on compositional distances for each tissue and each analytical strategy separately, Euclidian distances were calculated using the corresponding metabolome or ionone sub-dataset with zeros for undetected metabolomic features for the 52 accession samples with Multi Experiment viewer v 4.9. Then, accession sample clustering based on compositional data was performed under R (https://www.R-project.org/) using the Euclidian distance matrices calculated with Multi Experiment viewer and with the "complete" method for sample hierarchical clustering. In order to compare distances between accessions based on genetics with the ones based on composition, each compositional Euclidian distance matrix was restricted to the 44 accessions having both molecular and fruit metabolome or elemental data. Then, a Mantel test was done to measure the association between the molecular distance matrix and each metabolomic or elemental distance matrix by calculating the Pearson correlation between these two matrices, with its *p*-value calculated with 10000 Monte Carlo simulations using XLSTAT (v2019, Addinsoft, Paris, France). When the *p*-value was below 0.001, the two matrices were considered as significantly correlated.

To select annotated metabolic features that are relevant to classifying the *C. melo* accessions, random forest (RF) analysis was performed using the subset of annotated molecular features [27] from the combined data set (Table S4). Feature selection by RF technology was as described earlier [29] using hyperparameter optimization proposed by [75]. The R-package randomForest v 4.6–14 [76,77] was used for training classification trees. The hyperparameter tuning was performed using the tuneranger function from the tuneRanger package V 0.2 [75]. The optimized parameters were mtry, node size, and sample size. Other non-tuned RF parameter settings were ntree = 10,000, importance = TRUE, replace = TRUE. The decision tree was created by the functions rpart and rpart.plot from the rpart V 4.1–13 and rpart.plot V 2.2.0 packages, respectively, using the method "class". We selected the top 20 molecular features according to the mean decrease of accuracy from 10 repeated RF analyses. A decision tree was created, limiting the tree to only those branches that contain all members of a class. The averaged mean decrease of accuracy and mean decrease of the Gini index were correlated by Pearson's correlation coefficient (r^2) 0.931 assuming linear association.

All data were visualized by the respective R-packages and the Multi Experiment Viewer [74] in combination with Microsoft Excel and PowerPoint.

Supplementary Materials: The following are available online at http://www.mdpi.com/2218-1989/10/3/121/s1, Figure S1: Phylogenetic tree and alignments based on direct PCR sequencing of melon accessions not included in GBS, Figure S2: Fruit quality global measurements of 52 accession samples, Figure S3: Hierarchical clustering analysis of 52 accession samples performed per analytical platform for fruit flesh, Figure S4: Hierarchical clustering analysis of 52 accession samples performed per analytical platform for fruit rind, Table S1: Description of the melon accessions used in this study, Table S2: Genes and primers used for PCR-based genome sequencing and polymorphism identification, Table S3: Genetic distance matrix of the 44 GBS-sequenced accessions, Table S4: List of 605 provisionally-annotated molecular features from the combined data set with assignment of variable-importance obtained by RF technology.

Author Contributions: Conceptualization, A.A.S. and R.D.H.; sample preparation and biochemical analyses, A.A., J.W.A., B.B., J.B., M.H.B., C.D., A.E., R.G., T.H.H., D.R., M.M., S.M. (Sagit Meir), S.M. (Sonia Miller), R.M., I.R., J.K.S., R.C.H.d.V., J.L.W., E.Y.; genetic analyses, A.G., N.K., E.O., H.S.P., E.L., U.S., Y.T., Y.B., A.F., S.B.-D., G.T. data curation and mining, F.B., A.E., D.J., J.K., A.M.; writing and original draft preparation, A.E., R.D.H., J.K., A.M., H.S.P., A.A.S.; manuscript reviewing and editing, all authors; funding acquisition, R.D.H. All authors have read and agreed to the submitted version of the manuscript.

Funding: The data presented in this paper were generated within the META-PHOR project financed by the European Union Framework 6 programme [2006-FOODCT-036220].

Acknowledgments: A.E. and J.K. acknowledge technical and laboratory support of the GC-MS analyses by Ines Fehrle.

Conflicts of Interest: The authors declare no conflict of interest.

References

1. Pitrat, M. Melon. In *Handbook of Plant Breeding: Vegetable*; Prohens, J., Nuez, F., Eds.; Springer Science: New York, NY, USA, 2008; Volume I, pp. 283–316.
2. Burger, Y.; Paris, H.S.; Cohen, R.; Katzir, N.; Tadmor, Y.; Lewinsohn, E.; Schaffer, A.A. Genetic diversity of *Cucumis melo*. *Hortic. Rev.* **2010**, *36*, 165–198.
3. Pech, J.-C.; Bouzayen, M.; Latche, A. Climacteric fruit ripening: Ethylene-dependent and independent regulation of ripening pathways in melon fruit. *Plant Sci.* **2008**, *175*, 114–120. [CrossRef]
4. Kirkbride, J.H., Jr. *Biosystematic Monograph of the Genus Cucumis [Cucurbitaceae]*; Parkway Publishers: Boone, NC, USA, 1993.
5. Robinson, R.W.; Decker-Walters, D.S. *Cucurbits*; CAB International: Wallingford, Oxon, UK, 1997; pp. 65–70.
6. Pitrat, M.; Hanelt, P.; Hammer, K. Some comments on infraspecific classification of cultivars of melon. In *Proceedings of Cucurbitaceae 2000: The 7th Eucarpia Meeting on Cucurbit Genetics and Breeding, Ma'ale Ha Hamisha, Israel, 19–23 March 2000*; Acta Horticulturae No. 510; Katzir, N., Paris, H.S., Eds.; ISHS: Leuven, Belgium, 2000; pp. 29–36.
7. Decker-Walters, D.S.; Chung, S.M.; Staub, J.E.; Quemada, H.D.; López-Sesé, A.I. The origin and genetic affinities of wild populations of melon [*Cucumis melo*, Cucurbitaceae] in North America. *Plant Syst. Evol.* **2002**, *233*, 183–197. [CrossRef]
8. Zhao, G.; Lian, Q.; Zhang, Z.; Fu, Q.; He, Y.; Ma, S.; Ruggieri, V.; Monforte, A.J.; Wang, P.; Julca, I.; et al. A comprehensive genome variation map of melon identifies multiple domestication events and loci influencing agronomic traits. *Nat. Genet.* **2019**, *51*, 1607–1615. [CrossRef]
9. Hadfield, K.A.; Rose, J.K.C.; Bennett, A.B. The respiratory climacteric is present in Charentais [*Cucumis melo* cv. Reticulatus F1 Alpha] melons ripened on or off the plant. *J. Exp. Bot.* **1995**, *46*, 1923–1925. [CrossRef]
10. Aubert, C.; Bourger, N. Investigation of volatiles in Charentais cantaloupe melons [*Cucumis melo* Var. *cantalupensis]. Characterization of aroma constituents in some cultivars. J. Agric. Food Chem.* **2004**, *52*, 4522–4528. [CrossRef]
11. Renner, S.S.; Schaefer, H.; Kocyan, A. Phylogenetics of *Cucumis* [Cucurbitaceae]: Cucumber [*C. sativus*] belongs in an Asian/Australian clade far from melon [*C. melo*]. *BMC Evol. Biol.* **2007**, *7*, 58–69. [CrossRef]
12. Sebastian, P.; Schaefer, H.; Telford, I.R.H.; Renner, S.S. Cucumber [*Cucumis sativus*] and melon [*C. melo*] have numerous wild relatives in Asia and Australia, and the sister species of melon is from Australia. *Proc. Natl. Acad. Sci. USA* **2010**, *107*, 14269–14273. [CrossRef]
13. Paris, H.S.; Amar, Z.; Lev, E. Medieval emergence of sweet melons, *Cucumis melo* [Cucurbitaceae]. *Ann. Bot.* **2012**, *110*, 23–33. [CrossRef]
14. Paris, H.S.; Amar, Z.; Lev, E. Medieval history of the duda'im melon [*Cucumis melo*, Cucurbitaceae]. *Econ. Bot.* **2012**, *66*, 276–284. [CrossRef]
15. Janick, J.; Paris, H.S.; Parrish, D.C. The cucurbits of Mediterranean antiquity: Identification of taxa from ancient images and descriptions. *Ann. Bot.* **2007**, *100*, 1441–1457. [CrossRef]

16. Paris, H.S. Semitic-language records of snake melons [*Cucumis melo*, Cucurbitaceae] in the medieval period and the "piqqus" of the "faqqous". *Genet. Resour. Crop Evol.* **2012**, *59*, 31–38. [CrossRef]
17. Deleu, W.; Esteras, C.; Roig, C.; González-To, M.; Fernández-Silva, I.; Gonzalez-Ibeas, D.; Blanca, J.; Aranda, M.A.; Arús, P.; Nuez, F.; et al. A set of EST-SNPs for map saturation and cultivar identification in melon. *BMC Plant Biol.* **2009**, *9*, 90. [CrossRef]
18. Blanca, J.M.; Cañizares, J.; Ziarsolo, P.; Esteras, C.; Mir, G.; Nuez, F.; Garcia-Mas, J.; Picó, M.B. Melon transcriptome characterization: Simple sequence repeats and single nucleotide polymorphisms discovery for high throughput genotyping across the species. *Plant Genome* **2011**, *4*, 118–131. [CrossRef]
19. Endl, J.; Achigan-Dako, E.G.; Pandey, A.K.; Monforte, A.J.; Pico, B.; Schaefer, H. Repeated domestication of melon [*Cucumis melo*] in Africa and Asia and a new close relative from India. *Am. J. Bot.* **2018**, *105*, 1662–1671. [CrossRef]
20. Leida, C.; Moser, C.; Esteras, C.; Sulpice, R.; Lunn, J.E.; de Langen, F.; Monforte, A.J.; Picó, B. Variability of candidate genes, genetic structure and association with sugar accumulation and climacteric behavior in a broad germplasm collection of melon [*Cucumis melo* L.]. *BMC Genet.* **2015**, *16*, 28. [CrossRef]
21. McNair, J. Angiosperm phylogeny on a Chemical Basis. *Bull. Torrey Bot. Club* **1935**, *62*, 515–532. [CrossRef]
22. Waterman, P.G. The current status of chemical systematics. *Phytochemistry* **2007**, *68*, 2896–2903. [CrossRef]
23. van Treuren, R.; van Eekelen, H.D.; Wehrens, R.; de Vos, R.C. Metabolite variation in the lettuce gene pool: Towards healthier crop varieties and food. *Metabolomics* **2018**, *14*, 146. [CrossRef]
24. Bernillon, S.; Biais, B.; Deborde, C.; Maucourt, M.; Cabasson, C.; Gibon, Y.; Moing, A. Metabolomic and elemental profiling of melon fruit quality as affected by genotype and environment. *Metabolomics* **2013**, *9*, 57–77. [CrossRef]
25. Freilich, S.; Lev, S.; Gonda, I.; Reuveni, E.; Portnoy, V.; Oren, E.; Lohse, M.; Galpaz, N.; Bar, E.; Tzuri, G.; et al. Systems approach for exploring the intricate associations between sweetness, color and aroma in melon fruits. *BMC Plant Biol.* **2015**, *15*, 71. [CrossRef] [PubMed]
26. Esteras, C.; Rambla, J.L.; Sánchez, G.; López-Gresa, M.P.; González-Mas, M.C.; Fernández-Trujillo, J.P.; Bellés, J.M.; Granell, A.; Picó, M.B. Fruit flesh volatile and carotenoid profile analysis within the *Cucumis melo* L. species reveals unexploited variability for future genetic breeding. *J. Sci. Food Agric.* **2018**, *98*, 3915–3925. [CrossRef] [PubMed]
27. Moing, A.; Aharoni, A.; Biais, B.; Rogachev, I.; Meir, S.; Brodsky, L.; Allwood, J.W.; Erban, A.; Dunn, W.B.; Kay, L.; et al. Extensive metabolic cross talk in melon fruit revealed by spatial and developmental combinatorial metabolomics. *New Phytol.* **2011**, *190*, 683–696. [CrossRef]
28. Daub, C.O.; Kloska, S.; Selbig, J. MetaGeneAlyse: Analysis of integrated transcriptional and metabolite data. *Bioinformatics* **2003**, *19*, 2332–2333. [CrossRef]
29. Erban, A.; Fehrle, I.; Martinez-Seidel, F.; Brigante, F.; Lucini Más, A.; Baroni, V.; Wunderlin, D.; Kopka, J. Discovery of food identity markers by metabolomics and machine learning technology. *Sci. Rep.* **2019**, *9*, 9697. [CrossRef]
30. Wahyuni, Y.; Ballester, A.R.; Tikunov, Y.; de Vos, R.C.; Pelgrom, K.T.; Maharijaya, A.; Sudarmonowati, E.; Bino, R.J.; Bovy, A.G. Metabolomics and molecular marker analysis to explore pepper [*Capsicum* sp.] biodiversity. *Metabolomics* **2013**, *9*, 130–144. [CrossRef]
31. De Vos, R.C.H.; Schipper, B.; Hall, R.D. High-performance liquid chromatography–mass spectrometry analysis of plant metabolites in *Brassicaceae*. In *Plant Metabolomics*; Methods in Molecular Biology [Methods and Protocols]; Hardy, N., Hall, R., Eds.; Humana Press: Totowa, NJ, USA, 2011; Volume 860.
32. Liu, C.; Jiang, D.; Cheng, Y.; Deng, X.; Chen, F.; Fang, L.; Ma, Z.; Xu, J. Chemotaxonomic study of *Citrus, Poncirus* and *Fortunella* genotypes based on peel oil volatile compounds—Deciphering the genetic origin of Mangshanyegan [*Citrus nobilis* Lauriro]. *PLoS ONE* **2013**, *8*, e58411. [CrossRef]
33. Obando-Ulloa, J.M.; Moreno, E.; García-Mas, J.; Nicolai, B.; Lammertyn, J.; Monforte, A.J.; Fernández-Trujillo, J.P. Climacteric or non-climacteric behavior in melon fruit: 1. Aroma volatiles. *Postharvest Biol. Technol.* **2008**, *49*, 27–37. [CrossRef]
34. Jacquin, H. *Monographie Complète Du Melon*; Rousselon: Paris, France, 1832.
35. Paris, H.S. A proposed subspecific classification for *Cucurbita pepo*. *Phytologia* **1986**, *61*, 133–138.
36. Gong, L.; Paris, H.S.; Nee, M.H.; Stift, G.; Pachner, M.; Vollmann, J.; Lelley, T. Genetic relationships and evolution in *Cucurbita pepo* [pumpkin, squash, gourd] as revealed by simple sequence repeat polymorphisms. *Theor. Appl. Genet.* **2012**, *124*, 875–891. [CrossRef]

37. Gur, A.; Tzuri, G.; Meir, A.; Sa'ar, U.; Portnoy, V.; Katzir, N.; Schaffer, A.A.; Li, L.; Burger, J.; Tadmor, Y. Genome-wide linkage-disequilibrium mapping to the candidate gene level in melon [*Cucumis melo*]. *Sci. Rep.* **2017**, *7*, 9770. [CrossRef]
38. Cohen, S.; Itkin, M.; Yeselson, Y.; Tzuri, G.; Portnoy, V.; Harel-Baja, R.; Lev, S.; Sa'ar, U.; Davidovitz-Rikanati, R.; Baranes, N.; et al. The PH gene determines fruit acidity and contributes to the evolution of sweet melons. *Nat. Commun.* **2014**, *5*, 4026. [CrossRef]
39. Harel-Beja, R.; Tzuri, G.; Portnoy, V.; Lotan-Pompan, M.; Lev, S.; Cohen, S.; Dai, N.; Yeselson, L.; Meir, A.; Libhaber, S.E.; et al. A genetic map of melon highly enriched with fruit quality QTLs and EST markers. including sugar and carotenoid metabolism genes. *Theor. Appl. Genet.* **2010**, *121*, 511–533. [CrossRef]
40. Tzuri, G.; Zhou, X.; Chayut, N.; Yuan, H.; Portnoy, V.; Meir, A.; Sa'ar, U.; Baumkoler, F.; Mazourek, M.; Lewinsohn, E.; et al. A 'golden' SNP in CmOr governs the fruit flesh color of melon [*Cucumis melo*]. *Plant J.* **2015**, *82*, 267–279. [CrossRef]
41. Diaz, A.; Fergany, M.; Formisano, G.; Ziarsolo, P.; Blanca, J.; Fei, Z.; Staub, J.E.; Zalapa, J.E.; Cuevas, H.E.; Dace, G.; et al. A consensus linkage map for molecular markers and quantitative trait loci associated with economically important traits in melon [*Cucumis melo* L.]. *BMC Plant Biol.* **2011**, *11*, 111. [CrossRef]
42. Oren, E.; Tzuri, G.; Dafna, A.; Meir, A.; Kumar, R.; Katzir, N.; Elkind, Y.; Freilich, S.; Schaffer, A.A.; Tadmor, Y.; et al. High-density NGS-based map construction and genetic dissection of fruit shape and rind netting in *Cucumis melo*. *Theor. Appl. Genet.* **2020**, in press. [CrossRef] [PubMed]
43. Oren, E.; Tzuri, G.; Vexler, L.; Dafna, A.; Meir, A.; Faigenboim, A.; Kenigswald, M.; Portnoy, V.; Schaffer, A.A.; Levi, A.; et al. The multi-allelic APRR2 gene is associated with fruit pigment accumulation in melon and watermelon. *J. Exp. Bot.* **2019**, *70*, 3781–3794. [CrossRef]
44. Galpaz, N.; Gonda, I.; Shem-Tov, D.; Barad, O.; Tzuri, G.; Lev, S.; Fei, Z.; Xu, Y.; Mao, L.; Jiao, C.; et al. Deciphering genetic factors that determine melon fruit-quality traits using RNA-Seq-based high-resolution QTL and eQTL mapping. *Plant J.* **2018**, *94*, 169–191. [CrossRef]
45. Chayut, N.; Yuan, H.; Ohali, S.; Meir, A.; Sa'ar, U.; Tzuri, G.; Zheng, Y.; Mazourek, M.; Gepstein, S.; Zhou, X.; et al. Distinct mechanisms of the ORANGE protein in controlling carotenoid flux. *Plant Physiol.* **2017**, *173*, 376–389. [CrossRef]
46. Périn, C.; Gomez-Jimenez, M.; Hagen, L.; Dogimont, C.; Pech, J.C.; Latché, A.; Pitrat, M.; Lelièvre, J.M. Molecular and genetic characterization of a non-climacteric phenotype in melon reveals two loci conferring altered ethylene response in fruit. *Plant Physiol.* **2002**, *129*, 300–309. [CrossRef]
47. Ríos, P.; Argyris, J.; Vegas, J.; Leida, C.; Kenigswald, M.; Tzuri, G.; Troadec, C.; Bendahmane, A.; Katzir, N.; Picó, B.; et al. ETHQV6.3 is involved in melon climacteric fruit ripening and is encoded by a NAC domain transcription factor. *Plant J.* **2017**, *91*, 671–683. [CrossRef]
48. Eduardo, I.; Arús, P.; Monforte, A.J. Development of a genomic library of near isogenic lines [NILs] in melon [*Cucumis melo* L.] from the exotic accession PI161375. *Theor. Appl. Genet.* **2005**, *112*, 139–148. [CrossRef]
49. Vegas, J.; Garcia-Mas, J.; Monforte, A.J. Interaction between QTLs induces an advance in ethylene biosynthesis during melon fruit ripening. *Theor. Appl. Genet.* **2013**, *126*, 1531–1544. [CrossRef]
50. Saladié, M.; Cañizares, J.; Phillips, M.A.; Rodriguez-Concepcion, M.; Larrigaudière, C.; Gibon, Y.; Stitt, M.; Lunn, J.E.; Garcia-Mas, J. Comparative transcriptional profiling analysis of developing melon [*Cucumis melo* L.] fruit from climacteric and non-climacteric varieties. *BMC Genom.* **2015**, *16*, 440. [CrossRef]
51. Buttery, R.G.; Seifert, R.M.; Ling, L.C.; Soderstrom, E.L.; Ogawa, J.M.; Turnbaugh, J.G. Additional aroma components of honeydew melon. *J. Agric. Food Chem.* **1982**, *30*, 1208–1211. [CrossRef]
52. Gross, K.C.; Sams, C.E. Changes in cell wall neutral sugar composition during fruit ripening: A species survey. *Phytochemistry* **1984**, *23*, 2457–2461. [CrossRef]
53. Jetter, R.; Kunst, L.; Samuels, A.L. Composition of plant cuticular waxes. *Biol. Plant Cuticle* **2008**, *23*, 145–181.
54. Cohen, H.; Dong, Y.; Szymanski, J.; Lashbrooke, J.; Meir, S.; Almekias-Siegl, E.; Zeisler-Diehl, V.V.; Schreiber, L.; Aharoni, A. A multilevel study of melon fruit reticulation provides insight into skin ligno-suberization hallmarks. *Plant Physiol.* **2019**, *179*, 1486–1501. [CrossRef]
55. Dai, N.; Cohen, S.; Portnoy, V.; Tzuri, G.; Harel-Beja, R.; Pompan-Lotan, M.; Carmi, N.; Zhang, G.; Diber, A.; Pollock, S.; et al. Metabolism of soluble sugars in developing melon fruit: A global transcriptional view of the metabolic transition to sucrose accumulation. *Plant Mol. Biol.* **2011**, *76*, 1–18. [CrossRef]
56. Fang, C.; Luo, J. Metabolic GWAS-based dissection of genetic bases underlying the diversity of plant metabolism. *Plant J.* **2019**, *97*, 91–100. [CrossRef]

57. Elshire, R.J.; Glaubitz, J.C.; Sun, Q.; Poland, J.A.; Kawamoto, K.; Buckler, E.S.; Mitchell, S.E. A robust, simple genotyping-by-sequencing [GBS] approach for high diversity species. *PLoS ONE* **2011**, *6*, 1–10. [CrossRef]
58. Garcia-Mas, J.; Benjak, A.; Sanseverino, W.; Bourgeois, M.; Mir, G.; González, V.M.; Hénaff, E.; Câmara, F.; Cozzuto, L.; Lowy, E.; et al. The genome of melon [*Cucumis melo* L.]. *Proc. Natl. Acad. Sci. USA* **2012**, *109*, 11872–11877. [CrossRef]
59. Glaubitz, J.C.; Casstevens, T.M.; Lu, F.; Harriman, J.; Elshire, R.J.; Sun, Q.; Buckler, E.S. TASSEL-GBS: A high capacity genotyping by sequencing analysis pipeline. *PLoS ONE* **2014**, *9*, e90346. [CrossRef]
60. Bradbury, P.J.; Zhang, Z.; Kroon, D.E.; Casstevens, T.M.; Ramdoss, Y.; Buckler, E.S. TASSEL: Software for association mapping of complex traits in diverse samples. *Bioinformatics* **2007**, *23*, 2633–2635. [CrossRef]
61. Biais, B.; Allwood, J.W.; Deborde, C.; Xu, Y.; Mickael, M.; Beauvoit, B.; Dunn, W.B.; Jacob, D.; Goodacre, R.; Rolin, D.; et al. ^{1}H NMR, GC−EI-TOFMS, and data set correlation for fruit metabolomics: Application to spatial metabolite analysis in melon. *Anal. Chem.* **2009**, *81*, 2884–2894. [CrossRef]
62. Ward, J.L.; Forcat, S.; Beckmann, M.; Bennett, M.; Miller, S.J.; Baker, J.M.; Hawkins, N.D.; Vermeer, C.P.; Lu, C.; Lin, W.; et al. The metabolic transition during disease following infection of *Arabidopsis thaliana* by *Pseudomonas syringae* pv. *tomato*. *Plant J.* **2010**, *63*, 443–457. [CrossRef]
63. Allwood, J.W.; Erban, A.; de Koning, S.; Dunn, W.B.; Luedemann, A.; Lommen, A.; Kay, L.; Löscher, R.; Kopka, J.; Goodacre, R. Inter-laboratory reproducibility of fast gas chromatography–electron impact–time of flight mass spectrometry [GC–EI–TOF/MS] based plant metabolomics. *Metabolomics* **2009**, *5*, 479–496. [CrossRef]
64. Luedemann, A.; Strassburg, K.; Erban, A.; Kopka, J. TagFinder for the quantitative analysis of gas chromatography-mass spectrometry [GC-MS] based metabolite profiling experiments. *Bioinformatics* **2008**, *24*, 732–737. [CrossRef]
65. Strehmel, N.; Hummel, J.; Erban, A.; Strassburg, K.; Kopka, J. Retention index thresholds for compound matching in GC-MS metabolite profiling. *J. Chromatogr. B* **2008**, *871*, 182–190. [CrossRef]
66. Kopka, J.; Schauer, N.; Krueger, S.; Birkemeyer, C.; Usadel, B.; Bergmueller, E.; Doermann, P.; Weckwerth, W.; Gibon, Y.; Stitt, M.; et al. GMD@CSB.DB: The Golm Metabolome Database. *Bioinformatics* **2005**, *21*, 1635–1638. [CrossRef]
67. Verhoeven, H.A.; Jonker, H.; De Vos, R.C.H.; Hall, R.D. Solid Phase Micro-Extraction GC–MS Analysis of Natural Volatile components in Melon and Rice. In *Book Humana, Methods for Plant Metabolomics*; Hardy, N.G., Hall, R.D., Eds.; Springer–Humana: London, UK, 2012; pp. 85–99.
68. Lommen, A. MetAlign: Interface-driven, versatile metabolomics tool for hyphenated full-scan mass spectrometry data preprocessing. *Anal. Chem.* **2009**, *81*, 3079–3086. [CrossRef]
69. Tikunov, Y.M.; Laptenok, S.; Hall, R.D.; Bovy, A.; de Vos, R.C.H. MSClust: A tool for unsupervised mass spectra extraction of chromatography-mass spectrometry ion-wise aligned data. *Metabolomics* **2012**, *8*, 714–718. [CrossRef]
70. Moussaieff, A.; Rogachev, I.; Brodsky, L.; Malitsky, S.; Toal, T.W.; Belcher, H.; Yativ, M.; Brady, S.M.; Benfey, P.N.; Aharoni, A. High-resolution metabolic mapping of cell types in plant roots. *Proc. Natl. Acad. Sci. USA* **2013**, *110*, E1232–E1241. [CrossRef] [PubMed]
71. Smith, C.A.; Want, E.J.; O'Maille, G.; Abagyan, R.; Siuzdak, G. XCMS: Processing mass spectrometry data for metabolite profiling using nonlinear peak alignment, matching, and identification. *Anal. Chem.* **2006**, *78*, 779–787. [CrossRef]
72. Scholz, M.; Kaplan, F.; Guy, C.L.; Kopka, J.; Selbig, J. Non-linear PCA: A missing data approach. *Bioinformatics* **2005**, *21*, 3887–3895. [CrossRef]
73. Wehrens, R.; Hageman, J.A.; van Eeuwijk, F.; Kooke, R.; Flood, P.J.; Wijnker, E.; Keurentjes, J.J.; Lommen, A.; van Eekelen, H.D.; Hall, R.D.; et al. Improved batch correction in untargeted MS-based metabolomics. *Metabolomics* **2016**, *12*, 1–12. [CrossRef]
74. Saeed, A.I.; Sharov, V.; White, J.; Li, J.; Liang, W.; Bhagabati, N.; Braisted, J.; Klapa, M.; Currier, T.; Thiagarajan, M.; et al. TM4: A free, open-source system for microarray data management and analysis. *BioTechniques* **2003**, *34*, 374–378. [CrossRef]

75. Probst, P.; Wright, M.; Boulesteix, A.-L. Hyperparameters and tuning strategies for random forest. *WIREs Data Min. Knowl. Discov.* **2019**, *9*, e1301. [CrossRef]
76. Breiman, L. Random forests. *Mach. Learn.* **2001**, *45*, 5–32. [CrossRef]
77. Liaw, A.; Wiener, M. Classification and regression by randomForest. *R. News* **2002**, *2*, 18–22.

metabolites

MDPI

Article

Color Mutations Alter the Biochemical Composition in the San Marzano Tomato Fruit

Gabriella Dono [1], Jose Luis Rambla [2,3], Sarah Frusciante [4], Antonio Granell [2], Gianfranco Diretto [4] and Andrea Mazzucato [1,*]

1 DAFNE Dept. of Agriculture and Forest Sciences, University of Tuscia, Via S.C. de Lellis snc, 01100 Viterbo, Italy; gabriella.dono91@hotmail.it
2 IBMCP Institute for Plant Molecular and Cell Biology (CSIC-UPV), Carrer de l'Enginyer Fausto Elio, s/n, 46022 Valencia, Spain; jrambla@ibmcp.upv.es (J.L.R.); agranell@ibmcp.upv.es (A.G.)
3 Department of Agricultural and Environmental Sciences, Jaume I University, Av. Vicent Sos Baynat, s/n. 12071 Castellòn de la Plana, Spain
4 ENEA, Casaccia Research Center, Via Anguillarese 301, S. Maria di Galeria, 00123 Rome, Italy; sarah.frusciante@enea.it (S.F.); gianfranco.diretto@enea.it (G.D.)
* Correspondence: mazz@unitus.it; Tel.: +39-0761-357370

Received: 14 February 2020; Accepted: 12 March 2020; Published: 15 March 2020

Abstract: San Marzano (SM) is a traditional Italian landrace characterized by red elongated fruits, originating in the province of Naples (Italy) and cultivated worldwide. Three mutations, *yellow flesh* (*r*), *green flesh* (*gf*) and *colorless fruit epidermis* (*y*) were introduced into SM by backcross and the resulting introgression lines (ILs) produced the expected yellow, brown and pink fruit variants. In addition, ILs carrying double combinations of those mutations were obtained. The six ILs plus the SM reference were analyzed for volatile (VOC), non-polar (NP) and polar (P) metabolites. Sixty-eight VOCs were identified, and several differences evidenced in the ILs; overall *gf* showed epistasis over *r* and *y* and *r* over *y*. Analysis of the NP component identified 54 metabolites; variation in early carotenoids (up to lycopene) and chlorophylls characterized respectively the ILs containing *r* and *gf*. In addition, compounds belonging to the quinone and xanthophyll classes were present in genotypes carrying the *r* mutation at levels higher than SM. Finally, the analysis of 129 P metabolites evidenced different levels of vitamins, amino acids, lipids and phenylpropanoids in the ILs. A correlation network approach was used to investigate metabolite–metabolite relationships in the mutant lines. Altogether these differences potentially modified the hedonistic and nutritional value of the berry. In summary, single and combined mutations in *gf*, *r* and *y* generated interesting visual and compositional diversity in the SM landrace, while maintaining its original typology.

Keywords: fruit pigmentation; introgression lines; metabolomics; mass spectrometry; San Marzano landrace; tomato

1. Introduction

In tomato (*Solanum lycopersicum* L.), fruit color is one of the most important traits affecting consumer liking, and it is the result of combined effect of carotenoids, flavonoids and eventually chlorophylls. The red color of ripe fruit comes from the accumulation of all-trans-lycopene; mutants affected in the carotenoid pathway have an altered carotenoid composition, resulting in different fruit colors [1,2]. Besides carotenoids, flavonoids play a role in determining the color of tomato fruit, particularly at the epidermal level [3]. One of the most abundant flavonoids in tomato fruit peel is the yellow-colored naringenin chalcone, often preceding and paralleling the production of lycopene in the pericarp [4]. Lastly, chlorophylls can eventually have a role in defining the color of the fruit; although they are normally degraded during ripening, "stay green" (SGR) mutants exist, that maintain important chlorophyll levels in the ripe fruit. Color variations do not only change fruit pigmentation, but they also affect the entire set of tomato fruit attributes, adversely or positively impacting the organoleptic and nutritional properties. Indeed, ripening involves a number of physiological processes ranging from the chlorophyll breakdown to the consecutive carotenoid accumulation, but also of other compounds belonging to the secondary metabolism, which play functional roles in plants [5], and have a potential in preventing diseases and promoting health in animals and humans [6,7]. Among them, acids, tocochromanols, quinones, fatty acids, sugars and polyols, as well as glycoalkaloids vary during ripening, thus modifying the nutritional value and the antioxidant activity of the fruit [8]. Three tomato mutants, *yellow flesh* (*r*), *colorless fruit epidermis* (*y*) and *green flesh* (*gf*), which are representative of the main classes of color pigments, were chosen to be studied at metabolite level, with the aim of assessing the impact of these mutations on the fruit metabolome with emphasis on compounds that co-participate in defining fruit qualities.

Yellow-fruited tomatoes have been documented since the first introduction in Europe [4,9]. The yellow color is due to the *r* mutation, represented by loss-of-function alleles of phytoene synthase 1 (*Psy1*; *Psy1* catalyzes the first rate-limiting step in the carotenoid pathway, the condensation of two molecules of geranylgeranyl diphosphate in phytoene [10]), which results in the inhibition of the whole carotenoid biosynthesis [11]. Recently, *r* tomatoes have been meeting with an increasing success, for the color novelty and the peculiar organoleptic qualities [12].

The *y* fruit mutant was originally described as a monogenic recessive variant leading to the formation of a colorless fruit peel [13]. The mutation, mapped on the short arm of Chr1 [14], involves the *SlMYB12* transcription factor, causing the lack of naringenin chalcone, one of the major flavonoids in tomato fruit peel, which gives the yellow color and has been proposed to influence the characteristics and function of the cuticular layer [15,16]. The pink *y*-type fruit mutation has been identified in numerous cultivated varieties that are highly consumed in Asian countries.

Fruits of *gf* tomato mutant were described for their characteristic muddy brown color, resulting from the accumulation of lycopene coupled with the heterochronic presence of chlorophyll in the ripen fruit due to a lack of chlorophyll degradation [17]. In *S. lycopersicum*, the *Gf* locus maps on the long arm of Chr8. Further studies indicated that *gf* is a member of the SGR gene family, *SlSGR1*, a Mg-decheletase gene needed for chlorophyll catabolism [18,19]. Nowadays, many cultivated tomatoes, heirloom varieties but also modern hybrids, exhibiting the *gf* phenotype are commercially available; indeed, these varieties are appreciated and have been reported to be superior for taste-related compounds [20,21].

San Marzano (from now on referred to as SM) is a traditional variety grown in the region of Naples, Italy; it is considered an important model for fruit quality parameters, because of its intense and uniform red color, which revealed peculiar sensory profiles. Indeed SM fruits are covered by the leaves, which promotes the lycopene accumulation, but at the same time reduces the sugar content, giving birth to its typical bittersweet flavor; this is also mainly due to the ratio between the sugar content, where fructose and galactose are the main carbohydrates, and organic acids, mostly citric acid, a major factor in determining sourness, but also flavor intensity [22–24]. The effect of San Marzano extracts has also been the object of studies aimed at assessing nutraceutical properties [25].

The *r*, *y* and *gf* mutations, together with other variants affecting the tomato fruit phenotype, have been introgressed by backcrossing into the common genetic background SM; the single mutant lines have also been used to express two variants in double combinations [26].

In this work, we analyzed the effect of introducing color mutations in the metabolite complement of the SM fruit. Namely we compared a comprehensive set of fruit volatile compounds (VOCs), including those involved in flavor, and both non-polar (NP) and polar (P) specialized metabolites, including those involved in health properties, in the *r*, *y* and *gf* single mutant lines and their respective double mutants, to the ripe fruit of the SM genetic background. The double mutants expanded the effect on metabolism of the introduced single mutations by revealing additional additive or epistatic effects that could be further exploited for the improvement and diversification of the SM landrace by introducing innovation, while maintaining the characteristic SM typology.

2. Materials and Methods

2.1. Plant Material and Growth Conditions

Three single mutant lines harboring *r*, *y* and *gf* and the double mutants representing all the possible combinations in the SM background were studied in comparison with the corresponding wild-type (Table 1). Details on the backcross scheme used to obtain these introgression lines, as well as the growth conditions used were reported before [26]. Briefly, eight plants per accession were transplanted and cultivated in twin rows in an unheated tunnel following standard cultural practices for indeterminate tomatoes, using tutors and weekly removal of lateral shoots. Daily temperature was controlled by a ventilation system and plants were irrigated through a drop system. The trial was repeated with identical materials and methods for two consecutive years (2017 and 2018).

Table 1. Extended names, genetic symbols and description of the berry color of the seven studied genotypes, including San Marzano (SM), three single mutant and their respective double mutants.

Class of Material	Class of Variation	Name	Genetic Symbol	Fruit Color
Wild-type	-	San Marzano	SM	Red
San Marzano fruit variants	Chlorophyll	*green flesh*	*gf*	Muddy brown
	Carotenoids	*yellow flesh*	*r*	Yellow
	Flavonoids	*colorless fruit epidermis*	*y*	Pink
	Double mutants	*yellow flesh + green flesh*	*r_gf*	Light yellow
	-	*colorless fruit epidermis + green flesh*	*y_gf*	Wine-coloured
	-	*colorless fruit epidermis + yellow flesh*	*y_r*	Green

2.2. Fruit Sampling

Before sampling, berries were visually inspected and only intact and healthy tomatoes were collected. Two biological replicates for genotype, each represented by four fully ripe berries, were harvested over a period of three days, during the first week of August, for each year. A longitudinal pericarp wedge was excised from each of the four appropriately washed berries and cut into cubes; each replica, consisting of about 30 g of fresh material, was immediately frozen in liquid nitrogen and homogenized until a fine powder was obtained. Aliquots of about 10 g of this material were freeze-dried for the analysis of non-volatile secondary metabolites. All samples, both frozen and freeze-dried, were stored at −80 °C until analysis.

2.3. Volatile Detection and Quantification

For volatile analysis, two biological replicates and two technical replicates were processed and analyzed independently for the two year experiments. Prior to the analysis, frozen fruit powder (0.5 g fresh weight) from each sample was weighed in a 15 mL vial, closed and incubated at 37 °C for 10 min. Then, 1.1 g of $CaCl_2 \cdot 2H_2O$ and 500 µL of EDTA 100 mM (pH 7.5) were added; samples were then gently shaken and sonicated for 5 min. Then,1 mL of the homogenized mixture was transferred into a 10 mL screw cap headspace vial, where volatiles were collected by head space solid-phase microextraction as previously described [27]. A 65 µm PDMS/DVB SPME fiber (Supelco Analytical, Bellefonte, PA, USA) was used for all the analysis. Pre-incubation and extraction were performed at 50 °C for 10 and 20 min respectively. Desorption was performed for 1 min at 250 °C in splitless mode. Volatile extraction and injection were performed by means of a CombiPAL autosampler (CTC Analytics AG, Zwingen, Switzerland)). Separation and detection were performed by a 6890N gas chromatograph coupled to a 5975B mass spectrometer (Agilent Technologies Inc., Waldbronn, Germany) with DB-5ms fused silica capillary column (60 m, 0.25 mm, 1 µm) (J&W Scientific, Agilent Technologies Inc., Santa Clara, CA, USA). Oven temperature conditions were 40 °C for 2 min, 5 °C/min ramp until 250 °C and then held isothermally at 250 °C for 5 min. Helium was used as carrier gas at 1.2 mL/min constant flow. Ionization was performed by electron impact (ionization energy, 70 eV; source temperature 230 °C). Data acquisition was performed in scan mode (mass range *m/z* 35–250; 6.2 scans per second). Chromatograms and spectra were recorded and processed using the Enhanced ChemStation software (Agilent). Untargeted analysis of all the compounds in the chromatogram was performed by means of the MetAlign 3 software [28]. Compounds were unequivocally identified by comparison of both mass spectra and retention time to those of pure standards (SIGMA-Aldrich, St. Louis, MO, USA). For quantification, peak areas of selected specific ions were integrated for each compound and normalized by comparison with the peak area of the same compound in a reference sample injected regularly, in order to correct for variations in detector sensitivity and fiber aging. This reference sample consisted of a homogeneous mixture of all the samples analyzed. Data for each sample were expressed as the relative content of each metabolite compared to those in the SM reference.

2.4. Non-Volatile Detection and Quantification

For both non-volatile P and NP metabolites, two biological replicates and two technical replicates, for two years, were processed and analyzed independently. Prior to analysis, 10 mg of freeze-dried fruit powder from each sample were weighed and extracted (i) with 0.75 mL cold 75% (*v*/*v*) methanol with for 0.5 mg/L formononetin as internal standard (IS) for P metabolites, as previously described [29]; and with (ii) 0.25 mL cold 100% (*v/v*), 1 mL of chloroform spiked with 25 mg/L α-tocopherol acetate as internal standard and 0.25 mL 50 mM Tris buffer (pH 7.5, containing 1 M NaCl) as described in [30]. Liquid chromatography coupled to high-resolution mass spectrometry (LC-HRMS) conditions were as previously reported for, respectively, polar [31] and non-polar [32] metabolomes.

Metabolite identification was performed by comparing chromatographic and spectral/MS properties with authentic standards, if available, and reference spectra, and based on the *m*/*z* accurate masses as found in the Pubchem database for monoisotopic masses, or in the Metabolomics Fiehn Lab Mass Spectrometry Adduct Calculator for adduct ions. Quantification of each metabolite was carried by calculating the relative contents to the formononetin (P) and α-tocopherol acetate (NP) IS levels.

2.5. Statistical and Bioinformatics Analyses

Raw data were firstly inspected and manually curated for the presence of outliers (e.g., when % st.dev./avg exceed 30%). For principal component analysis (PCA), the complete dataset after $\log_2$ transformation and including all replicates was considered. Untargeted analysis of VOC, and NP and P metabolomes was carried out as previously reported using, respectively, MetAlign and the SIEVE software (v2.2, ThermoFisher Scientific, Waltham, MA, USA; [33]).

As untargeted analysis revealed a consistent year effect, the "Gen*Year" interaction was investigated by two-way multivariate analysis of variance (MANOVA) on those metabolites than presented less than 30% missing values. The analysis was performed with the PROC GLM procedure and the MANOVA statement implemented in the SAS software package (v9.4M6, SAS® University Edition, SAS Institute Inc., North Carolina State, USA). Since "Gen*Year" interaction was found to be the less consistent source of variation, allowance was made for the existing interaction, data were mediated over the two years and all genotypes were presented with a single mean value. PCA was performed with SIMCA-P version 11 (Umetrics, Umea, Sweden) with Unit Variance normalization. The differences between each line and the SM reference were assessed using Student's *t*-test at the 5% significance level ($p < 0.05$). Graphs were elaborated with Excel (Microsoft Office 2013, Microsoft Corporation, Washington, DC, USA).

Venn diagrams were generated using Venny software (https://bioinfogp.cnb.csic.es/tools/venny/index.html, v2.1). Correlation networks were generated using average values over the two years under study, as previously described [34,35]. To better evaluate most robust metabolite-metabolite associations (e.g., significant correlations), the MCODE Cytoscape plugin was used [36].

3. Results

3.1. Untargeted Analysis of Volatile, Non-Polar and Polar Metabolites

Untargeted metabolomics aims to gather information on global metabolic profiles by retrieving, in an unsupervised way, as many metabolites are detectable in a GC-MS/LC-HRMS chromatograms. The comparison of the entire VOC, NP and P metabolome between the SM control and the six mutated lines detected the total features of their metabolic profile differences, setting the stage for a more specific targeted metabolomics study later in this paper. By using this approach, 263 VOCs, 746 NP and 110 P compounds were identified in the samples, many of which were differentially accumulated in, at least, one pairwise comparison (Table S1). For VOCs, the first two principal components explained over 51% of total variance; PC1 separated *gf* and *y* from the other lines, while PC2 kept *r* plus all the combinations harboring the *y* mutation distinct from the others (Figure 1a). In Figure 1b, PCA of the first two components for the NP compounds explained almost 53% of the total variance; SM and two *green flesh* genotypes were clearly separated from *y* mutants plus *r* by the PC1, and *r* mutants were grouped in the lower quadrants by the PC2. For P metabolites, the first two PCA components explained the over 55% of total variance (Figure 1c); PC1 kept more clearly SM and the lines carrying *r* separated from *y* mutants, and *y_r* was positioned exactly halfway between its parental lines. PC2 separated lines carrying *yellow flesh*, plus *y* from SM, which was in the upper side of the graph together with *gf* and *y_gf*. Overall, the untargeted metabolomic analysis revealed that the mutations mostly affected NP rather than VOC and P metabolome, with the *r* mutant showing the highest extent of changes, including 112 NP compounds (106 down- and 6 up-regulated; Table S1). In addition, *r* also displayed the highest epistatic attitude towards *y* (in the VOC and NP fractions) and *gf* (P metabolome), while, notably, *y* was epistatic to *gf* in the NP untargeted metabolome.

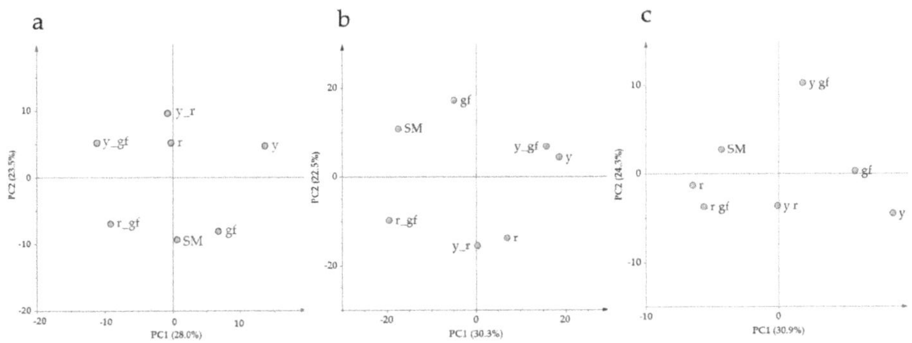

Figure 1. PC1 × PC2 score plots of the six mutated lines plus San Marzano (SM) according to relative values of 263 VOCs (**a**), 746 NP (**b**) and 110 P (**c**) metabolites measured by GC-LS and LC-HRMS. Line symbols are explained in Table 1.

3.2. Estimation of "Gen*Year" Interaction in the Quantification of Targeted Metabolites

Analysis of untargeted metabolites revealed the presence of a consistent Year effect (Figure S1). Seventy-eight VOCs, 33 NP and 69 P metabolites were independently subjected to multivariate ANOVA; the interaction was not significant for VOCs, but it was highly significant for NP and significant for P compounds (Table S2). "Gen*Year" interaction was found to be the least consistent source of variation; therefore, data were mediated over the two years and all genotypes were presented with a single mean value in targeted analyses.

3.3. Targeted Analysis of Volatile Compounds

In order to give a more specific characterization of the flavor, volatile composition of each of the mutated lines in comparison with the wild-type SM was carried out. The selected analytical strategy allowed the relative quantification of 68 VOCs unequivocally identified by both mass spectra and retention index with those of authentic standards (Table S3). Overall, eight compounds were related to benzenoids (B), ten to branched-chain amino acid-relatives (BCAA), nine to apocarotenoids (C), two to esters (E), twenty-four to fatty acids derivatives (L), four to phenylalanine derivatives (Phe), two to sulfur compounds (S) and six to monoterpenoids (T).

PCA of the volatile composition revealed that the first two components explained about the 54% of the total variance; the score plot showed the position of the double mutants, related to their parental lines, and with respect to SM (Figure 2a). Indeed, *r* and *y_r* were co-located in the same dial, in agreement with the VOC untargeted metabolome plot; conversely, *r_gf* placed halfway between *r* and *gf* according to PC1 (Figure 2a). Moreover, PC1 kept the mutants *r* and *y_r* separated from the other lines. PC2 placed *y_r* and its parental lines in the upper side of the graph, together with SM. The corresponding loading plot was able to identify groups of metabolites, often belonging to the same metabolic pathways, as apocarotenoids (C, in red) and lipids (L, in light blue), terpenoids (T, in green blue) and branched-chain amino acid derivatives (BCAA, in blue; Figure 2b). A comparison between the score and the loading plots revealed the overall compositional differences between the mutated lines.

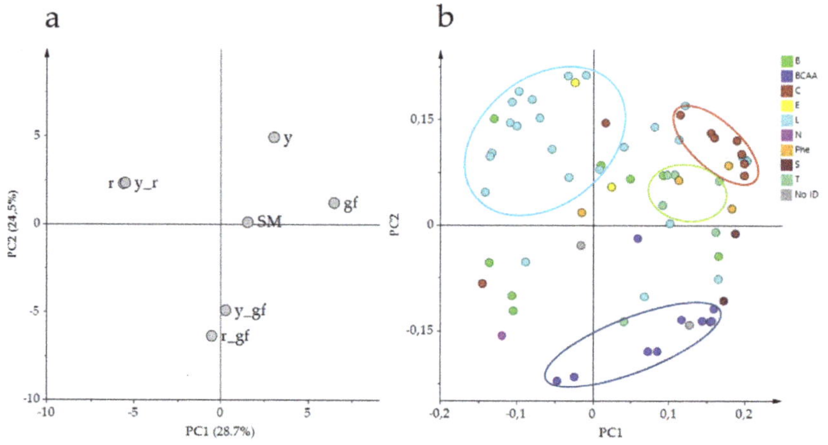

Figure 2. PCA of log2 values of 68 volatile compounds measured by a solid-phase micro-extraction gas-chromatography coupled to mass spectrometry (HS-SPME/GC-MS). (**a**) PC1 × PC2 score plot of the six mutated lines plus San Marzano (SM). (**b**) PC1 X PC2 loading plot. Line (a) and metabolite class (b) abbreviations are explained in Tables 1 and 2 respectively.

Indeed, one of the most obvious features was that all the lines harboring the *yellow flesh* mutation (*r*, *y_r* and *r_gf*) and, in a lesser extent, also the double mutant *y_gf*, were characterized by producing lower levels of volatile apocarotenoids, which was particularly dramatic in the case of some linear apocarotenoids such as 6-methyl-5-hepten-2-one (Figure 3). In the case of the mutants for *yellow flesh* this was in accordance with the scarcity of their carotenoid precursors. These lines also showed lower levels of several phenylalanine derivatives. A characteristic feature of the lines *y*, *r* and *y_r* was the high production of fatty acid derivatives together with low levels of branched-chain amino acid-related volatiles, conversely to double mutant lines *y_gf* and *r_gf*, which showed the opposite pattern. Finally, *y* and *gf* lines were characterized by higher levels of apocarotenoids and terpenoids (Figure 2b).

Table 2. Number of compounds in the different categories and classes of metabolites that are significantly different from San Marzano (SM) in each of the six lines under study.

Metabolomics Fraction	Metabolic Class	Abbreviation	No. of Compounds	No. of Differentially Accumulated Compounds Over SM					
				gf	*r*	*y*	*r_gf*	*y_gf*	*y_r*
Volatile	Benzenoids	B	8	2	2	3	2		3
	Branched-chain amino acid derivatives	BCAA	10	3	3	2	4		1
	Apocarotenoids	C	9	3	5	3	6	3	5
	Esters	E	2						
	Fatty acid derivatives	L	24	3	4	6	1		4
	Others	Phe, S, T, No ID *	15	6	4	8	4		6
Total		**VOCs**	68	17	18	22	17	3	19
Non-polar	Carotenoids	CAR	15	2	5	3	7	5	7
	Chlorophylls	CHL	8	2	1		2	3	
	Fatty acids	FA	14	1				1	
	Phospholipids	PHO	1			1	1		
	Quinones	QUI	6	1	2		2		3
	Tocopherols	TOC	5	1	1	1		1	2
	Others	STE, No ID *	5	1	1		1		
Total		**NP**	54	8	10	5	13	10	12
Polar	Amino acids	AA	19	2	6	3	4	3	6
	Acids	AC	17	5	5	6	5	4	4
	Amines	AM	4			1	1		1
	Sugars and polyols	SAP	15	3	3	3	2	4	2
	Alkaloids	ALK	11	1	4	2	3	1	2
	Phenylpropanoids	PHE	55	6	13	24	18	18	16
	Vitamins	VIT	3	1	2				1
	Others	A, NU, LI, No ID *	5			1		2	1
Total		**P**	129	18	33	40	33	32	33
Gran total			251	43	61	67	63	45	64

* Undefined compounds.

To further investigate the volatile compounds content of each of the mutated line in comparison with their original parental SM, a *t*-test analysis was performed. Out of 68 VOCs identified, the line with the highest number of compounds significantly different from SM was *y*, mainly because of differences in fatty acid derivatives, among other VOCs (Table 2). The lines containing the *r* mutation strongly differed for apocarotenoid volatiles, a group of metabolites considered to be involved in tomato flavor [37], representing the metabolic pathways most notably altered in these ILs (Figure 3). Some volatiles, such as the benzenoid eugenol, had higher levels in all mutant lines (Figure 3a). Mutants carrying *yellow flesh* had lower levels of the apocarotenoid β-ionone (Figure 3b), as well as 6-methyl-5-hepten-2-one, where *y_gf* acquired lower levels too (Figure 3c). Furthermore, *r*, *y* and their double mutant had higher levels of many fatty acid derivatives, such as *(E)*-2-pentenal (Figure 3d) and 1-penten-3-one (Figure 3e). Lastly, *y* was enriched in phenylalanine derivatives, such as 2-phenylethanol (Figure 3f).

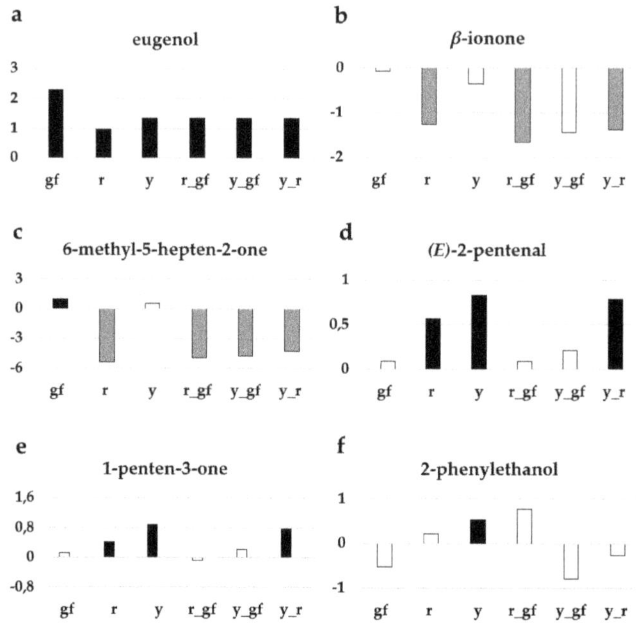

Figure 3. Relative log2 variation in selected volatiles involved in tomato flavor: (**a**) eugenol, (**b**) β-ionone, (**c**) 6-methyl-5-hepten-2-one, (**d**) *(E)*-2-pentenal, (**e**) 1-penten-3-one, (**f**) 2-phenylethanol in the six fruit mutant lines in San Marzano (SM) background compared with the recurrent parent. Line symbols are reported in Table 1. Bars colored in grey and black indicate means significantly lower and higher than SM for $p \leq 0.05$ after Student's *t*-test, respectively.

3.4. Targeted Analysis of Non-Polar Metabolites

To investigate changes at the NP specialized metabolome, LC-HRMS was used to determine the level of 54 known and previously validated compounds. They were divided in different metabolic classes, including 14 fatty acids (FA), one phospholipid (PHO), two sterols (STE), 15 carotenoids (CAR), eight chlorophylls (CHL), six quinones (QUI), and five tocochromanols (TOC; Table S4). The score plot of the first two PCA components explained about the 57% of the total variance, with double mutants differently spaced from their parental lines (Figure 4a). Indeed, PC1 kept the mutants carrying *r* separated from SM and the other lines. PC2 clearly separated *y_r* and its parental lines from the three mutants carrying *gf* plus SM, partially confirming untargeted metabolomics results. The loading plot grouped metabolites belonging to the same metabolic pathways (Figure 4b). A comparison between the score and the loading plots revealed the compositional differences between the mutated lines. Indeed, the chlorophyll group in the upper side of PC2 characterized the *green flesh* genotypes, while two quinones were in the lower side of PC2, in correspondence of the *yellow flesh* genotypes. The carotenoid group was split into early carotenoids (up to lycopene) and xanthophylls at the opposite sides of PC1; they were respectively decreased and increased in the *yellow flesh* mutant lines (Figure 4b).

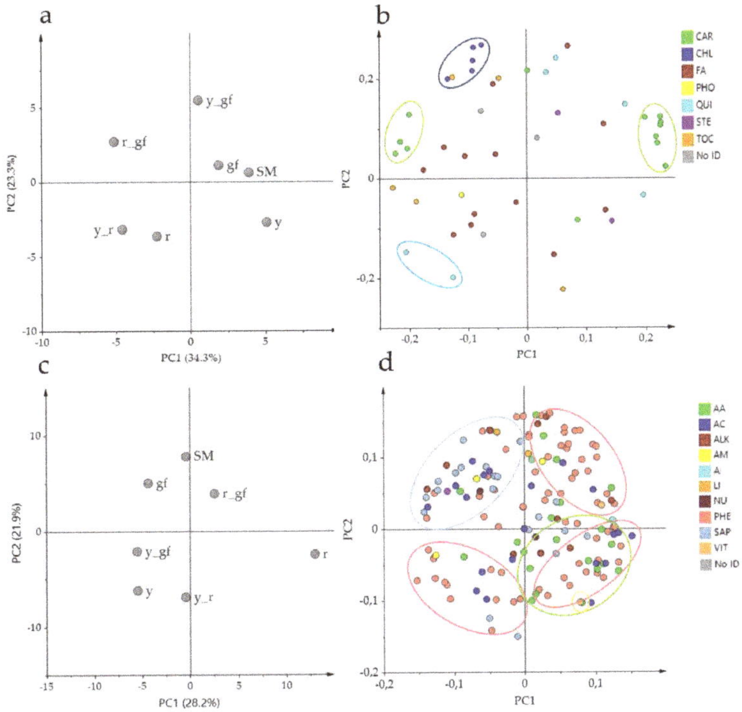

Figure 4. PCA of log2 values of 54 NP (**a**,**b**) and 128 P (**c**,**d**) metabolites measured by LC-HRMS. (**a**–**c**) PC1 × PC2 score plot of the six mutated lines plus San Marzano (SM). (**b**–**d**) PC1 × PC2 corresponding loading plots. Line (**a**,**c**) and metabolite class (**b**,**d**) abbreviations are explained in Tables 1 and 2 respectively.

To further investigate the metabolic changes of each mutated line in comparison with the original parental SM, a *t*-test analysis was performed (Table 2; Table S4), and we particularly focused on metabolites with sensorial (color, taste) and health-related properties. The lines carrying *r* showed the highest number of NP compounds different from SM, mainly due, as expected, to differences in carotenoids (Table 2). The lower number of differences was shown by the single mutant *y*, indicating that this genotype is more similar to SM for NP targeted compounds.

r, *r_gf* and *y_r* were characterized by levels of phytoene (Figure 5a), β-carotene (Figure 5b) and lycopene (Figure 5c) lower than SM and the other lines, in agreement with previous reports; *r*, *r_gf*, *y_gf*, and *y_r* reported higher levels of the xanthophylls all-trans-neoxanthin (Figure 5d) and luteoxanthin (Table S4). At chlorophyll metabolism level, *gf* was characterized by higher contents of both chlorophyll a (Table S4) and b, with the latter also higher in *y_gf* and *r_gf* (Figure 5e). Moreover, *r_gf* showed higher levels of pheophytin a and pheophorbide a (Table S4). Drawing the attention on quinones, plastoquinone increased in lines carrying *r* (Figure 5f) and plastoquinol-9 in *r* and *y_r* (Table S4). Lastly, δ-tocopherol amount was higher than SM only in *r_gf*, while γ-tocopherol and β-tocopherol enhanced in *r_gf* and *y_gf* (Figure 5g,h).

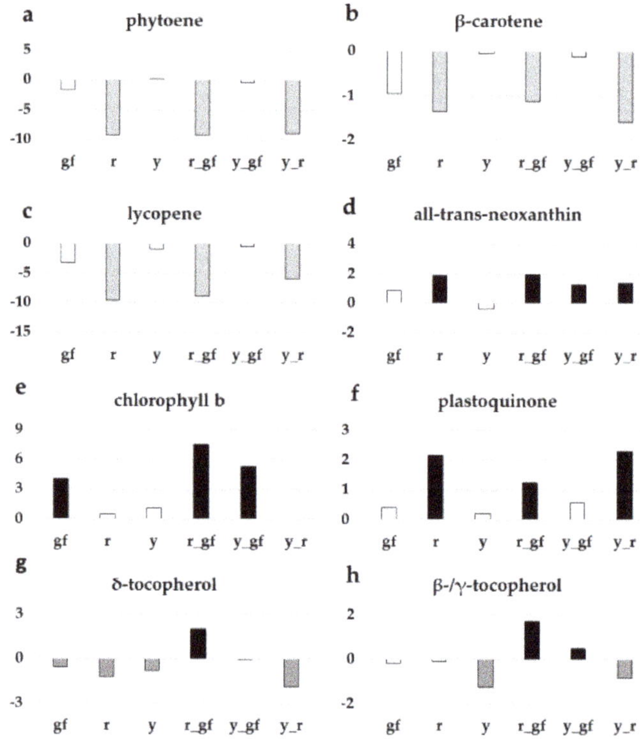

Figure 5. Relative log2 variation in (**a**) phytoene, (**b**) β-carotene, (**c**) lycopene, (**d**) all-trans-neoxanthin, (**e**) chlorophyll b, (**f**) plastoquinone, (**g**) δ-tocopherol, (**h**) β-/γ-tocopherol, of six fruit mutant lines in San Marzano (SM) background compared with the recurrent parent. Line symbols are reported in Table 1. Bars colored in grey and black indicate means significantly lower and higher than SM for $p \leq 0.05$ after Student's *t*-test, respectively.

3.5. Targeted Analysis of Polar Metabolites

The relative quantification of 128 polar metabolites allowed to complete the metabolomics characterization of the six mutants under study. P metabolites were divided into different metabolic classes, including 19 amino acids (AA), 17 acids (AC), four amines (AM), two lipids (LI), one nucleic acid (NU), 15 sugars and polyols (SAP), 11 alkaloids (ALK), 55 phenylpropanoids (PHE) and three vitamins (VIT; Table S5). The score plot of the first two components explained about the 52% of the total variance, with PC1 that particularly separated *r* and *r_gf* from all the other lines, and PC2 that identified a group including *gf* and *r_gf* together with SM (Figure 4c). Interestingly, P untargeted and targeted metabolomes differently separated the mutants under study, providing clues about a large extent of distinct metabolic components contributing to their chemical profiles. The loading plot grouped metabolites belonging to the same metabolic pathway (Figure 4d). By the comparison of the score and the loading plots, the position of some metabolites in relation to the lines studied was highlighted; indeed, many kaempferols and quercetins were in the PHE group corresponding to the area of the *y* mutants. On the contrary, many naringenins grouped in the opposite side. Most AAs were grouped together, matching with *r* and in opposition to the SAP group (Figure 4d).

To investigate the P metabolite content of each of the mutated lines in comparison with SM, a *t*-test analysis was performed (Table 2; Table S5), giving emphasis to nutritional- and sensorial attribute-related molecules. The line with the highest number of differentially accumulated polar compounds was the *y* single mutant, with a preponderance of down-regulated metabolites, as expected; the PHE group mostly contributed to this diversity (Table 2). On the contrary, *gf* was the line more like SM. Notably, lines containing *r* showed a higher number of AA over SM (Table 2).

As already highlighted in the corresponding loading plot (Figure 4d), lines carrying *y* had lower levels of the PHE naringenin chalcone glucoside (Figure 6a), and conversely higher levels of kaempferol-hexose deoxyhexose-pentose compared to SM, a biochemical phenotype also observed in *gf* and *r_gf* too for the latter (Figure 6b). Regarding the ALK group, calystegine B1 resulted statistically lower in all the lines, with exception of *gf* and *y_gf*, which however displayed reduced amounts compared to SM (Figure 6c). In addition, a series of primary metabolites were characterized by higher levels in the mutants under study: for example, the AA proline in *r* and *y_r* (Figure 6d) and the VIT nicotinamide in *gf*, *r* and *y_r* (Figure 6e). Similarly, the SAP glucoheptulose was higher in *gf*, whereas all the other lines were more similar to SM (Figure 6f). The LI phosphocoline displayed lower levels in lines carrying the *y* mutation (Figure 6g), while the AC sinapinic acid was higher in all lines, with the only exception of *r* (Figure 6h).

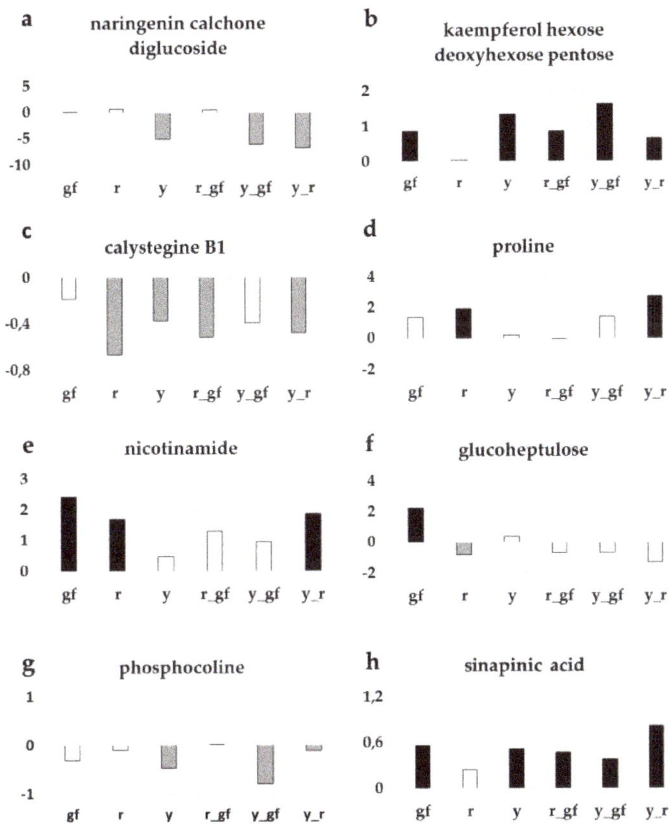

Figure 6. Relative log2 variation in (**a**) naringenin chalcone glucoside, (**b**) kaempferol-hexose-deoxyhexose-pentose, (**c**) calystegine B1, (**d**) proline, (**e**) nicotinamide, (**f**) glucoheptulose, (**g**) phosphocoline, (**h**) sinapinic acid, of six fruit mutant lines in San Marzano (SM) background compared with the recurrent parent. Line symbols are reported in Table 1. Bars colored in grey and black indicate means significantly lower and higher than SM for $p \leq 0.05$ after Student's *t*-test, respectively.

3.6. Bioinformatics to Investigate Metabolite-Metabolite Relationships

Bioinformatic approaches, including Venn diagram visualization and correlation network analysis, were used in order to achieve a deeper understanding of the biochemical perturbations and relationships occurring in the SM mutants under study. Venn diagrams showed the degree of overlap for VOC, NP and P metabolites in each double mutant and its two parental single mutants. For line-specific metabolites in each group, all lines, except *y_gf* and *y_r*, showed a higher number of significantly up-regulated metabolites (Figure 7). When the overlaps between single mutants and the respective double combinations were considered, the epistasis of *r* over *gf* (29 metabolites were in common between *r* and *r_gf*) and *y* (28 metabolites in common between *r* and *y_r*) and of *y* over *gf* (21 metabolites in common between *y* and *y_gf*) were found (Figure 7).

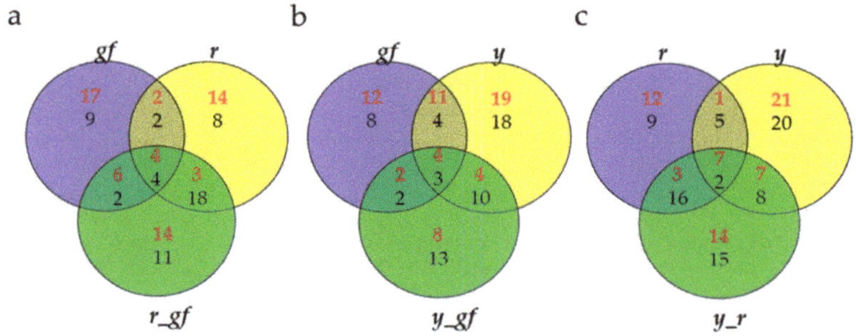

Figure 7. Venn diagrams for (**a**) *gf*, *r* and *r_gf*; (**b**) *gf*, *y* and *y_gf*; (**c**) *r*, *y* and *y_r*. Red and black numbers correspond to up- and down-regulated metabolites, respectively.

Furthermore, we used a correlation network approach to investigate mutation-induced alteration at VOC, NP and P metabolome levels. To this purpose, three networks were built by integrating all differentially accumulated metabolites in, at least, each single mutant and the corresponding double mutants (Figure 8). Overall, the three force-directed networks allowed the achievement of specific topologies according to the distribution of the significant correlation networks existing in each metabolite-metabolite interaction; notably, a high extent of conservation was observed in each network, either in the direct (PHE in the *y*- and CAR in the *r*-yielding networks) or not direct (L volatiles in *r*, *y* and *r_y*; and AA in *r*, *gf* and *r_gf*) targets of the mutations. In order to evidence the more robust and strongest correlations, the MCODE Cytoscape plugin was applied to each of the three networks (Figures S2–S4). In this way, it was possible to identify the highly interconnected regions, which resulted particularly abundant in the *r*/*gf* mutants, with a lower number of greatly dense clusters including primary (SAP, AA) and secondary (isoprenoids as CAR, CHL, QUI, T and PHE) compounds; on the contrary, the *y*/*r* mutants were characterized by a higher number of clusters with lower density, and with a high representation of amino and organic acids, besides CAR and PHE.

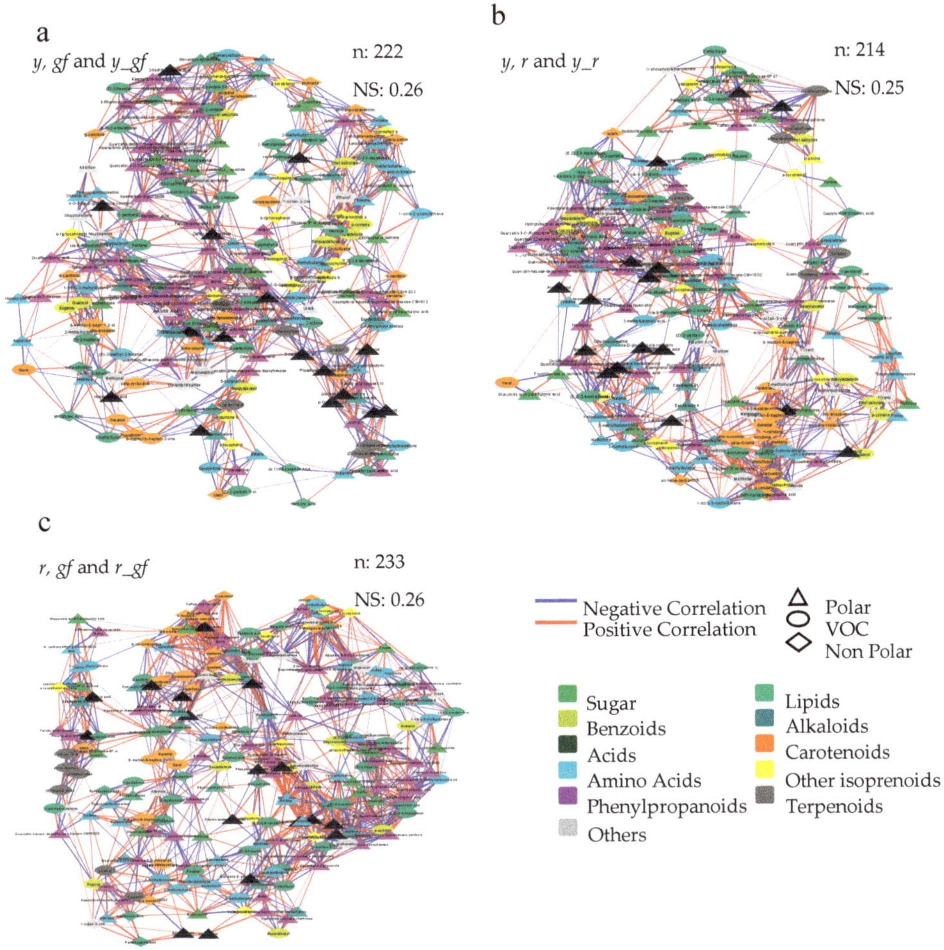

Figure 8. Correlation networks of metabolomics data in (**a**) *y*, *gf* and *y_gf*; (**b**) *y*, *r* and *y_r*; (**c**) *r*, *gf* and *r_gf* mutants. All the data from volatile (VOC), non-polar (NP) and polar (P) analyses were used as fold change to the San Marzano (SM) reference. Network topology was directed by the force of Pearson correlation coefficient index (Table S6). Each node represents VOC (turquoise circle), a NP (diamond) or a P (triangle) metabolite. Lines joining the nodes represent positive (red) or negative (blue) correlations. Node sizes are proportional to the respective node strengths, which are shown in Tables S7–S9. Node color is depending on the metabolic class of each compound as indicated in the figure. Number of nodes (n) and network strength (NS) are shown at the top of each network. Only correlations with $|\rho| > 0.95$ are shown ($p \leq 0.05$).

4. Discussion

This study focused on the analysis of the volatile and non-volatile compounds of six tomato lines introgressing mutations that have been selected as representative for affecting the main classes of fruit pigments. Interest towards these mutations is proved by the increasing introduction of cultivars with yellow, pink and brown fruits in the fresh tomato market and by scientific studies addressing the properties of single [38,39] and multiple [40] mutants. The characterization of the biochemical effects of these mutations was based on the comparison with the SM original parent, with the final aim of defining whether the lines showing aesthetic novelty were also diversified for organoleptic and nutritional characteristics.

When the studied compounds were considered all together, *y* showed the highest level of variation. Venn diagrams indicated epistasis of *r* over both *gf* and *y*, as highlighted by the univariate analysis. Interestingly, when integrated and subjected to a correlation network analysis, differentially accumulated metabolites in any of the six single and double mutant ILs exhibited a strong level of coordination: indeed, irrespectively to the metabolic class object of the mutation, metabolites acting in the same pathway clustered together, thus indicating a great conservative capacity of the fruit metabolism in its mutation-derived reorganization. This finding is consistent with previous reports showing, either in tomato and grape, a general phenotypic-metabolic plasticity in response to genetic or environmental changes [24,30,41]. However, looking at the sub-clusters generated by each network, distinct and specific relationships were unraveled, with the mutants carrying the *y* and *r* mutations involving the largest number of metabolites belonging to highly diversified pathways.

4.1. *Biochemical Changes in Fruits of Yellow Flesh and of Its Combinations with y and gf*

Mutants containing *r* were clearly separated from the other lines by NP targeted metabolites, mainly due to the compositional differences of carotenoids. The double mutant *r_gf* mapped in a distinct position compared with *r* and *y_r*, indicating partially additive and partially synergistic effects of this combination. The primary effect of the *r* mutation in the carotenoid pathway was to strongly reduce phytoene, and the colored carotenoids β-carotene and lycopene, as expected [42]. Phytofluene was also reduced in *r* genotypes, as previously described in tomato accessions [43] and in ripening fruits of *Psy1* knockout lines [44]. In the latter case, a concomitant decrease in the volatile apocarotenoids was also observed [44]. In literature, the study of *r* metabolites was often limited to the analysis of carotenoids, including the xanthophyll lutein, whose levels, in line with our results, did not vary [11]. Intriguingly, we evidenced that all-trans-neoxanthin, the last xanthophylls in the carotenoid pathway, in *r* mutants showed levels higher than SM; it is possible that this represents a cellular strategy implemented to compensate for the decrease of other carotenoids. Indeed, xanthophylls act in flowing the energy through the photosynthetic apparatus and protecting organisms against damage caused by photosynthesis itself [45]. In addition, both violaxanthin and neoxanthin are key substrates in ABA biosynthesis, and their variation can influence fruit ripening and the attitude towards abiotic stresses [46]. In this context, enhanced levels of xanthophylls could provide enough metabolic flux to guarantee an adequate ABA production. The same hypothesis can be extended to the analysis of quinones, such as plastoquinone and plastoquinol-9, whose levels increased in *yellow flesh* mutants. The importance of quinones in basic metabolic processes, such as respiration and photosynthesis, has been established [47] and an increase in their content is plausible, in a scenario where it is necessary to compensate for the lack of carotenoids.

Important metabolic changes occurred in the *r* mutants for P compounds; several amino acids positively varied, as proline in the single mutant and in combination with *y*, and valine/norvaline in the single mutant or combined with *gf*. Proline and valine have significant functions in plant cells as plant stress sensor, ABA and polyamines interactors, and precursors of BCAA-derived volatiles [48,49]. Similarly, the vitamin nicotinamide, that plays a primary role in pyridine metabolism [50], also increased in *r*, alone or with *y*.

Carotenoid-derived volatiles have an important role in tomato flavor, as their levels positively correlate with tomato acceptability [51]. Loss-of-function of *Psy1* in *yellow flesh* mutants led to the lack of substrates for apocarotenoid production [52], thus justifying the low amount of C VOCs such as β-ionone [53] and 6-methyl-5-hepten-2-one [54]. Conversely, the L 1-penten-3-one and *(E)*-2-pentenal increased in *r* and *r_y*, suggesting a role for the lipoxygenase–linoleate (LOX) enzyme that catalyzes the oxidation of polyunsaturated fatty acids by molecular oxygen with the formation of unstable hydroperoxides which in turn oxidize carotenoid pigments [55].

In summary, compensating the depletion of the principal carotenoids, lines carrying the *r* mutation offer peculiar nutraceutical properties, for an increased content of amino acids, vitamins, xanthophylls and quinones. Xanthophylls have been highlighted in recent studies for their positive contribution towards total dietary carotenoid intake [56], whereas quinones showed positive properties in treating cardiovascular diseases, chronic gingivitis and periodontitis and a favorable impact on cancer treatment and human reproductive health [57]. Differently, tomato *r* lines will probably show a lower score for the aroma, considering that some of the apocarotenoids providing floral or fruity notes are less represented in yellow fruits.

4.2. Biochemical Changes in Fruits of Colorless Fruit Epidermis and of Its Combinations with r and gf

Multivariate analysis of targeted P compounds tightly grouped lines carrying the *y* mutation, an expected result as *y* is characterized by substantial modifications of the class of phenylpropanoids, mainly due to the lack of yellow pigment naringenin chalcone [15]. As also observed in the case of the NP fraction, untargeted and targeted metabolomes did not perfectly matched, suggesting the existence of larger, still unexplored, metabolic changes. The decrease or absence of naringenins also characterized *y* and its two double mutants, leading to the conclusion that *y* is epistatic on *gf* and *r* for this class of compounds. Compared to red SM fruits, kaempferol-hexose-deoxyhexose-pentose was higher in *y* mutants, even if many other kaempferols showed no differences. Similarly, several quercetins showed higher levels in lines carrying *y*. Variations in the PHE pathways could explain the reduced phosphocholine levels detected in the three *y* mutants, since increased activity of pathways generating cuticular lipids in tomato fruit peel precedes that of phenylpropanoid and flavonoid biosynthesis pathways [58].

No substantial differences in carotenoids were found between *y* and SM, except the higher all-trans-neoxanthin levels in *y_gf* and *y_r*. Conversely, tocochromanol metabolism was altered in *y* mutant berries as there was a decrease in tocopherols, mainly affecting *y* and *y_r*, confirming previous reports [15,59].

Among VOCs, differences were found in compounds derived from phenylalanine, such as eugenol and 2-phenylethanol, in line with the known alterations of *y* in the phenylpropanoid branch of phenylalanine catabolism [54]. Indeed, a significant number of VOCs derived from amino acids, are considered to have an influence on tomato flavor and liking, such as guaiacol and eugenol. In addition, 2-phenylethanol is a main contributor to tomato flavor, increasing floral aroma and the perception of sweetness [60,61], it is plausible that increase of these class of volatiles exerts a positive effect on the aroma in lines carrying the *y* mutation. In addition, *y* mutants were characterized by higher production of many fatty acid derivatives, which are the most abundant volatiles produced in the tomato fruit [62]. These VOCs include several C_5 compounds such as 1-penten-3-one or (E)-2-pentenal, and C_6 volatiles such as 1-hexanol, (*Z*)-3-hexenal, (*E*)-2-hexenal, or hexanal, among others, that are classified as "green leaf" volatiles due to their 'green' characteristic, with a fresh aroma of cut grass. Although some studies suggested a reduced impact on tomato flavor and no effect on consumer preference of these compounds [63,64], others claim their potential impact on overall flavor intensity and liking [37].

In addition to variation of the aroma, *y* mutants are endowed of higher kaempferol and quercetin levels, but also present the drawback of lower naringenin and tocopherols. Since all tocochromanols are potent lipid-soluble antioxidants and are essential dietary nutrients for mammals as vitamin E [65], this aspect represents a negative trait in *y* tomato lines.

Therefore, the *y* mutation is to be considered as a source of more aesthetic (color) and organoleptic (aroma) than nutritional novelty. Although no clear relationship between colorless epidermis and fruit shelf-life, peculiar mechanical properties of the *y* epicarp were manifested by the fact that the peel of the mutant fruit was richer of lignin [15,66]. Intriguingly, the SM lines carrying *y*, alone or in combination, showed a higher resistance to storage, indicating that pigment variation in the peel implicates different mechanical properties and post-harvest behavior of the fruit [26].

4.3. Biochemical Changes in Fruits of Green Flesh and of Its Combinations with r and y

The group of *green flesh* mutants showed no changes in carotenoids, but as expected, it was primarily affected in the content of chlorophylls. *gf* belongs to the so-called cosmetic-type SGR mutants, that have been described in several plant species to retain substantial amount of chlorophylls in fruits during ripening, maintaining other ripening-related metabolites, such as lycopene, unchanged [67]. Accordingly, all the SM lines containing *gf* showed fruits with higher amounts of chlorophylls. Among other NP metabolites, several tocopherols increased in *gf* single and double mutants, confirming previous data that showed higher tocopherol levels in *gf* [68,69] and *r_gf* (A. Mazzucato and G.P. Soressi, unpublished results) mutants.

For P metabolites, *gf* was similar to SM, with few notable exceptions involving flavonoids, such as kaempferols, and sugars, such as glucoheptulose.

No substantial variations were found for *gf* mutants in VOCs except for a high eugenol content that was paralleled by the other lines.

Because of the positive effects of chlorophylls and chlorophyll-related metabolites on cellular inflammation and as anti-mutagen and anti-carcinogen agent [70,71], and of the antioxidant function of vitamin E-producing metabolites, that reduce free-radical damage to membrane lipids by scavenging peroxyl radicals [72,73], tomato lines carrying *gf* may be considered an alternative to red tomatoes offering added nutraceutical value. In addition, *gf* may play important roles in enhancing organoleptic qualities. Although a direct, positive effect of *gf* on soluble solids content has not clearly been established, an SlGLK2-enhanced chlorophyll content in immature green fruits led to an increment in total soluble solid content in ripe fruits, possibly by the positive regulation of sugar metabolism enzyme-encoding genes [69]. This effect will be additive to that of other taste-related compounds, such as glutamate [20]. In our *gf* lines, glutamic acid was not significantly different from SM, but, considering only data in one year, it was substantially higher in all lines. All these properties justify the interest towards breeding new *gf* tomato lines and the present success in the market promises a further development in the near future.

5. Conclusions

This work was based on the study of the effects of mutations with an emerging commercial interest, compared within the traditional Italian tomato variety San Marzano. This biochemical and bioinformatics characterization has given further insights on the effect of each mutation on fruit aesthetic, flavor and nutritional composition. The analysis of the respective three double mutants offered an added value, making it possible to establish epistatic or synergistic effects between each pair of mutations and represented a starting point for breeding new tomato lines with different phenotypes. Although *r* and *y* cause the decrement of the most important classes of health-related pigments (carotenoids and flavonoids), the compensating increase of other metabolites with nutraceutical (xanthophylls, tocopherols, amino acids) or flavor-related (phenylalanine and fatty acid derivatives) positive properties make the studied lines worthy of attention for breeding novel and better tomatoes. Novel phenotypes could take advantage of new breeding techniques as genome editing to recapitulate the original mutations in different tomato backgrounds, or in different species, and obtain the wanted variation directly avoiding backcrossing and linkage drag effects. Ultimately, advanced breeding programs will convert the new lines into novel elite varieties of commercial interest.

Supplementary Materials: The following are available online at http://www.mdpi.com/2218-1989/10/3/110/s1, Figure S1: PC1 × PC2 score plots of the six mutated lines plus San Marzano (SM), sampled in two different years, according to relative values of 263 volatile (**a**), 746 non-polar (**b**) and 110 polar (**c**) metabolites measured by a solid-phase micro-extraction gas-chromatography coupled with a mass spectrometry instrument (HS-SPME-GC-MS), Figure S2: Subnetworks generated from the *y*, *gf* and *y_gf* global network of Figure 8, using the MCODE Cytoscape plugin. Each subnetwork highlights the highest densely connected regions present in the global network. Each node represents a VOC (turquoise circle), non-polar (diamond) or a polar metabolite (triangle). Lines joining the nodes represent correlations; direct and inverse correlations are shown in red and blue, respectively. Node sizes are proportional to the corresponding node strengths (nss), which are shown in

Supplemental Table S7. Node color is depending on the metabolic class of each compound as indicated in the figure. Number of nodes (n) and network strength (NS) are shown on top of each network. Only correlations with $|\rho| > 0.95$ are shown (p-value 0.05), Figure S3: Subnetworks generated from the *y*, *r* and *y_r* global network of Figure 8 using the MCODE Cytoscape plugin. Each subnetwork highlights the highest densely connected regions present in the global network. Each node represents a VOC (turquoise circle), non-polar (diamond) or a polar metabolite (triangle). Lines joining the nodes represent correlations; direct and inverse correlations are shown in red and blue, respectively. Node sizes are proportional to the corresponding node strengths (nss), which are shown in Supplemental Table S7. Node color is depending on the metabolic class of each compound as indicated in the figure. Number of nodes (n) and network strength (NS) are shown on top of each network. Only correlations with $|\rho| > 0.95$ are shown (p-value 0.05), Figure S4: Subnetworks generated from the *r*, *gf* and *r_gf* global network of Figure 8 using the MCODE Cytoscape plugin. Each subnetwork highlights the highest densely connected regions present in the global network. Each node represents a VOC (turquoise circle), non-polar (diamond) or a polar metabolite (triangle). Lines joining the nodes represent correlations; direct and inverse correlations are shown in red and blue, respectively. Node sizes are proportional to the corresponding node strengths (nss), which are shown in Supplemental Table S7. Node color is depending on the metabolic class of each compound as indicated in the figure. Number of nodes (n) and network strength (NS) are shown on top of each network. Only correlations with $|\rho| > 0.95$ are shown (p-value 0.05), Table S1: Number of volatile (VOC), polar (P) and non-polar (NP) untargeted metabolites significantly different from Sam Marzano (SM) in the six introgression lines. Line symbols are explained in Table 1, Table S2: Fisher's F test (F) and relative degree of significance (P) after two-way multivariate analysis of variance (MANOVA) for three main categories of metabolites analyzed, Table S3: Mean (M) and standard deviation (SD) of volatile compounds in two years for the fold change of individual values over the San Marzano value and significance of Student's t test between each mutant line and San Marzano. Line symbols are explained in Table 1, Table S4: Mean (M) and standard deviation (SD) of non-polar compounds in two years for the fold change of individual values over the San Marzano value and significance of Student's t test between each mutant line and San Marzano. Line symbols are explained in Table 1, Table S5: Mean (M) and standard deviation (SD) of polar compounds in two years for the fold change of individual values over the San Marzano value and significance of Student's t test between each mutant line and San Marzano. Line symbols are explained in Table 1, Table S6: Metabolite-metabolite significant correlations (Pearson coefficient; $r > |0.95|$, p-value < 0.05) in tomato single mutant and the corresponding double mutants (*r*, *y* and *r_y*; *r*, *gf* and *r_gf*; *y*, *gf* and *y_gf*), Table S7: Node table of y, *gf* and *y_gf* double mutant Network. For each node the corresponding node strength (ns) is reported, Table S8: Node table of *y*, r and *y_r* double mutant Network. For each node the corresponding node strength (ns) is reported, Table S9: Node table of gf, r and *r_gf* double mutant Network. For each node the corresponding node strength (ns) is reported.

Author Contributions: A.M., G.D. (Gianfranco Diretto) and A.G. conceptualized and designed the experiments, G.D. (Gabriella Dono) G.D. (Gianfranco Diretto) and J.L.R. performed the experiments, G.D. (Gabriella Dono), G.D. (Gianfranco Diretto), S.F. and J.L.R. performed statistical analysis, A.M. and G.D. (Gabriella Dono) drafted the manuscript. All the authors reviewed and edited the manuscript. All authors have read and agreed to the published version of the manuscript.

Funding: This work was supported by the European Commission through-H2020 SFS-7a-2014 TRADITOM (634561) and the COST Actions ROXy CA18210 and EUcarotene CA15136 for networking. The Italian Ministry for Education, University and Research (MIUR) Law 232/216, Department of Excellence and the Spanish Ministry project BIO2016-78601-R are also acknowledged.

Acknowledgments: The authors acknowledge M.E. Picarella and the Metabolomics Lab at IBMCP for expert technical assistance.

Conflicts of Interest: The authors declare no conflict of interest.

References

1. Lewinsohn, E.; Sitrit, Y.; Bar, E.; Azulay, Y.; Ibdah, M.; Meir, A.; Yosef, E.; Zamir, D.; Tadmor, Y. Not just colors—carotenoid degradation as a link between pigmentation and aroma in tomato and watermelon fruit. *Trends Food Sci. Technol.* **2005**, *16*, 407–415. [CrossRef]
2. Gascuel, Q.; Diretto, G.; Monforte, A.J.; Fortes, A.M.; Granell, A. Use of natural diversity and biotechnology to increase the quality and nutritional content of tomato and grape. *Front. Plant Sci.* **2017**, *8*, 652. [CrossRef]
3. Bovy, A.G.; Gómez-Roldán, V.; Hall, R.D. Strategies to optimize the flavonoid content of tomato fruit. *Recent Adv. Polyphen. Res.* **2010**, *2*, 138–162.
4. Hunt, G.M.; Baker, E.A. Phenolic constituents of tomato fruit cuticles. *Phytochemistry* **1980**, *19*, 1415–1419. [CrossRef]
5. Dixon, R.A. Natural products and plant disease resistance. *Nature* **2001**, *411*, 843–847. [CrossRef]
6. Sands, D.C.; Morris, C.E.; Dratz, E.A.; Pilgeram, A. Elevating optimal human nutrition to a central goal of plant breeding and production of plant-based foods. *Plant Sci.* **2009**, *177*, 377–389. [CrossRef] [PubMed]

7. Martin, C.; Li, J. Medicine is not health care, food is health care: Plant metabolic engineering, diet and human health. *New Phytol.* **2017**, *216*, 699–719. [CrossRef] [PubMed]
8. Martínez-Valverde, I.; Periago, M.J.; Provan, G.; Chesson, A. Phenolic compounds, lycopene and antioxidant activity in commercial varieties of tomato (*Lycopersicum esculentum*). *J. Sci. Food Agric.* **2002**, *82*, 323–330. [CrossRef]
9. Jenkins, J.A.; Mackinney, G. Carotenoids of the apricot tomato and its hybrids with yellow and tangerine. *Genetics* **1955**, *40*, 715.
10. Rodríguez-Villalón, A.; Gas, E.; Rodríguez-Concepción, M. Phytoene synthase activity controls the biosynthesis of carotenoids and the supply of their metabolic precursors in dark-grown Arabidopsis seedlings. *Plant J.* **2009**, *60*, 424–435. [CrossRef]
11. Kachanovsky, D.E.; Filler, S.; Isaacson, T.; Hirschberg, J. Epistasis in tomato color mutations involves regulation of phytoene synthase 1 expression by cis-carotenoids. *Proc. Natl. Acad. Sci. USA* **2012**, *109*, 19021–19026. [CrossRef] [PubMed]
12. Rocha, M.D.C.; Deliza, R.; Corrêa, F.M.; do Carmo, M.G.; Abboud, A.C. A study to guide breeding of new cultivars of organic cherry tomato following a consumer-driven approach. *Food Res.* **2013**, *51*, 265–273. [CrossRef]
13. Lindstrom, E.W. Inheritance in tomato. *Genetica* **1925**, *10*, 305–317.
14. Rick, C.M.; Butler, L. Cytogenetics of the tomato. *Adv. Genet.* **1956**, *8*, 267–382.
15. Adato, A.; Mandel, T.; Mintz-Oron, S.; Venger, I.; Levy, D.; Yativ, M.; Domìnguez, E.; Wang, Z.; De Vos, R.C.; Jetter, R.; et al. Fruit-surface flavonoid accumulation in tomato is controlled by a SlMYB12-regulated transcriptional network. *PLoS Genet.* **2009**, *5*, e1000777. [CrossRef]
16. Ballester, A.R.; Molthoff, J.; de Vos, R.; Hekkert, B.T.; Orzaez, D.; Fernández-Moreno, J.P.; Tripodi, P.; Grandillo, S.; Martin, C.; Heldens, J.; et al. Biochemical and molecular analysis of pink tomatoes: Deregulated expression of the gene encoding transcription factor SlMYB12 leads to pink tomato fruit color. *Plant Physiol.* **2010**, *152*, 71–84. [CrossRef]
17. Kerr, E.A. Green flesh, gf. *Rep. Tomato Genet. Coop.* **1956**, *6*, 17.
18. Barry, C.S.; McQuinn, R.P.; Chung, M.Y.; Besuden, A.; Giovannoni, J.J. Amino acid substitutions in homologs of the STAY-GREEN protein are responsible for the green-flesh and chlorophyll retainer mutations of tomato and pepper. *Plant Physiol.* **2008**, *147*, 179–187. [CrossRef]
19. Barry, C.S.; Cornelius, S.; Pandey, P. A survey of cultivated heirloom tomato varieties identifies four new mutant alleles at the green-flesh locus. *Mol. Breed.* **2009**, *24*, 269–276. [CrossRef]
20. Bortolotti, S.; Boggio, S.B.; Delgado, L.; Orellano, E.G.; Valle, E.M. Different induction patterns of glutamate metabolizing enzymes in ripening fruits of the tomato mutant green flesh. *Physiol. Plant.* **2003**, *119*, 384–391. [CrossRef]
21. Cocaliadis, M.F.; Fernández-Muñoz, R.; Pons, C.; Orzaez, D.; Granell, A. Increasing tomato fruit quality by enhancing fruit chloroplast function. A double-edged sword? *J. Exp. Bot.* **2014**, *65*, 4589–4598. [CrossRef] [PubMed]
22. Loiudice, R.; Impembo, M.; Laratta, B.; Villari, G.; Voi, A.L.; Siviero, P.; Castaldo, D. Composition of San Marzano tomato varieties. *Food Chem.* **1995**, *53*, 81–89. [CrossRef]
23. Ercolano, M.R.; Carli, P.; Soria, A.; Cascone, A.; Fogliano, V.; Frusciante, L.; Barone, A. Biochemical, sensorial and genomic profiling of traditional Italian tomato varieties. *Euphytica* **2008**, *164*, 571–582. [CrossRef]
24. D'Esposito, D.; Ferriello, F.; Dal Molin, A.; Diretto, G.; Sacco, A.; Minio, A.; Barone, A.; Di Monaco, R.; Cavella, S.; Tardella, L.; et al. Unravelling the complexity of transcriptomic, metabolomic and quality environmental response of tomato fruit. *BMC Plant Biol.* **2017**, *17*, 66. [CrossRef] [PubMed]
25. Barone, D.; Cito, L.; Tommonaro, G.; Abate, A.A.; Penon, D.; De Prisco, R.; Penon, A.; Forte, I.M.; Benedetti, E.; Cimini, A.; et al. Antitumoral potential, antioxidant activity and carotenoid content of two Southern Italy tomato cultivars extracts: San Marzano and Corbarino. *J. Cell Physiol.* **2018**, *233*, 1266–1277. [CrossRef]
26. Dono, G.; Picarella, M.E.; Pons, C.; Santangelo, E.; Monforte, A.; Granell, A.; Mazzucato, A. Characterization of a repertoire of tomato fruit genetic variants in the San marzano genetic background. *Sci. Hortic.* **2020**, *261*, 108927. [CrossRef]
27. Rambla, J.L.; Medina, A.; Fernández-del-Carmen, A.; Barrantes, W.; Grandillo, S.; Cammareri, M.; Lopez-Casado, G.; Rodrigo, G.; Alonso, A.; Garcìa-Martinez, S. Identification, introgression, and validation of fruit volatile QTLs from a red-fruited wild tomato species. *J. Exp. Bot.* **2017**, *68*, 429–442. [CrossRef]

28. Tikunov, Y.; Lommen, A.; De Vos, C.R.; Verhoeven, H.A.; Bino, R.J.; Hall, R.D.; Bovy, A.G. A novel approach for nontargeted data analysis for metabolomics. Large-scale profiling of tomato fruit volatiles. *Plant Physiol.* **2005**, *139*, 1125–1137. [CrossRef]
29. Fasano, C.; Diretto, G.; Aversano, R.; D'Agostino, N.; Di Matteo, A.; Frusciante, L.; Giuliano, G.; Carputo, D. Transcriptome and metabolome of synthetic *Solanum* autotetraploids reveal key genomic stress events following polyploidization. *New Phytol.* **2016**, *210*, 1382–1394. [CrossRef]
30. Rambla, J.L.; Trapero-Mozos, A.; Diretto, G.; Rubio-Moraga, A.; Granell, A.; Gómez-Gómez, L.; Ahrazem, O. Gene-metabolite networks of volatile metabolism in Airen and Tempranillo grape cultivars revealed a distinct mechanism of aroma bouquet production. *Front. Plant Sci.* **2016**, *7*, 1619. [CrossRef]
31. Diretto, G.; Rubio-Moraga, A.; Argandoña, J.; Castillo, P.; Gómez-Gómez, L.; Ahrazem, O. Tissue-specific accumulation of sulfur compounds and saponins in different parts of garlic cloves from purple and white ecotypes. *Molecules* **2017**, *22*, 1359. [CrossRef] [PubMed]
32. Sulli, M.; Mandolino, G.; Sturaro, M.; Onofri, C.; Diretto, G.; Parisi, B.; Giuliano, G. Molecular and biochemical characterization of a potato collection with contrasting tuber carotenoid content. *PLoS One* **2017**, *12*, e0184143. [CrossRef] [PubMed]
33. Cappelli, G.; Giovannini, D.; Basso, A.L.; Demurtas, O.C.; Diretto, G.; Santi, C.; Girelli, G.; Bacchetta, L.; Mariani, F. A *Corylus avellana* L. extract enhances human macrophage bactericidal response against *Staphylococcus aureus* by increasing the expression of anti-inflammatory and iron metabolism genes. *J. Funct. Foods* **2018**, *45*, 499–511. [CrossRef]
34. Ahrazem, O.; Argandoña, J.; Fiore, A.; Aguado, C.; Luján, R.; Rubio-Moraga, Á.; Marro, M.; Araujo-Andrade, C.; Loza-Alvarez, P.; Diretto, G.; et al. Transcriptome analysis in tissue sectors with contrasting crocins accumulation provides novel insights into apocarotenoid biosynthesis and regulation during chromoplast biogenesis. *Sci. Rep.* **2018**, *8*, 1–17. [CrossRef] [PubMed]
35. Diretto, G.; Frusciante, S.; Fabbri, C.; Schauer, N.; Busta, L.; Wang, Z.; Matas, A.J.; Fiore, A.; Rose, J.K.C.; Fernie, R.A. Manipulation of β-carotene levels in tomato fruits results in increased ABA content and extended shelf life. *Plant Biotechnol. J.* **2019**, in press. [CrossRef] [PubMed]
36. Bader, G.D.; Hogue, C.W. An automated method for finding molecular complexes in large protein interaction networks. *BMC Bioinf.* **2003**, *4*, 2. [CrossRef] [PubMed]
37. Tieman, D.; Zhu, G.; Resende, M.F.; Lin, T.; Nguyen, C.; Bies, D.; Rambla, J.L.; Stephanie Ortiz Beltran, K.; Taylor, M. A chemical genetic roadmap to improved tomato flavor. *Science* **2017**, *355*, 391–394. [CrossRef]
38. Rodríguez-Burruezo, A.; Prohens, J.; Roselló, S.; Nuez, F. "Heirloom" varieties as sources of variation for the improvement of fruit quality in greenhouse-grown tomatoes. *J Hort. Sci. Biotechnol.* **2005**, *80*, 453–460. [CrossRef]
39. Flores, P.; Sánchez, E.; Fenoll, J.; Hellín, P. Genotypic variability of carotenoids in traditional tomato cultivars. *Food Res. Int.* **2017**, *100*, 510–516. [CrossRef]
40. Kang, S.I.; Hwang, I.; Goswami, G.; Jung, H.J.; Nath, U.K.; Yoo, H.J.; Lee, J.M.; Nou, I.S. Molecular insights reveal *Psy1*, *SGR*, and *SlMYB12* genes are associated with diverse fruit color pigments in tomato (*Solanum lycopersicum* L.). *Molecules* **2017**, *22*, 2180. [CrossRef]
41. Dal Santo, S.; Fasoli, M.; Negri, S.; D'Incà, E.; Vicenzi, N.; Guzzo, F. Plasticity of the berry ripening program in a white grape variety. *Front. Plant Sci.* **2016**, *7*, 970. [CrossRef] [PubMed]
42. LeRosen, A.L.; Went, F.W.; Zechmeister, L. Relation between genes and carotenoids of the tomato. *Proc. Natl. Acad. Sci. USA* **1941**, *27*, 236. [CrossRef] [PubMed]
43. Kang, B.; Gu, Q.; Tian, P.; Xiao, L.; Cao, H.; Yang, W. A chimeric transcript containing Psy1 and a potential mRNA is associated with yellow flesh color in tomato accession PI 114490. *Planta* **2014**, *240*, 1011–1021. [CrossRef] [PubMed]
44. Gady, A.L.; Vriezen, W.H.; Van de Wal, M.H.; Huang, P.; Bovy, A.G.; Visser, R.G.; Bachem, C.W. Induced point mutations in the *phytoene synthase 1* gene cause differences in carotenoid content during tomato fruit ripening. *Mol. Breed.* **2012**, *29*, 801–812. [CrossRef]
45. Demmig-Adams, B.; Adams Iii, W.W. Photoprotection and other responses of plants to high light stress. *Ann. Rev. Plant Biol.* **1992**, *4*, 599–626. [CrossRef]
46. Parry, A.D.; Babiano, M.J.; Horgan, R. The role of cis-carotenoids in abscisic acid biosynthesis. *Planta* **1990**, *182*, 118–128. [CrossRef]

47. Nohl, H.; Jordan, W.; Youngman, R.J. Quinones in biology: Functions in electron transfer and oxygen activation. *Free Radic. Biol. Med.* **1986**, *2*, 211–279. [CrossRef]
48. Ghorbanli, M.; Gafarabad, M.; Amirkian, T.; Bahareh, A.M. Investigation of proline, total protein, chlorophyll, ascorbate and dehydroascorbate changes under drought stress in Akria and Mobil tomato cultivars. *Iran. J. Plant Physiol.* **2013**, *3*, 651–658.
49. Pál, M.; Tajti, J.; Szalai, G.; Peeva, V.; Végh, B.; Janda, T. Interaction of polyamines, abscisic acid and proline under osmotic stress in the leaves of wheat plants. *Sci. Rep.* **2018**, *8*, 1–12. [CrossRef]
50. Alferez, F.M.; Gerberich, K.M.; Li, J.L.; Zhang, Y.; Graham, J.H.; Mou, Z. Exogenous nicotinamide adenine dinucleotide induces resistance to citrus canker in citrus. *Front. Plant Sci.* **2018**, *9*, 1472. [CrossRef]
51. Vogel, J.T.; Tieman, D.M.; Sims, C.A.; Odabasi, A.Z.; Clark, D.G.; Klee, H.J. Carotenoid content impacts flavor acceptability in tomato (*Solanum lycopersicum*). *J. Sci. Food Agric.* **2010**, *90*, 2233–2240. [CrossRef] [PubMed]
52. Fray, R.G.; Grierson, D. Identification and genetic analysis of normal and mutant phytoene synthase genes of tomato by sequencing, complementation and co-suppression. *Plant Mol. Biol.* **1993**, *22*, 589–602. [CrossRef] [PubMed]
53. Vogel, J.T.; Tan, B.C.; McCarty, D.R.; Klee, H.J. The carotenoid cleavage dioxygenase 1 enzyme has broad substrate specificity, cleaving multiple carotenoids at two different bond positions. *J. Biol. Chem.* **2008**, *283*, 11364–11373. [CrossRef]
54. Koeduka, T.; Fridman, E.; Gang, D.R.; Vassão, D.G.; Jackson, B.L.; Kish, C.M.; Orlova, I.; Spassova, S.M.; Lewis, N.G.; Noel, J.P.; et al. Eugenol and isoeugenol, characteristic aromatic constituents of spices, are biosynthesized via reduction of a coniferyl alcohol ester. *Proc. Natl. Acad. Sci. USA* **2006**, *103*, 10128–10133. [CrossRef] [PubMed]
55. Borrelli, G.M.; Troccoli, A.; Di Fonzo, N.; Fares, C. Durum wheat lipoxygenase activity and other quality parameters that affect pasta color. *Cereal Chem.* **1999**, *76*, 335–340. [CrossRef]
56. Biehler, E.; Alkerwi, A.A.; Hoffmann, L.; Krause, E.; Guillaume, M.; Lair, M.L.; Bohn, T. Contribution of violaxanthin, neoxanthin, phytoene and phytofluene to total carotenoid intake: Assessment in Luxembourg. *J. Food Compos. Anal.* **2012**, *25*, 56–65. [CrossRef]
57. Liu, M.; Lu, S. Plastoquinone and ubiquinone in plants: Biosynthesis, physiological function and metabolic engineering. *Front. Plant Sci.* **2016**, *7*, 1898. [CrossRef]
58. Mintz-Oron, S.; Mandel, T.; Rogachev, I.; Feldberg, L.; Lotan, O.; Yativ, M.; Jetter, R.; Venger, I.; Adato, A.; Aharoni, A. Gene expression and metabolism in tomato fruit surface tissues. *Plant Physiol.* **2008**, *147*, 823–851. [CrossRef]
59. Fernandez-Moreno, J.P.; Tzfadia, O.; Forment, J.; Presa, S.; Rogachev, I.; Meir, S.; Orzaez, D.; Aharoni, A.; Granell, A. Characterization of a new pink-fruited tomato mutant results in the identification of a null allele of the SlMYB12 transcription factor. *Plant Physiol.* **2016**, *171*, 1821–1836. [CrossRef]
60. Tieman, D.; Taylor, M.; Schauer, N.; Fernie, A.R.; Hanson, A.D.; Klee, H.J. Tomato aromatic amino acid decarboxylases participate in synthesis of the flavor volatiles 2-phenylethanol and 2-phenylacetaldehyde. *Proc. Natl. Acad. Sci. USA* **2006**, *103*, 8287–8292. [CrossRef]
61. Baldwin, E.A.; Goodner, K.; Plotto, A. Interactions of volatiles, sugars, and acids on perception of tomato aroma and flavor descriptors. *J. Food Sci.* **2008**, *73*, S294–S307. [CrossRef] [PubMed]
62. Rambla, J.L.; Tikunov, Y.M.; Monforte, A.J.; Bovy, A.G.; Granell, A. The expanded tomato fruit volatile landscape. *J. Exp. Bot.* **2014**, *65*, 4613–4623. [CrossRef] [PubMed]
63. Chen, G.; Hackett, R.; Walker, D.; Taylor, A.; Lin, Z.; Grierson, D. Identification of a specific isoform of tomato lipoxygenase (TomloxC) involved in the generation of fatty acid-derived flavor compounds. *Plant Physiol.* **2004**, *136*, 2641–2651. [CrossRef] [PubMed]
64. Tieman, D.; Bliss, P.; McIntyre, L.M.; Blandon-Ubeda, A.; Bies, D.; Odabasi, A.Z.; Rodriguez, G.R.; van der Knaap, E.; Taylor, M.G.; Goulet, C. The chemical interactions underlying tomato flavor preferences. *Curr. Biol.* **2012**, *22*, 1035–1039. [CrossRef]
65. Schneider, C. Chemistry and biology of vitamin E. *Mol. Nutr. Food Res.* **2005**, *49*, 7–30. [CrossRef]
66. Domínguez, E.; España, L.; López-Casado, G.; Cuartero, J.; Heredia, A. Biomechanics of isolated tomato (*Solanum lycopersicum*) fruit cuticles during ripening: The role of flavonoids. *Funct. Plant Biol.* **2009**, *36*, 613–620. [CrossRef]

67. Akhtar, M.S.; Goldschmidt, E.E.; John, I.; Rodoni, S.; Matile, P.; Grierson, D. Altered patterns of senescence and ripening in *gf*, a stay-green mutant of tomato (*Lycopersicon esculentum* Mill.). *J. Exp. Bot.* **1999**, *50*, 1115–1122. [CrossRef]
68. Almeida, J.; Azevedo, M.D.S.; Spicher, L.; Glauser, G.; vom Dorp, K.; Guyer, L.; del Valle Carranza, A.; Asis, R.; de Souza, A.P. Down-regulation of tomato PHYTOL KINASE strongly impairs tocopherol biosynthesis and affects prenyllipid metabolism in an organ-specific manner. *J. Exp. Bot.* **2016**, *67*, 919–934. [CrossRef]
69. Lupi, A.C.D.; Lira, B.S.; Gramegna, G.; Trench, B.; Alves, F.R.R.; Demarco, D.; Peres, L.E.P.; Purgatto, E.; Freschi, L.; Rossi, M. *Solanum lycopersicum* GOLDEN 2-LIKE 2 transcription factor affects fruit quality in a light- and auxin-dependent manner. *PLoS ONE* **2019**, *14*, e0212224. [CrossRef]
70. Lin, C.Y.; Lee, C.H.; Chang, Y.W.; Wang, H.M.; Chen, C.Y.; Chen, Y.H. Pheophytin a inhibits inflammation via suppression of LPS-induced nitric oxide synthase-2, prostaglandin E2, and interleukin-1β of macrophages. *Int. J. Mol.* **2014**, *15*, 22819–22834. [CrossRef]
71. Mishra, V.K.; Bachet, R.K.; Husen, A. Medicinal uses of chlorophyll: A critical overview. In *Chlorophyll: Structure, Function and Medicinal Uses*; Le, H., Salcedo, E., Eds.; Nova Science Publishers, Inc.: Hauppauge, NY, USA, 2011.
72. Brigelius-Flohé, R.; Kelly, F.J.; Salonen, J.T.; Neuzil, J.; Zingg, J.M.; Azzi, A. The European perspective on vitamin E: Current knowledge and future research. *Am. J. Clin. Nutr.* **2002**, *76*, 703–716. [CrossRef] [PubMed]
73. Niki, E.; Traber, M.G. History of vitamin E. *Ann. Nutr. Metab.* **2012**, *61*, 207–212. [CrossRef] [PubMed]

metabolites

MDPI

Article

Placenta, Pericarp, and Seeds of Tabasco Chili Pepper Fruits Show a Contrasting Diversity of Bioactive Metabolites

Felipe Cervantes-Hernández, Paul Alcalá-González, Octavio Martínez and José Juan Ordaz-Ortiz *

Unidad de Genómica Avanzada, Centro de Investigación y de Estudios Avanzados del Instituto Politécnico Nacional (CINVESTAV), Km. 9.6, Libramiento Norte Carretera Irapuato-León, Irapuato, Gto. 36824, Mexico; felipe.cervantes@cinvestav.mx (F.C.-H.); paulpascual92@hotmail.com (P.A.-G.); octavio.martinez@cinvestav.mx (O.M.)

* Correspondence: jose.ordaz.ortiz@cinvestav.mx

Received: 26 August 2019; Accepted: 23 September 2019; Published: 28 September 2019

Abstract: Chili pepper (*Capsicum* spp.) is one of the most important horticultural crops worldwide, and its unique organoleptic properties and health benefits have been established for centuries. However, there is little knowledge about how metabolites are distributed throughout fruit parts. This work focuses on the use of liquid chromatography coupled with high resolution mass spectrometry (UHPLC-ESI-HRMS) to estimate the global metabolite profiles of the pericarp, placenta, and seeds of Tabasco pepper fruits (*Capsicum frutescens* L.) at the red mature stage of ripening. Our main results putatively identified 60 differential compounds between these tissues and seeds. Firstly, we found that pericarp has a higher content of glycosides, showing on average a fold change of 5 and a fold change of 14 for terpenoids when compared with other parts of the fruit. While placenta was the richest tissue in capsaicinoid-related compounds, alkaloids, and tocopherols, with a 35, 3, and 7 fold change, respectively. However, the seeds were richer in fatty acids and saponins with fold changes of 86 and 224, respectively. Therefore, our study demonstrates that a non-targeted metabolomic approach may help to improve our understanding of unexplored areas of plant metabolism and also may be the starting point for a detailed analysis in complex plant parts, such as fruits.

Keywords: *Capsicum frutescens* L.; non-targeted metabolomics; secondary metabolism; Liquid Chromatography coupled to Mass Spectrometry (LC-MS)

1. Introduction

Chili pepper (*Capsicum* spp.) is one of the most important crops worldwide. It is used as a main ingredient for many dishes in different cultures, such as Asian, Latin-American, and Mediterranean cultures, due to its organoleptic properties [1]. There are 40 accepted chili species but only five are considered domesticated: *C. annuum*, *C. chinense* Jacq, *C. frutescens*, *C. baccatum*, and *C. pubescens* [2]. In 2017, 36 million tons of chili peppers were produced globally, with Mexico being the second largest producer [3]. Wild pepper populations of Tabasco pepper (*C. frutescens* L.) and *C. annuum* can be found in some states of Mexico, increasing the relevance for collecting and characterizing these species as resources for the breeding of cultivated peppers [4,5]. Previous studies have extensively described metabolite diversity in *C. annuum* [6,7] but very little is known on *C. chinense* and *C. frutescens*. Most of these studies have undertaken a targeted approach, where the main focus has been to quantitate for specific metabolites, such as capsiate, dihydrocapsiate, capsaicin, dihydrocapsaicin, carotenoids, fatty acids, and amino acids [8,9]. To date, there are no studies with a comprehensive global profiling of tissue specific in *C. frutescens* and *C. chinense*.

Capsicum species are known to be rich in compounds such as capsaicinoids, capsinoids, carotenoids, flavonoids, vitamins, essential oils, and other phytochemicals, which provide a unique taste, aromatic properties, and health benefits [10–14]. Capsaicinoids consist mainly of two congeners, capsaicin and dihydrocapsaicin. Capsinoids also have two major analogues, capsiate and dihydrocapsiate. However, there are structural differences: Capsaicinoids are fatty acid amides linked with vanillylamine, whereas capsinoids are fatty acid esters linked to vanillyl alcohol [15,16]. Several compounds identified in the *Capsicum* species have been studied, because the medicinal potential examples of these compounds are icariside E5, capsosides, and capsianosides [17]; hydroxycinnamic derivatives, O-glycosides of quercetin, luteolin, and chrysoeriol [7]. The reported medicinal benefits are related to areas, such as anti-inflammatory [18], anti-cancer [19], anti-microbial, antioxidant properties [1,20,21], and those with weight-loss properties [22]. In addition, some epidemiological studies of a number of these antioxidants reported that they possess anti-atherosclerotic, antitumor, antimutagenic or anticarcinogenic activity [23–25]. It may be that these properties have the ability to help us to address the identification, isolation, and production of nutraceutical compounds or new natural medicinal compounds.

Nevertheless, the location and relative abundance of these metabolites and their precursors in different parts of the chili pepper fruit, such as the pericarp, placenta and seed, remain unclear. It is known that some compounds are synthesized and accumulated into specific tissues in the *Capsicum* genus [1,26]; for example, capsaicin is synthesized mainly in the placenta [27,28], while anthocyanins are described as being accumulated in pericarp during fruit development [29]. Materska, reported a placental and pericarp comparison in chili pepper fruit, where placenta was the richest in flavonoids, while the pericarp presented a larger diversity in glycosylated compounds. Despite this, little is known about the tissue specific spatial-temporal location of other classes of compounds and products of the secondary metabolism in *Capsicum* fruits [30]. Investigating the metabolic diversity on fruit tissues is essential in order to gain a comprehensive understanding of the function of specific parts of the fruit at the fundamental level. Consequently, this will enable the possible exploitation of natural products either for pharmaceutical or food products. In the past, these have been done with histochemical methods, by staining tissues sections with various chemical to reveal the presence of specific compounds either by visual or microscope inspection [31].

Metabolomics is defined as the comprehensive analysis of all low molecular weight organic compounds (<1500 Da) in a biological system [32]. Mass spectrometry has become the most widely applied platform for metabolomics, due to the wide range of molecules that can be analyzed on a single run [33]. Global profiling or non-targeted mass spectrometry-based metabolomics have gained importance in the study of crop species and have been applied to investigate potato, tomato, rice, wheat, strawberry, cucumber, and tobacco [34–37]. In the field of plant metabolomics, liquid chromatography coupled with electrospray ionization high resolution mass spectrometry (UPLC-ESI-HRMS) has emerged as the technique of choice for the putative identification of metabolites in complex matrices. This technique has been widely used, due to its sensitivity, selectivity, and analysis capability [38,39]. Nevertheless, metabolite identification for unknown compounds still remains a big challenge to overcome. In that respect, the recommendations by the Metabolomics Standards Initiative (MSI) recognize five different levels for metabolite confidence annotations. Level 0 requires the full compound 3D structure and stereochemistry information. Levels which are more common include: Level 1 identifications need a confirmation by two orthogonal parameters such as retention time and MS/MS spectrum, normally with match reference standards; and Level 2 requires at least two orthogonal pieces of information, including evidence that excludes all other candidates. Data for Level 2 should describe probable structure and be matched to literature data or databases by diagnostic evidence [40].

Combining existing bioinformatic tools with high resolution mass spectrometry data can reveal unclear relationships of metabolites and their possible function at a spatial-temporal distribution level. As a first attempt to construct the chili fruit metabolome, we produced a hand curated dataset, that contains 60 putative identified metabolites, which include alkaloids and terpenoids that are unreported in *Capsicum*, with significant differences and relative abundances between three sections (pericarp, placenta, and seed) of Tabasco pepper fruit at the mature red stage, using an UHPLC-QTOF-HRMS platform, combined with the use of Progenesis QI for small molecules, as a tool for the pre-identification of unknown metabolites. Our results underline the global metabolic differences in complexity, mainly based on the secondary metabolism of these fruit parts.

2. Results

2.1. Non-Targeted Metabolomic Analysis

Our data for two tissues and the seed of Tabasco chili fruit comprised a total of 87 files or chromatograms per ionization mode (positive and negative polarities) as shown in Figure 1 and were uploaded into Progenesis QI. The dataset was first aligned (retention time): Each chromatogram was aligned against each other and automatically compared to a reference profile selected by the software that contained the highest number of features (potential compounds). Then, peak picking was performed by default parameters. A total of 1980 features were detected in the aqueous phase and 1481 were detected in the diethyl ether extract, for both ionization modes. Figure 1 shows a typical chromatographic profile in positive and negative ionization mode of placenta tissue, displaying some representative compounds of chili pepper fruit.

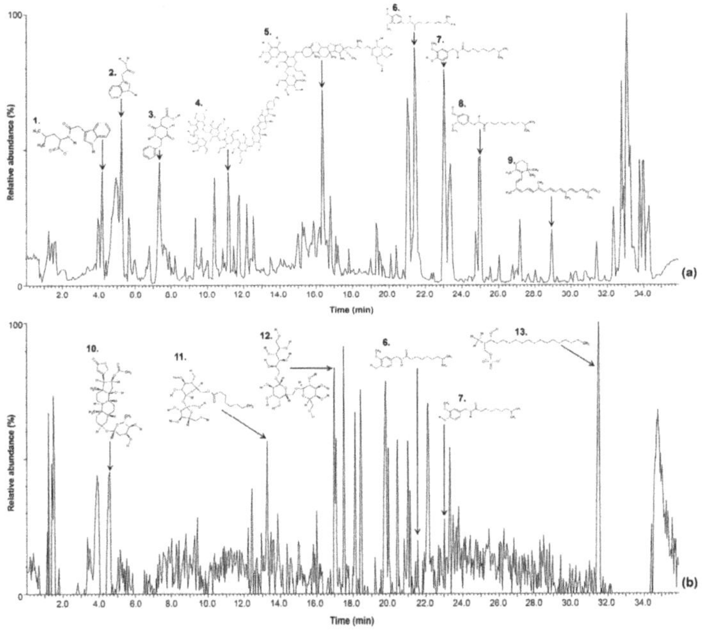

Figure 1. (**a**) Base peak intensity chromatographic profile of placenta tissue from chili pepper fruit on a Charged Surface Hybrid (CSH) C18 column obtained with Electrospray Ionization (ESI) positive mode on a mass range from 100 to 1500. 1. 2-[(1H-Indol-3-ylacetyl) amino]-4-methylpentanoate; 2. Indole-3-acetamide; 3. L-cis-Cyclo (aspartylphenylalanyl); 4. Capsicosin; 5. Yamogenintetroside B; 6. Capsaicin; 7. Dihydrocapsaicin; 8. Homodihydrocapsaicin; 9. β-Carotinal. (**b**) Base peak intensity chromatographic profile of placenta tissue from chili pepper fruit on a CSH C18 column obtained with ESI negative mode. 10. Oleandrigenin monodigitoxoside; 11. β-D-fructofuranosyl 6-O-octanoyl-α-D-glucopyranoside; 12. β-(1->6)-galactotriitol; 13. (2S,3R)-2-Azaniumyl-3-hydroxyoctadecyl phosphate.

Distribution of the features between parts was compared using principal component analysis (PCA) of loading and score plots describing a significant grouping by part for both extraction phases (Figure 2). In addition, quality controls (QC samples) were also considered, as shown in Figure 2. The QC samples cluster tightly in comparison to the total variance in the projection, suggesting a dataset deemed to be of high quality. Tables 1 and 2 describe loadings that mostly contribute to principal components for each extraction. Putative organic compounds catalogued as saponins (SPNS), such as Tuberoside J, Matesaponin 5, and Asparagoside B significantly contributed to component 1 in the aqueous phase, while other important features for component 2 belong to flavonoid class (FLV) and SPNS compounds. Furthermore, both components in the organic phase were mainly composed of glycerolipids (GL) and terpenoids (TER), as well as a putative carotenoid, (5cis,5′cis,9cis,11′cis)-1,2,7,7′,8,8′-Hexahydro-1,2-epoxy-ψ, ψ-carotene were important for contribution of component 1 in organic phase.

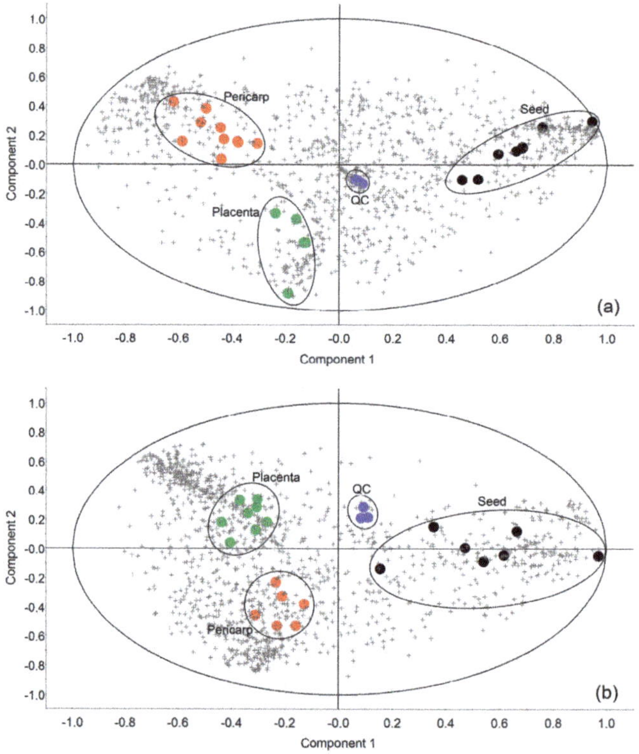

Figure 2. (**a**) Principal Component Analysis (PCA) Bi-plot of loadings (features: crosses) obtained in ESI positive mode and scores (samples: colored circles) extracted with methanol:water phase (component 1:33.10%; component 2: 15.72%; loadings = 1294 features; n = 26); (**b**) PCA Bi-plot of loadings (features: crosses) and scores (samples: colored circles) extracted with diethyl ether phase (component 1:37.15%; component 2: 15.14%; loadings = 1391 features; n = 24).

Table 1. Loadings most contributing to principal components for the aqueous phase.

Putative Identification	Class	PC1
Tuberoside J	SPNS	0.2012
Asparagoside B	SPNS	0.1939
Matesaponin 5	SPNS	0.1871
Oleanolic acid 3-O-[O-β-D-glucopyranosyl-(1->4)-O-β-D-glucopyranosyl-(1->3)-O-α-L-rhamnopyranosyl- (1->2)-α -L-arabinopyranoside]	SPNS	0.1822
Capsicosin	SPNS	0.1650
		PC2
(3″-Apiosyl-6″-malonyl) astragalin	FLV	0.1401
Pratenol B	BZD	0.0881
Asparagoside B	SPNS	0.0750
Matesaponin 5	SPNS	0.0673
Kaempferol 3-xylosylglucoside	FLV	0.0658

PC1: Principal Component 1; PC2: Principal Component 2; BZD: benzoyl derivate; FLV: flavonoid; SPNS: saponin.

Table 2. Loading most contributing to principal components for the organic phase.

Putative Identification	Class	PC1
Abietane	TER	0.1450
(5cis,5′cis,9cis,11′cis)-1,2,7,7′,8,8′-Hexahydro-1,2-epoxy-ψ, ψ-carotene	CARO	0.1429
Lycoperoside D	SPNS	0.1324
Phyllohydroquinone	TER	0.1302
2-Capryloolemyristin	GL	−0.0796
		PC2
α,α′-Trehalose 6-mycolate	GL	0.1505
2-Capryloolemyristin	GL	0.1226
MG(14:0/0:0/0:0)	GL	−0.1284
Uralenneoside	BZD	−0.0757
Abietane	TER	0.0719

PC1: Principal Component 1; PC2: principal Component 2; BZD: benzoyl derivate; Caro: carotenoid; GL: Glycerolipids; SPNS: saponin; TER: terpenoid.

2.2. *Level 1 and 2 Metabolomic Identification Analysis*

We putatively identified approximately 270 compounds and classified them in 52 compound classes (Table S1). Putative identifications were taken into consideration with a high match score (>90%). Terpenoids, fatty acids, and glycosylated compounds were the most abundant groups. As was predicted, different capsaicinoid compounds were also detected with a high match score. Alkaloids, carotenoids, saponins, and phenolic compounds were also detected in our study.

Capsaicin and dihydrocapsaicin were validated by matching their retention times and MS/MS spectra with those of the analytical standard (Level 1 identification). Furthermore, commonly reported compounds, such as carotenoids and capsaicin related compounds, were detected and putatively identified in the same manner in our samples (Figure 3).

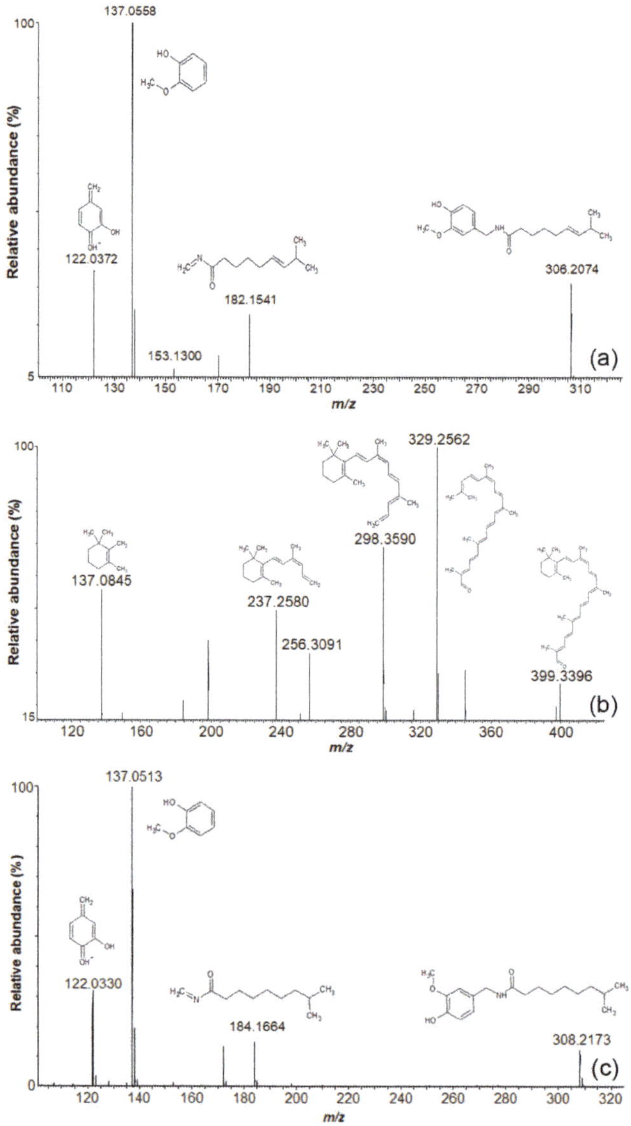

Figure 3. Mass spectrum of most common compounds in Tabasco chili pepper using Ultra High Pressure Liquid Chromatography MSMS Quadrupole Time of Flight (UHPLC-MS^2 Q-TOF; collision energy ramp: 20–40 eV) ESI positive ionization mode of placenta tissue, as putatively identified by Progenesis QI for small molecules. (**a**) Capsaicin; (**b**) β-carotinal; and (**c**) dihydrocapsaicin.

2.3. Different Metabolomic Profiles in Capsicum Sections

Based on results from volcano plot comparisons, we found a total of 60 putative compounds with significant differences between the parts of the Tabasco pepper fruit. Figure 4 shows the distribution of features between placenta and pericarp tissues. As was expected [28], capsaicinoids were more abundant in placenta than pericarp. In contrast, pericarp was richer in glycosylated compounds and terpenoids such as acalyphin, capsiate, and capsidiol. Some significant ions remain unknown that still need to be identified.

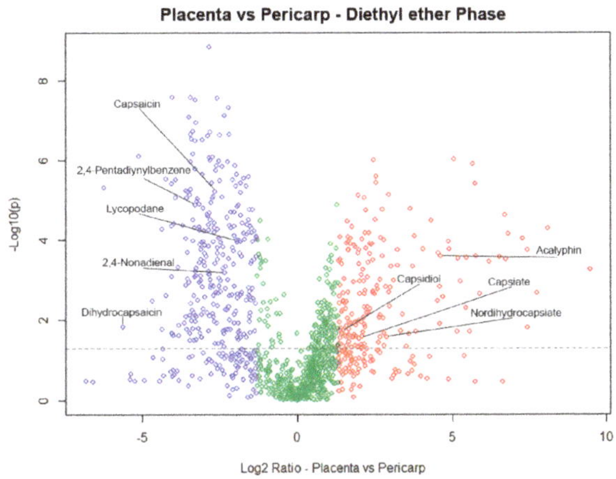

Figure 4. Volcano plot comparison of relative abundance between tissues of 1394 features in ESI positive ionization mode: Placenta (left), and pericarp (right), unchanged (green); one-way ANOVA $p = 0.05$ (dotted line) Y axis: p value, X axis: fold change.

After feature screening and putative identification, Venn diagrams were generated (Figure 5) to show similarities and differences between fruit parts, according to the fold change values obtained in Table S1. Noticeably, around 30% of the complete dataset of identification was shared by all three fruit parts in almost all the extraction solvents, the exception being for the organic phase in the negative ionization mode, which only share 9.32% similarities. As can be seen in the Venn diagrams, a greater number of putative metabolites were identified in the positive ionization mode. Those easily detected in the positive ionization mode were the compound classes alkaloids, carotenoids, fatty acids, glycosylated compounds, terpenoids, and saponins; while in the negative ionization mode amino acid-derivate compounds, sphingolipids, and phospholipids were detected.

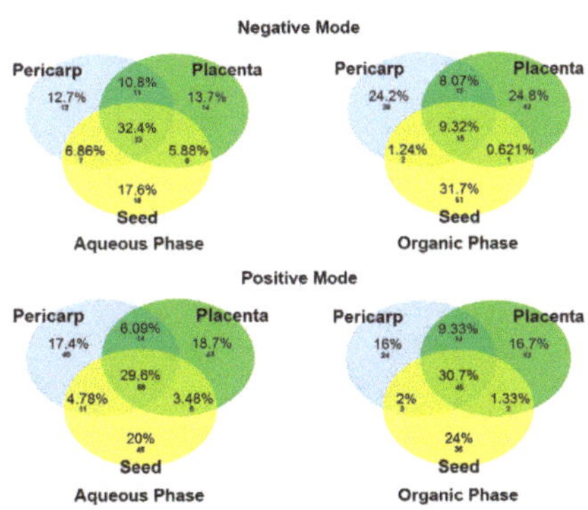

Figure 5. Venn diagrams of the complete dataset of putative metabolites in different fruit parts at the red mature stage of Tabasco chili pepper (*C. frutescens*); labels are in percent and number of metabolites.

Listed in Table 3 is the putative identification that presents a differential abundance between fruit parts, including the fragmentation pattern (MS^2) and adducts for putative annotation. Fold change values are shown in Table S1.

Table 3. Differential putative identifications in parts of Tabasco pepper fruit by UHPLC-MS^2 in both ESI modes.

Compound Name	Formula	Class	Adduct	Precursor (*m/z*)	Fragments (*m/z*)
α-campholenaldehyde	$C_{10}H_{16}O$	TER	[M + H − H_2O] +	135.1180	109.1021(3.7)
Jasmolone	$C_{11}H_{16}O_2$	JASM	[M + H − $2H_2O$] +	145.1027	133.1026(4.0), 121.1024(3.3), 107.0864(3.5)
2,4-Pentadiynylbenzene	$C_{11}H_8$	BZD	[2M + NH4] +	298.1669	177.0684(2.8), 145.0399(1.6), 117.0428(1.9)
Uralenneoside	$C_{12}H_{14}O_8$	BZD	[M + H] +	287.0755	287.0741(63.6), 285.0609(0.4), 257.0637(2.6), 203.0493(0.5), 153.0300(1.45), 135.0542(0.1)
Synephrine acetonide	$C_{12}H_{17}NO_2$	BZD	[2M + FA − H] -	459.2565	208.2805(8.9)
Cuscohygrine	$C_{13}H_{24}N_2O$	AK	[M + H] +	225.1977	197.1340(5.5), 183.1184 (5.6)
Acalyphin	$C_{14}H_{20}N_2O_9$	GC	[M + Na] +	383.1044	325.0952(1.3), 299.0774(4.6), 165.0311(0.3)
Pratenol B	$C_{15}H_{12}O_7$	BZD	[M + H − H_2O] +	287.0546	153.0195(2.8), 131.0512 (2.3)
Lycopodane	$C_{15}H_{25}N$	AK	[M + H − $2H_2O$] +	220.3782	184.1841(5.8)
Pedalitin	$C_{16}H_{12}O_7$	FLV	[M + H − H_2O] +	299.0570	299.0568(7.3), 165.0197(0.5)
Nordihydrocapsaicin	$C_{17}H_{27}NO_3$	CAPS	[M+H] +	294.2055	285.2240(3.6), 257.2282(2.8), 189.1653(3.9)
Nerolidyl acetate	$C_{17}H_{28}O_2$	TER	[M + H − $2H_2O$] +	229.1966	161.134 (12.7)
Capsaicin	$C_{18}H_{27}NO_3$	CAPS	[M + H] +	306.2075	182.1559(0.2), 137.0605(15.4), 122.0371(5.8)
Dihydrocapsaicin	$C_{18}H_{29}NO_3$	CAPS	[M + H] +	308.2240	9137.061 (5.5)
Artocarbene	$C_{19}H_{18}O_4$	PPN	[M + H] +	311.1301	175.0771(2.5), 169.0756(3.5), 163.0764(0.9), 160.0537(0.7), 137.0614(2.2), 131.0511(2.0)
1-(4-hydroxyphenyl)-7-phenyl-(6E)-6-hepten-3-ol	$C_{19}H_{22}O_2$	PPN	[M + Cl] −	317.1345	131.0808 (0.3)
Sterculynic acid	$C_{19}H_{30}O_2$	FAT	[M + H − H_2O] +	273.2235	273.2220(60.9), 255.2121(53.2), 173.1339(6.3), 163.0616(4.3), 161.1336(28.8), 147.1183(8.0)
Kaempferol 3-O-arabinoside	$C_{20}H_{18}O_{10}$	FLV	[M + H − $2H_2O$] +	383.0783	325.0730(4.0), 299.0568(7.3), 165.0197(0.5)
all-trans-3,4-Didehydroretinoate	$C_{20}H_{26}O_2$	PRN	[M + H] +	281.1929	181.1024(21.3), 165.0731(19.1), 157.1027(23.0), 155.0870(37.3), 145.1027(27.6), 128.0636(66.9)
Cinncassiol C	$C_{20}H_{28}O_7$	TER	[M + H − H_2O] +	363.1781	332.1368(0.6), 314.1253(0.5), 222.1141(0.6), 136.0677(2.1), 135.0456(0.2), 119.0495(0.8)
Isopimaric acid	$C_{20}H_{30}O_2$	TER	[M + H − $2H_2O$] +	285.2239	284.2974(0.7), 257.2282(2.8)
2′-Hydroxyisoorientin	$C_{21}H_{20}O_{12}$	FLV	[M + H] +	465.1051	303.0512 (7.4)
5,7,3′-trihydroxy-3,5′-dimethoxy-2′-(3′-methylbut-2-enyl)flavone	$C_{22}H_{22}O_7$	FLV	[M + H] +	399.1472	381.1379 (0.9)
Vestitone 7-glucoside	$C_{22}H_{26}O_9$	PPN	[M + ACN + H] +	417.1577	221.0831 (2.6)
6-O-Acetylaustroinulin	$C_{22}H_{36}O_4$	TER	[M + ACN + Na] +	787.5307	733.4879 (7.0)
xi-8-Acetonyldihydrosanguinarine	$C_{23}H_{19}NO_5$	AK	[M + H − H_2O] +	372.1245	344.1288(0.8), 149.0352(0.3)
Quercetin 3-(6″-malonyl-glucoside)	$C_{24}H_{22}O_{15}$	FLV	[M + H] +	551.1061	303.0514 (23.3)
12′-apo-β-carotenal	$C_{25}H_{34}O$	TER	[2M + FA − H] −	745.5259	685.5227(231.2), 539.4294(47.0)
Kaempferol 3-xylosylglucoside	$C_{26}H_{28}O_{15}$	FLV	[M + H] +	581.1525	341.2486(1.6), 287.0557(49.5), 153.0195(2.8), 131.0512(2.3)
11′-Carboxy-α-tocopherol	$C_{26}H_{42}O_4$	TPHE	[M + H] +	419.3222	177.1023 (0.5)
β-tocopherol	$C_{28}H_{48}O_2$	TPHE	[2M − H] -	831.7267	417.6959(90.2)
Amarogentin	$C_{29}H_{30}O_{13}$	GC	[M + NH4] +	604.2051	325.073(4.0), 299.0568(7.3), 165.0197(0.5)
Rhamnazin 3-rutinoside	$C_{29}H_{34}O_{16}$	FLV	[M + 2Na − H] +	683.1485	303.0514 (23.3)
Myrciacitrin V	$C_{30}H_{30}O_{13}$	FLV	[M + ACN + H] +	640.2098	151.0407 (5.4)
Bryononic acid	$C_{30}H_{46}O_3$	CBN	[M + ACN + Na] +	518.3645	358.1972(0.5), 342.2300(1.5), 320.2464(1.8), 222.1338(0.3), 196.1848(0.3)
3,7-Dihydroxy-25-methoxycucurbita-5,23-dien-19-al	$C_{31}H_{50}O_4$	STR	[M + Cl] −	521.3404	485.7277(21.0)
Capsianoside I	$C_{32}H_{52}O_{14}$	TER	[M + Na] +	683.3298	683.3291(4.4), 365.1088(1.6), 363.0929(1.6), 271.2444(7.5)

Table 3. *Cont.*

Compound Name	Formula	Class	Adduct	Precursor (m/z)	Fragments (m/z)
Diosgenin 3-O-beta-D-glucoside	$C_{33}H_{52}O_8$	TER	[M + H] +	577.3759	468.2101(67.0), 441.1756(9.8), 415.3230(23.2), 397.3135(4.5), 397.1857(3.8), 271.0622(4.8)
Kidjoranin-3-O-β-digitoxopyranoside	$C_{36}H_{48}O_{10}$	SPNS	[M − H_2O − H] −	621.3029	621.3011(167.6), 579.2889(8.2), 285.1144(46.0), 255.0975(5.6)
Feruloyl-β-sitosterol	$C_{39}H_{58}O_4$	TER	[2M + Hac − H] −	1239.9012	1239.893(28.0), 887.5754(11.0)
Ubiquinol-6	$C_{39}H_{60}O_4$	PRN	[M + Na] +	615.4544	394.3743(20.0), 322.2779(2.0), 310.3341(3.8), 134.1078(2.0)
Fistuloside A	$C_{39}H_{62}O_{13}$	SPNS	[M + H] +	739.4309	577.3766(32.5), 468.2101(67.0), 441.1756(9.8), 415.3230(23.2), 397.3135 (4.5),271.0622(4.8)
Nigroxanthin	$C_{40}H_{54}O_2$	TER	[M + 2Na − H] +	611.3841	467.2684(1.0), 449.3285(2.3), 305.2134(4.5), 287.2032(2.6), 269.1927(1.9)
Ursolic acid 3-[glucosyl-(1->4)-xyloside]	$C_{41}H_{66}O_{12}$	TER	[M + Na] +	773.4399	773.4389(4.5), 686.3793(2.1), 611.3844(5.0), 449.3285(2.3), 305.2134(4.5), 287.2032(2.6)
Melilotoside B	$C_{41}H_{68}O_{12}$	TER	[M + H] +	753.4203	267.1773 (0.4)
Licoricesaponin C2	$C_{42}H_{62}O_{15}$	TER	[M + Na] +	829.3952	829.3948(32.4), 723.3561(3.4),624.3786(4.6),310.1940(1.3), 250.1564(8.8), 146.0618(1.3)
Tuberoside L	$C_{51}H_{84}O_{23}$	SPNS	[M + H] +	1065.5627	670.3848(1.4), 611.3847(1.6), 449.3291(1.6), 432.3226(0.5)
Yamogenintetroside B	$C_{52}H_{86}O_{22}$	SPNS	[M + 2Na − H] +	1107.5319	854.4602(7.0), 762.4256(11.1), 559.4917(1.7), 541.4820(2.5), 426.3396(5.1), 309.1197(1.9)
Oleanolic acid 3-O-[O-β-D-glucopyranosyl-(1->4)-O-β-D-glucopyranosyl-(1->3)-O-α-L-rhamnopyranosyl-(1->2)-α-L-arabinopyranoside]	$C_{53}H_{86}O_{21}$	TER	[M + Na] +	1081.5464	773.4401(1.3), 611.3846(3.4), 449.3298(3.2), 153.0195(6.5)
Tragopogonsaponin F	$C_{56}H_{80}O_{21}$	SPNS	[M + CH_3OH + H] +	1121.5515	786.4322 (43.4)
Trigofoenoside G	$C_{56}H_{92}O_{27}$	SPNS	[M + H] +	1197.5995	1197.5994(31.3), 829.3948(32.4), 723.3561(3.4), 624.3786(4.6), 338.1889(1.5), 250.1564(8.8)
Hovenoside D	$C_{57}H_{92}O_{26}$	TER	[M + CH_3OH + H] +	1225.6028	1210.6307(124.9), 1064.5708(55.2), 870.4542(25.2), 442.3347(11.2), 325.1173(14.4), 301.0726(19.0)
Capsicosin	$C_{57}H_{94}O_{29}$	TER	[M + H] +	1243.6144	595.3883(18.9), 433.3333(18.6), 415.3237(9.2), 289.2185(18.0), 271.2091(10.8), 161.1340(12.7)
Eleutheroside L	$C_{59}H_{96}O_{25}$	SPNS	[M + Na] +	1227.6129	932.493(12.4), 399.3288(8.4), 285.2599(2.2)
β-L-arabinose 1-phosphate(2-)	$C_5H_9O_8P$-2	GC	[M + Cl] −	262.9693	262.9688(2.6), 218.9505(4.6)
Capsicoside A	$C_{63}H_{106}O_{35}$	SPNS	[M + H − $2H_2O$] +	1387.6618	901.4882(5.1), 739.4320(24.5), 577.3766(32.5), 468.2101(67.0), 441.1756(9.8), 415.3230(23.2)
Matesaponin 5	$C_{65}H_{106}O_{31}$	SPNS	[M + Na] +	1405.6713	757.4389(195.6), 595.3839(200.6), 451.2716(56.9), 289.2162(151.2), 271.2075(66.5), 253.1970(56.8)
Pyridoxamine	$C_8H_{12}N_2O_2$	PYR	[M+H−H_2O] +	151.0872	135.0247 (0.3)
3-[3,4-Dihydroxy-2-(hydroxymethyl)-1-pyrrolidinyl]propanamide	$C_8H_{16}N_2O_4$	AK	[M + H − H_2O] +	187.1093	175.1117(1.3), 155.0443(0.2), 116.0711(1.3), 112.0767(0.5), 109.0296(0.9)
2,4-Nonadienal	$C_9H_{14}O$	CBN	[M + K] +	177.0683	169.1144(11.7), 157.1133(15.3), 155.0982(17.7), 153.0822(13.1), 142.0889(25.6), 128.0724(24.6)

AK: Alkaloids; BZD: benzoyl derivate; CAPS: capsaicinoids; CBN: carbonyl derivate; FAT: fatty acids; FLV: flavonoids; GC: glycoside compounds; Jasm: jasmones; PPN: phenylpropanoids; PYR: pyrazines; SPNS: saponins; TER: terpenoids; TPHE: tocopherols. Relative abundance values of fragments ions are in brackets.

3. Discussion

The global metabolic comparison between the tissues and seeds of *C. frutescens* showed several feature differences between the pericarp, placenta, and seed. The Level 1 and 2 confidence metabolite annotations allowed us to assign a putative identification to these ions. Around 30% of metabolites were shared between all three parts. Compounds related to the primary metabolism showed few significant differences, they included amino acid related compounds, fatty acids, and phospholipids. As shown in the Venn diagram (Figure 5) and Table S1, placenta and pericarp have the biggest compound class diversity. Significantly, the seeds presented a higher number of putative identifications, and these were primarily saponins, terpenes, and fatty acids.

Pericarp compound classes were mainly composed of glycosylated compounds and terpenoids. Complementary to these findings, Materska demonstrated that chili pepper pericarp is abundant in glycosylated compounds [30]. Likewise, terpenoids were distributed in the whole fruit but pericarp showed a slightly higher proportion of them. These compounds are highly abundant in spices and herbs and give a wide range on the aroma and flavor spectrum [1]. Similarly, terpenoids have been described as showing antibiotic properties [41] and have been used in fragrances [42].

Placental tissue showed a large number of previously reported compounds with bioactivity, mainly capsaicin- and capsinoids-related compounds. In addition, alkaloids and tocopherols were present, a fact that is in agreement with current literature [28,30]. Found to be abundant in this fruit compartment were 6-O-acetylaustroinulin (terpenoid) and Myrciacitrin V (flavonoid) and they have not been reported in the *Capsicum* fruit.

The compounds found in chili seeds were predominantly fatty acids (3,16-Dihydroxypalmitate, Sterculynic acid) and saponins (Capsicoside A, Eleutheroside L and Tragopogonsaponin F) where they function as reserve nutrients for embryo development and propagation [43]. Moreover, seeds showed the presence of terpenes (Ursolic acid 3-[glucosyl-(1->4)-xyloside]) which are known to function as a natural promoter of predation and, as a consequence, a seed disperser [44]. Ritota et al. (2010) reported an abundance of fatty acids in sweet pepper species by nuclear magnetic resonance spectroscopy, in which polyunsaturated fatty acids were easily detected and pre-identified [6].

Our results were consistent with previously reported findings regarding the large diversity of secondary metabolites in fruits of *Capsicum* species and the non-targeted metabolomics profiling of Solanaceae [10,26,39,45–48]. Furthermore, new compounds, such as Myrciacitrin V, Feruloyl-β-sitosterol, 6-O-acetylaustroinulin and others, were putatively annotated as statistically significant in specific fruit parts.

Capsaicin, Dihydrocapsaicin, and capsaicinoids derivatives mainly accumulate in the placenta, as previously reported [27,49]. This class of compounds represents the most described and abundant metabolite in this genus and are predominantly known as being responsible for the pungency. Different bioactivity assays have been developed, demonstrating properties of capsaicinoids over different cell lines and metabolism, including as an analgesic and for weight-loss [22,50]. Large abundance of capsaicinoids in the chili fruit placenta was proposed by Tewksbury and Nabhan (2001), who suggest that capsaicin selectively discourages vertebrate predators (capsaicin has been found to repel or poison mammals) without deterring more effective and important seed dispersers, such as birds. [51].

A variety of new compounds in *Capsicum* genus, also reported in different species, were detected in pericarp, including isopimaric acid [41], which is a terpenoid with bioactive properties. Additionally, other compounds such as Pedalitin [52], Xi-8-acetonyldihydrosanguinarine [53], Pratenol B [54], Uralenneoside [55] were detected in pericarp and have been previously reported as bioactive compounds. Quercetin 3-(6″-malonyl-glucoside) is an anthocyanin-related compound that has not been reported in pepper fruit, but this compound class is well known to be localized mainly in pericarp [56,57], due to its involvement as a protection system against solar damage in plants and to attract potential pollinators [29].

New putative compounds in placental tissue, such as Lycopodane [58], 2,4-Pentadiynylbenzene [59], Myrciacitrin V [60], Cinncassiol C [61], and 6-O-acetylaustroinulin [62] have been reported as bioactive compounds, supporting the nutraceutical properties of chili pepper against metabolic disorders.

The existence of terpenes in seeds may result in different aromas that have been shown to firstly attract birds to mature fruits during the day [63] and secondly, to promote the dispersal of seeds. This function supports the ecological relationship between birds and chili pepper fruits, attracting the most beneficial vertebrate predators [51].

In summary, the non-targeted LC-MS metabolomics method that was developed in this study is shown to be a powerful tool for the putative identification of tissue-specific secondary metabolites at the red mature stage of chili pepper fruit. The use of databases available online gave rise to a faster comprehensive elucidation of global characteristics of a complex matrix than more traditional phytochemical studies. Nutraceutical, aroma, flavor, and new compounds that have not been reported before were putatively identified and related to pericarp, placenta or seeds of *C. frutescens*. As presented here, some of these compounds have been reported with bioactivity properties, supporting empirical properties of pepper fruit that have been known for centuries. The procedure developed here will be utilized for further studies in our laboratory, including to enable the exploration of comparisons between wild cultivars of chili pepper fruit with their cultivated counterparts and for the further understanding of secondary metabolism in this crop. We recommend that complementary analysis should be carried out to confirm structural elucidation. In addition, compound isolation and bioactivity properties should be considered in future studies.

4. Materials and Methods

4.1. Plant Material and Dissection of Tissues and Seed

Seeds of Tabasco pepper (*C. frutescens* L.) were treated with 3% hypochlorite solution. Plants were grown in optimum conditions (30–32 °C), at greenhouse facilities between June and September of 2016. Fruits from different plants were collected at 60 DPA (red ripe stage), washed with deionized water and immediately frozen with liquid nitrogen and stored at −80 °C until dissection and analysis.

Five biological replicates (plants) were considered for the experiment and three fruits per plant were collected. Each fruit was first placed into dry ice to facilitate hand dissection into pericarp, placenta, and seed using a sterile scalpel. All fruit parts were ground using a ball mill (Retsch MM301) under cold conditions and applying liquid nitrogen.

4.2. Chemicals, Reagents, and Standards

All chemicals and reagents were purchased from AccesoLab S.A. de C.V. (Mexico, Mexico). Capsaicin and dihydrocapsaicin analytical standards, formic acid, methanol, acetonitrile were HPLC grade and purchased from Sigma–Aldrich (Mexico, Mexico).

4.3. Sample Extraction and UHPLC-MS Analysis

For metabolite extraction, the method employed was adapted from Matyash [64] as follows: Methanol, 1.5 mL, was added to 100 mg of sample in a test tube and vortexed for 1 min, then, 5 mL of diethyl ether was added. The mixture was incubated with gentle stirring for one hour at room temperature. Next, 1.5 mL of ultra-pure water (18 Ω, milli-Q system) was added and mixed vigorously for a further minute then kept at room temperature for 10 min to allow phase separation. After that, the sample was centrifuged at 1000 × g for 10 min. Aqueous and organic layers were recovered separately and vacuum dried (miVac®, Genevac) at 30 °C for 30 min and finally kept at −80 °C until further analysis.

Three quality control (QC) samples were prepared to account for instrument drift and system calibration during analysis in UHPLC-QTOF-HRMS; each QC sample was prepared by mixing homogenously all sample extracts into a new single vial, in both separated phases containing polar and non-polar compounds. QC samples were distributed at the beginning, middle, and end of the injection run list. Analytical standards of capsaicin and dihydrocapsaicin were injected under the same conditions as samples. Extraction blanks were also considered during the experiment.

For LC-MS analysis, all samples (including QC, analytical standards and blank extraction) were resuspended in 1 mL of acetonitrile/ultra-pure water 50:50 (*v/v*) and filtered through a membrane of 0.2 μm (PTFE, Agilent Technologies, Santa Clara, USA). Samples were injected according to a randomized list order on an UPLC®(Acquity class I, Waters, Milford, CA, USA) coupled with an orthogonal QTOF (SYNAPT G1 HDMS, Waters, Milford, CA, USA) mass spectrometer. Chromatographic separation was achieved on a reversed phase CSH C18 column (2.1 mm × 150 mm, 1.7 μm, Waters, Milford, USA) maintained at 30 °C during chromatographic separation. Auto-sampling of 10 μL per sample was injected. Compounds were eluted using ultra-pure water with 0.1% (*v/v*) formic acid (solvent A) and acetonitrile with 0.1% (*v/v*) formic acid (solvent B) with a flow rate of 0.3 mL/min with the following gradient program: From 0.5 to 30 min, 1–75% B; 30 to 31 min, 75% B; 31 to 31.5 min, 75–100% B; 31.5 to 34.5, 100% B; 34.5 to 34.6, 100-1% B; 34.6 to 36 min, 1% B. The mass spectrometer mass range was set from 50 to 1500 Da. Both ionization modes were injected separated. For negative electrospray ionization (ESI) mode, the conditions were set as follows: Capillary voltage 2 kV; cone voltage 40 V; source temperature 150 °C; cone gas flow 20 L/h; desolvation temperature 350 °C; desolvation gas flow 600 L/h. For the positive ESI mode: Capillary voltage 3 kV; cone voltage 40 V; source temperature 130 °C; desolvation temperature 350 °C; desolvation gas flow 700 L/h. Leucine-Enkephalin (2 ng/mL) was infused as LockSpray reference internal mass calibrant at a flow rate of 5 μL/min and its signal was monitored every 10 s. The data format was collected in a continuum mode with a MS scan time of 1.5 s. In both the positive and negative ionization mode, data were acquired in MS^E experiments; using Argon as the collision gas with a collision energy in the trap region of 6 eV (Function 1, low energy) and ranged from 20–40 eV (Function 2, high voltage).

4.4. Data Analysis

Raw data was imported to Progenesis QI for small molecules software (Non-Linear Dynamics, Waters, Milford, MA, USA) for automatic alignment, normalization, deconvolution, and compound pre-identification over all samples separating the aqueous and organic phases. The RT range was limited from 0.5 to 35 min for pre-identification method. Pre-identification was performed using Chemspider Databases (PlantCyc, Plant Metabolic Network, KEGG, HMDB and ChEBI) and with an in-house database with a minimum match of 90% for precursor ions, MS/MS data and isotope distribution was included for increasing match score values. Statistics and graphics were performed using EZinfo 3.0 (Waters, Milford, MA, USA) and R (3.3.3v, Vienna, Austria) [65] software. Compounds were grouped according to their compound classes. The resulting data was first mean centered and scaled to Pareto and then submitted to a principal component analysis (PCA) using the first three components. Results were analyzed using one-way ANOVA and q-values were established using the false discovery rate (FDR < 0.01) to correct multiple comparisons by the Benjamini–Hochberg procedure [66].

Supplementary Materials: The following are available online at http://www.mdpi.com/2218-1989/9/10/206/s1, Table S1: Putative identification.xlsx.

Author Contributions: Conceptualization, J.J.O.-O.; data curation, F.C.-H. and P.A-G.; formal analysis, F.C.-H. and P.A.-G.; funding acquisition, J.J.O.-O.; methodology, F.C.-H. and P.A.-G.; supervision, O.M. and J.J.O.-O.; writing—original draft, F.C.-H.; writing—review and editing, O.M. and J.J.O.-O.

Funding: This work was supported by Consejo Nacional de Ciencia y Tecnología (CONACYT Mexico) for grant number 264354 and scholarship: 587878.

Acknowledgments: We thank Fernando Hernández-Godinez (Computational Biology Lab), María Esperanza Anaya-Gil (Metabolomics Lab), and Neftalí Ochoa-Alejo for their assistance, support, and suggestions during the experiment.

Conflicts of Interest: The funders had no role in the design of the study; in the collection, analyses, or interpretation of data; in the writing of the manuscript, or in the decision to publish the result.

References

1. Guzman, I.; Bosland, P.W.; O'Connell, M.A. Heat, Color, and Flavor Compounds in *Capsicum* Fruit. In *The Biological Activity of Phytochemicals*; Springer: New York, NY, USA, 2011; pp. 109–126. [CrossRef]
2. Bebeli, P.J.; Mazzucato, A. The Solanaceae—A Review of Recent Research on Genetic Resources and Advances in the Breeding of Tomato, Pepper and Eggplant. *Eur. J. Plant Sci. Biotechnol.* **2008**, *2*, 3–30.
3. FAOSTAT. Available online: http://www.fao.org/faostat/en/#data/QC (accessed on 2 November 2018).
4. Velázquez-Ventura, J.C.; de la Cruz-Lázaro, E.; Osorio-Osorio, R.; Preciado-Rangel, P. Morphological Variation of Wild Peppers (*Capsicum* spp.) from the State of Tabasco, Mexico. *Emir. J. Food Agric.* **2018**, *30*, 115–121. [CrossRef]
5. Hayano-Kanashiro, C.; Gámez-Meza, N.; Medina-Juárez, L.Á. Wild Pepper *Capsicum annuum* L. var. *glabriusculum*: Taxonomy, Plant Morphology, Distribution, Genetic Diversity, Genome Sequencing, and Phytochemical Compounds. *Crop Sci.* **2016**, *56*, 1–11. [CrossRef]
6. Ritota, M.; Marini, F.; Sequi, P.; Valentini, M. Metabolomic Characterization of Italian Sweet Pepper (*Capsicum annum* L.) by Means of HRMAS-NMR Spectroscopy and Multivariate Analysis. *J. Agric. Food Chem.* **2010**, *58*, 9675–9684. [CrossRef] [PubMed]
7. Marín, A.; Ferreres, F.; Tomás-Barberán, F.A.; Gil, M.I. Characterization and Quantitation of Antioxidant Constituents of Sweet Pepper (*Capsicum annuum* L.). *J. Agric. Food Chem.* **2004**, *52*, 3861–3869. [CrossRef]
8. Fayos, O.; Savirón, M.; Orduna, J.; Barbero, G.F.; Mallor, C.; Garcés-Claver, A. Quantitation of Capsiate and Dihydrocapsiate and Tentative Identification of Minor Capsinoids in Pepper Fruits (*Capsicum* spp.) by HPLC-ESI-MS/MS(QTOF). *Food Chem.* **2019**, *270*, 264–272. [CrossRef] [PubMed]
9. Ananthan, R.; Subhash, K.; Longvah, T. Capsaicinoids, Amino Acid and Fatty Acid Profiles in Different Fruit Components of the World Hottest Naga King Chilli (*Capsicum chinense* Jacq). *Food Chem.* **2018**, *238*, 51–57. [CrossRef] [PubMed]
10. Aranha, B.C.; Hoffmann, J.F.; Barbieri, R.L.; Rombaldi, C.V.; Chaves, F.C. Untargeted Metabolomic Analysis of *Capsicum* spp. by GC-MS. *Phytochem. Anal.* **2017**, *28*, 439–447. [CrossRef] [PubMed]
11. Fayos, O.; De Aguiar, A.C.; Jiménez-Cantizano, A.; Ferreiro-González, M.; Garcés-Claver, A.; Martínez, J.; Mallor, C.; Ruiz-Rodríguez, A.; Palma, M.; Barroso, C.G.; et al. Ontogenetic Variation of Individual and Total Capsaicinoids in Malagueta Peppers (*Capsicum frutescens*) during Fruit Maturation. *Molecules* **2017**, *22*, 736. [CrossRef]
12. Kumar, O.A.; Tata, S.S. Ascorbic Acid Contents in Chili Peppers (*Capsicum* L.). *Not. Sci. Biol.* **2009**, *1*, 50–52. [CrossRef]
13. Loizzo, M.R.; Pugliese, A.; Bonesi, M.; Menichini, F.; Tundis, R. Evaluation of Chemical Profile and Antioxidant Activity of Twenty Cultivars from *Capsicum annuum*, *Capsicum baccatum*, *Capsicum chacoense* and *Capsicum chinense*: A Comparison between Fresh and Processed Peppers. *LWT-Food Sci. Technol.* **2015**, *64*, 623–631. [CrossRef]
14. Nagy, Z.; Daood, H.; Koncsek, A.; Molnár, H.; Helyes, L. The Simultaneous Determination of Capsaicinoids, Tocopherols, and Carotenoids in Pungent Pepper Powder. *J. Liq. Chromatogr. Relat. Technol.* **2017**, *40*, 199–209. [CrossRef]
15. Kobata, K.; Tate, H.; Iwasaki, Y.; Tanaka, Y.; Ohtsu, K.; Yazawa, S.; Watanabe, T. Isolation of Coniferyl Esters from *Capsicum baccatum* L., and Their Enzymatic Preparation and Agonist Activity for TRPV1. *Phytochemistry* **2008**, *69*, 1179–1184. [CrossRef]
16. Huang, W.; Cheang, W.S.; Wang, X.; Lei, L.; Liu, Y.; Ma, K.Y.; Zheng, F.; Huang, Y.; Chen, Z.-Y. Capsaicinoids but Not Their Analogue Capsinoids Lower Plasma Cholesterol and Possess Beneficial Vascular Activity. *J. Agric. Food Chem.* **2014**, *62*. [CrossRef] [PubMed]
17. Iorizzi, M.; Lanzotti, V.; De Marino, S.; Zollo, F.; Blanco-Molina, M.; Macho, A.; Muñoz, E. New Glycosides from *Capsicum annuum* L. Var. *acuminatum*. Isolation, Structure Determination, and Biological Activity. *J. Agric. Food Chem.* **2001**, *49*, 2022–2029. [CrossRef] [PubMed]
18. Spiller, F.; Alves, M.K.; Vieira, S.M.; Carvalho, T.A.; Leite, C.E.; Lunardelli, A.; Poloni, J.A.; Cunha, F.Q.; de Oliveira, J.R. Anti-Inflammatory Effects of Red Pepper (*Capsicum baccatum*) on Carrageenan- and Antigen-Induced Inflammation. *J. Pharm. Pharmacol.* **2008**, *60*, 473–478. [CrossRef]

19. Anandakumar, P.; Kamaraj, S.; Jagan, S.; Ramakrishnan, G.; Devaki, T. Capsaicin Provokes Apoptosis and Restricts Benzo(a)Pyrene Induced Lung Tumorigenesis in Swiss Albino Mice. *Int. Immunopharmacol.* **2013**, *17*, 254–259. [CrossRef] [PubMed]
20. Jolayemi, A.; Ojewole, J. Comparative Anti-Inflammatory Properties of Capsaicin and Ethyl Acetate Extract of *Capsicum frutescens* Linn [Solanaceae] in Rats. *Afr. Health Sci.* **2013**, *13*, 357–361. [CrossRef]
21. M, S.; Gaur, R.; Sharma, V.; Chhapekar, S.S.; Das, J.; Kumar, A.; Yadava, S.K.; Nitin, M.; Brahma, V.; Abraham, S.K.; et al. Comparative Analysis of Fruit Metabolites and Pungency Candidate Genes Expression between Bhut Jolokia and Other *Capsicum* Species. *PLoS ONE* **2016**, *11*, e0167791. [CrossRef]
22. Varghese, S.; Kubatka, P.; Rodrigo, L.; Gazdikova, K.; Caprnda, M.; Fedotova, J.; Zulli, A.; Kruzliak, P.; Büsselberg, D. Chili Pepper as a Body Weight-Loss Food. *Int. J. Food Sci. Nutr.* **2017**, *68*, 392–401. [CrossRef]
23. Cai, Y.; Luo, Q.; Sun, M.; Corke, H. Antioxidant Activity and Phenolic Compounds of 112 Traditional Chinese Medicinal Plants Associated with Anticancer. *Life Sci.* **2004**, *74*, 2157–2184. [CrossRef] [PubMed]
24. Jackson, J.K.; Higo, T.; Hunter, W.L.; Burt, H.M. The Antioxidants Curcumin and Quercetin Inhibit Inflammatory Processes Associated with Arthritis. *Inflamm. Res.* **2006**, *55*, 168–175. [CrossRef]
25. Liu, R.H. Potential Synergy of Phytochemicals in Cancer Prevention: Mechanism of Action. *J. Nutr.* **2004**, *134*, 3479S–3485S. [CrossRef] [PubMed]
26. Wahyuni, Y.; Ballester, A.-R.; Sudarmonowati, E.; Bino, R.J.; Bovy, A.G. Metabolite Biodiversity in Pepper (*Capsicum*) Fruits of Thirty-Two Diverse Accessions: Variation in Health-Related Compounds and Implications for Breeding. *Phytochemistry* **2011**, *72*, 1358–1370. [CrossRef] [PubMed]
27. Gamboa-Becerra, R.; Ramírez-Chávez, E.; Molina-Torres, J.; Winkler, R. MSI.R Scripts Reveal Volatile and Semi-Volatile Features in Low-Temperature Plasma Mass Spectrometry Imaging (LTP-MSI) of Chilli (*Capsicum annuum*). *Anal. Bioanal. Chem.* **2015**, *407*, 5673–5684. [CrossRef] [PubMed]
28. Nugroho, L.H. Red Pepper (*Capsicum* spp.) Fruit: A Model for the Study of Secondary Metabolite Product Distribution and Its Management. *AIP Conf. Proc.* **2016**, *1744*, 020034. [CrossRef]
29. Aza-González, C.; Herrera-Isidrón, L.; Núñez-Palenius, H.G.; Martínez De La Vega, O.; Ochoa-Alejo, N. Anthocyanin Accumulation and Expression Analysis of Biosynthesis-Related Genes during Chili Pepper Fruit Development. *Biol. Plant.* **2013**, *57*, 49–55. [CrossRef]
30. Materska, M. Bioactive Phenolics of Fresh and Freeze-Dried Sweet and Semi-Spicy Pepper Fruits (*Capsicum annuum* L.). *J. Funct. Foods* **2014**, *7*, 269–277. [CrossRef]
31. Li, B.; Bhandari, D.R.; Janfelt, C.; Römpp, A.; Spengler, B. Natural products in *Glycyrrhiza glabra* (licorice) rhizome imaged at the cellular level by atmospheric pressure matrix-assisted laser desorption/ionization tandem mass spectrometry imaging. *Plant J.* **2014**, *80*, 161–171. [CrossRef]
32. Viant, M.R.; Kurland, I.J.; Jones, M.R.; Dunn, W.B. How Close Are We to Complete Annotation of Metabolomes? *Curr. Opin. Chem. Biol.* **2017**, *36*, 64–69. [CrossRef]
33. Kamphorst, J.J.; Lewis, I.A. Editorial Overview: Recent Innovations in the Metabolomics Revolution. *Curr. Opin. Biotechnol.* **2017**, *43*, 4–7. [CrossRef] [PubMed]
34. Frank, T.; Engel, K.-H. 8-Metabolomic Analysis of Plants and Crops. In *Metabolomics in Food and Nutrition*; Woodhead Publishing Limited: Sawston, UK, 2013; pp. 148–191. [CrossRef]
35. Schauer, N.; Fernie, A.R. Plant Metabolomics: Towards Biological Function and Mechanism. *Trends Plant Sci.* **2006**, *11*, 508–516. [CrossRef] [PubMed]
36. Christ, B.; Pluskal, T.; Aubry, S.; Weng, J.-K. Contribution of Untargeted Metabolomics for Future Assessment of Biotech Crops. *Trends Plant Sci.* **2018**, *23*, 1047–1056. [CrossRef] [PubMed]
37. Fang, C.; Fernie, A.R.; Luo, J. Exploring the Diversity of Plant Metabolism. *Trends Plant Sci.* **2019**, *24*, 83–98. [CrossRef] [PubMed]
38. Bijttebier, S.; Van der Auwera, A.; Foubert, K.; Voorspoels, S.; Pieters, L.; Apers, S. Bridging the Gap between Comprehensive Extraction Protocols in Plant Metabolomics Studies and Method Validation. *Anal. Chim. Acta* **2016**, *935*, 136–150. [CrossRef]
39. Matsuda, F.; Yonekura-Sakakibara, K.; Niida, R.; Kuromori, T.; Shinozaki, K.; Saito, K. MS/MS Spectral Tag-Based Annotation of Non-Targeted Profile of Plant Secondary Metabolites. *Plant J.* **2009**, *57*, 555–577. [CrossRef]
40. Blaženović, I.; Kind, T.; Ji, J.; Fiehn, O. Software Tools and Approaches for Compound Identification of LC-MS/MS Data in Metabolomics. *Metabolites* **2018**, *8*, 31. [CrossRef]

41. Smith, E.; Williamson, E.; Zloh, M.; Gibbons, S. Isopimaric Acid from *Pinus nigra* Shows Activity against Multidrug-Resistant and EMRSA Strains of *Staphylococcus aureus*. *Phyther. Res.* **2005**, *19*, 538–542. [CrossRef]
42. Bakkali, F.; Averbeck, S.; Averbeck, D.; Idaomar, M. Biological Effects of Essential Oils—A Review. *Food Chem. Toxicol.* **2008**, *46*, 446–475. [CrossRef]
43. Baud, S.; Dubreucq, B.; Miquel, M.; Rochat, C.; Lepiniec, L. Storage Reserve Accumulation in *Arabidopsis*: Metabolic and Developmental Control of Seed Filling. *Arab. B.* **2008**, e0113. [CrossRef] [PubMed]
44. Rodríguez, A.; Alquézar, B.; Peña, L. Fruit Aromas in Mature Fleshy Fruits as Signals of Readiness for Predation and Seed Dispersal. *New Phytol.* **2013**, *197*, 36–48. [CrossRef] [PubMed]
45. Rabara, R.C.; Tripathi, P.; Rushton, P.J. Comparative Metabolome Profile between Tobacco and Soybean Grown under Water-Stressed Conditions. *BioMed Res. Int.* **2017**, *2017*, 3065251. [CrossRef] [PubMed]
46. Yasir, M.; Sultana, B.; Anwar, F. LC—ESI—MS/MS Based Characterization of Phenolic Components in Fruits of Two Species of Solanaceae. *J. Food Sci. Technol.* **2017**, *55*, 2370–2376. [CrossRef] [PubMed]
47. López-Gresa, M.P.; Lisón, P.; Campos, L.; Rodrigo, I.; Rambla, J.L.; Granell, A.; Conejero, V.; Bellés, J.M. A Non-Targeted Metabolomics Approach Unravels the VOCs Associated with the Tomato Immune Response against *Pseudomonas Syringae*. *Front. Plant Sci.* **2017**, *8*, 01188. [CrossRef] [PubMed]
48. Heinig, U.; Aharoni, A. Analysis of Steroidal Alkaloids and Saponins in Solanaceae Plant Extracts Using UPLC-QTOF Mass Spectrometry. In *Methods in Molecular Biology*; Rodríguez-Concepción, M., Ed.; Springer: New York, NY, USA, 2014; Volume 1153, pp. 171–185. [CrossRef]
49. Aza-González, C.; Núñez-Palenius, H.G.; Ochoa-Alejo, N. Molecular Biology of Capsaicinoid Biosynthesis in Chili Pepper (*Capsicum* spp.). *Plant Cell Rep.* **2011**, *30*, 695–706. [CrossRef] [PubMed]
50. Reyes-Escogido, D.M.; Gonzalez-Mondragon, E.G.; Vazquez-Tzompantzi, E. Chemical and Pharmacological Aspects of Capsaicin. *Molecules* **2011**, *16*, 1253–1270. [CrossRef] [PubMed]
51. Tewksbury, J.J.; Nabhan, G.P. Seed Dispersal: Directed Deterrence by Capsaicin in Chillies. *Nature* **2001**, *412*, 403–404. [CrossRef]
52. Yamazaki, K.; Iwashina, T.; Kitajima, J.; Gamou, Y.; Yoshida, A.; Tannowa, T. External and Internal Flavonoids from Madagascarian Uncarina Species (Pedaliaceae). *Biochem. Syst. Ecol.* **2007**, *35*, 743–749. [CrossRef]
53. Furuya, T.; Ikuta, A.; Syōno, K. Alkaloids from Callus Tissue of *Papaver somniferum*. *Phytochemistry* **1972**, *11*, 3041–3044. [CrossRef]
54. Yoshihara, T.; Sakuma, T.; Ichihara, A. Yellow Fluorescent Stress Compounds, Pratenols A and B, from Red Clover (*Trifolium pratense*) Infected by *Kahatiella Caulivora*. *Biosci. Biotechnol. Biochem.* **1992**, *56*, 1955–1958. [CrossRef]
55. Jia, S.S.; Ma, C.M.; Li, Y.H.; Hao, J.H. Glycosides of Phenolic Acid and Flavonoids from the Leaves of *Glycyrrhiza uralensis* Ficsh. *Yao Xue Xue Bao* **1992**, *27*, 441–444. [PubMed]
56. Li, Q.; Somavat, P.; Singh, V.; Chatham, L.; Gonzalez de Mejia, E. A Comparative Study of Anthocyanin Distribution in Purple and Blue Corn Coproducts from Three Conventional Fractionation Processes. *Food Chem.* **2017**, *231*, 332–339. [CrossRef] [PubMed]
57. Wahyuni, Y.; Stahl-Hermes, V.; Ballester, A.R.; de Vos, R.C.H.; Voorrips, R.E.; Maharijaya, A.; Molthoff, J.; Zamora, M.V.; Sudarmonowati, E.; Arisi, A.C.M.; et al. Genetic Mapping of Semi-Polar Metabolites in Pepper Fruits (*Capsicum* sp.): Towards Unravelling the Molecular Regulation of Flavonoid Quantitative Trait Loci. *Mol. Breed.* **2014**, *33*, 503–518. [CrossRef] [PubMed]
58. Takayama, H.; Katakawa, K.; Kitajima, M.; Yamaguchi, K.; Aimi, N. Ten New Lycopodium Alkaloids Having the Lycopodane Skeleton Isolated from Lycopodium Serratum THUNB. *Chem. Pharm. Bull. Tokyo* **2003**, *51*, 1163–1169. [CrossRef] [PubMed]
59. Pappas, R.S.; Sturtz, G. Unusual Alkynes Found in the Essential Oil of *Artemisia dracunculus* L. Var. *dracunculus* from the Pacific Northwest. *J. Essent. Oil Res.* **2001**, *13*, 187–188. [CrossRef]
60. Bahmani, M.; Golshahi, H.; Saki, K.; Rafieian-Kopaei, M.; Delfan, B.; Mohammadi, T. Medicinal Plants and Secondary Metabolites for Diabetes Mellitus Control. *Asian Pac. J. Trop. Dis.* **2014**, *4*, S687–S692. [CrossRef]
61. Nohara, T.; Nishioka, I.; Tokubuchi, N.; Miyahara, K.; Kawasaki, T. Cinncassiol C1, a Novel Type of Diterpene from Cinnamomi Cortex. *Chem. Pharm. Bull. Tokyo* **1980**, *28*, 1969–1970. [CrossRef]
62. Novillo, F.; Velasco, E.; Delgado, G. Tri- and Diterpenoids from Some Species of Euphorbiaceae. Evaluation of Their Antiinflammatory and Cytotoxic Properties. *Planta Med.* **2015**, *81*, PM_106. [CrossRef]
63. Borges, R.M.; Ranganathan, Y.; Krishnan, A.; Ghara, M.; Pramanik, G. When Should Fig Fruit Produce Volatiles? Pattern in a Ripening Process. *Acta Oecol.* **2011**, *37*, 611–618. [CrossRef]

64. Matyash, V.; Liebisch, G.; Kurzchalia, T.V.; Shevchenko, A.; Schwudke, D. Lipid Extraction by Methyl-Tert-Butyl Ether for High-Throughput Lipidomics. *J. Lipid Res.* **2008**, *49*, 1137–1146. [CrossRef] [PubMed]
65. R Core Team. *R: A Language and Environment for Statistical Computing;* R Foundation for Statistical Computing: Vienna, Austria, 2018. Available online: https://www.R-project.org/ (accessed on 3 November 2018).
66. Benjamini, Y.; Yekutieli, D. The control of the false discovery rate in multiple testing under dependency. *Ann. Stat.* **2001**, *29*, 1165–1188.

metabolites

MDPI

Article

Phytohormone and Transcriptomic Analysis Reveals Endogenous Cytokinins Affect Kiwifruit Growth under Restricted Carbon Supply

Simona Nardozza [1,*], Janine Cooney [2], Helen L. Boldingh [2], Katrin G. Hewitt [2], Tania Trower [2], Dan Jones [1], Amali H. Thrimawithana [1], Andrew C. Allan [1,3] and Annette C. Richardson [4]

1 The New Zealand Institute for Plant and Food Research Limited (PFR), 1142 Auckland, New Zealand; dan.jones@plantandfood.co.nz (D.J.); amali.thrimawithana@plantandfood.co.nz (A.H.T.); andrew.allan@plantandfood.co.nz (A.C.A.)

2 The New Zealand Institute for Plant and Food Research Limited (PFR), 3240 Hamilton, New Zealand; janine.cooney@plantandfood.co.nz (J.C.); helen.boldingh@plantandfood.co.nz (H.L.B.); kati.hewitt@plantandfood.co.nz (K.G.H.); tania.trower@plantandfood.co.nz (T.T.)

3 School of Biological Sciences, University of Auckland, Private Bag 92019, 1142 Auckland, New Zealand

4 The New Zealand Institute for Plant and Food Research Limited (PFR), 0294 Kerikeri, New Zealand; annette.richardson@plantandfood.co.nz

* Correspondence: simona.nardozza@plantandfood.co.nz

Received: 16 December 2019; Accepted: 2 January 2020; Published: 4 January 2020

Abstract: Following cell division, fruit growth is characterized by both expansion through increases in cell volume and biomass accumulation in cells. Fruit growth is limited by carbon starvation; however, the mechanism controlling fruit growth under restricted carbohydrate supply is poorly understood. In a previous study using red-fleshed kiwifruit, we showed that long-term carbon starvation had detrimental effects on carbohydrate, anthocyanin metabolism, and fruit growth. To elucidate the mechanisms underlying the reduction in fruit growth during kiwifruit development, we integrated phytohormone profiling with transcriptomic and developmental datasets for fruit under high or low carbohydrate supplies. Phytohormone profiling of the outer pericarp tissue of kiwifruit showed a 6-fold reduction in total cytokinin concentrations in carbon-starved fruit, whilst other hormones were less affected. Principal component analysis visualised that cytokinin composition was distinct between fruit at 16 weeks after mid bloom, based on their carbohydrate supply status. Cytokinin biosynthetic genes (IPT, CYP735A) were significantly downregulated under carbon starvation, in agreement with the metabolite data. Several genes that code for expansins, proteins involved in cell wall loosening, were also downregulated under carbon starvation. In contrast to other fleshy fruits, our results suggest that cytokinins not only promote cell division, but also drive fruit cell expansion and growth in kiwifruit.

Keywords: cytokinin; fruit expansion; kiwifruit; phytohormone

1. Introduction

Actinidia Lindl. spp. (kiwifruit) fruit growth and development is characterised by a rapid growth phase, where cell numbers in the ovary tissue rapidly increase by cell division (up to four weeks after full bloom [1] in the outer pericarp), and a cell expansion phase where cell volume increases and starch accumulates in cells [2]. Later in development, cell expansion and fruit growth slow down, with fruit progressing through maturity and ripening [3]. Carbohydrate supply manipulation of girdled canes (removal of a bark strip around the cane containing the phloem tissue) during *Actinidia chinensis* var. *deliciosa* 'Hayward' fruit development significantly increased fruit weight, whilst fruit weight was not affected in ungirdled canes due to the ability of kiwifruit to redistribute photosynthates within the vine [4]. In addition to 'Hayward', *A. chinensis* var. *chinensis* 'Zes006' fruit weight was also affected when carbohydrate supply was manipulated in girdled shoots [5–7].

The dynamic interplay of phytohormones regulates the growth and development of fleshy fruit [8]. Cell division is primarily driven by cytokinins, whilst auxins and gibberellins are major players in cell expansion [9,10]. Abscisic acid increases as the fruit approaches maturity and together with ethylene is responsible for fruit ripening [11]. Studies on endogenous hormone profiles during kiwifruit development are limited, with most of the studies focused on the effects of exogenous hormone applications on fruit growth [12,13]. In kiwifruit, increases in endogenous cytokinin concentrations have been associated with cell division [14], fruit maturity, and ripening [14,15]. It has also been shown that exogenous cytokinin applications to developing kiwifruit after cell division had ceased increased both cell and fruit expansions [12]. In plants, cell expansion is the result of turgor pressure stress caused by an increase in cell osmotic potential and cell wall relaxation mediated by low-pH activated expansin proteins [16]. Expansin genes have been reported to be amongst genes induced by cytokinin [17], but their role in cytokinin-driven growth has not been elucidated.

Sugar and hormone signalling networks have been extensively studied in vegetative tissues (reviewed by Ljung et al. [18]) and to a lesser extent in reproductive tissues [19]. Cytokinins, in particular, have been implicated in the control of source–sink relationships of carbon compounds [20], and a role for them in controlling the availability of sugars in sink tissues has been proposed [21,22]. In Arabidopsis, high quantities of photosynthetically generated sugars have an effect on plant growth and on de novo cytokinin biosynthesis through the upregulation of two genes coding for key enzymes catalysing active cytokinin precursors synthesis: isopentenyltransferase (*AtIPT3*) and cytokinin hydroxylase (*CYP735A2*) [23]. Arabidopsis type-B response regulators, part of the MYB transcription factor family, mediate cytokinin transcriptional regulation [24]. A meta-analysis of published transcriptomic data identified a number of primary metabolic genes that are regulated by cytokinins including trehalose 6-phosphate pathway genes [17] and the *bZIP11* transcription factor. These genes are involved in the growth regulatory network including the sucrose non-fermenting-1 related protein kinase 1, which controls carbohydrate metabolism [25].

Kiwifruit carbon starvation mainly occurs when the leaf-to-fruit ratio on vines is low, but it can also be caused by environmental factors such as high temperature or low light during critical fruit growth phases reducing the photosynthetic carbon supply to fruit. In this study, we hypothesized that fruit growth under restricted carbohydrate supply was controlled at the phytohormonal level. To test this, we integrated phytohormone profiles with transcriptomic and developmental datasets for fruit grown under high or low carbohydrate supplies. Of all the phytohormones, cytokinin concentration positively correlated with fruit weight and the carbohydrate supply treatment. Cytokinin biosynthetic genes were also significantly downregulated under carbon starvation. As several expansin genes were also significantly downregulated under carbon starvation, we propose that cytokinin signalling affects fruit weight via regulation of cell expansion through expansins.

2. Results

2.1. Fruit Weight and Phytohormone Correlations

Carbon starvation during fruit development significantly restricted kiwifruit fruit growth by 16 weeks after mid bloom (WAMB; Figure 1a). Low carbohydrate supply fruit (carbon starved) were 25% smaller than untreated control fruit from ungirdled shoots, and 45% smaller than high carbohydrate supply fruit.

Phytohormones from the following classes were quantified in kiwifruit tissues: cytokinins, gibberellins, auxins, jasmonates, salicylates, and abscisates.

To ascertain if a particular phytohormone class was correlated with fruit weight, Pearson correlation analysis was performed. Phytohormones were measured from the fruit outer pericarp as this is the major fruit tissue, representing 60% of fruit tissue proportion [26,27]. Fruit weight was positively and highly correlated with total cytokinin concentration ($r = 0.91$, $p < 0.0001$, Figure 1b). Fruit weight was also negatively and significantly correlated with total gibberellins ($r = -0.61$, $p < 0.001$, Figure 1c), positively and significantly correlated with abscisic acid ($r = 0.65$, $p < 0.0001$, Figure 1d), and negatively and significantly correlated with jasmonic acid ($r = -0.55$, $p < 0.001$, Figure 1e). Fresh weight was not significantly correlated with other phytohormones such as salicylic acid and auxin (data not shown).

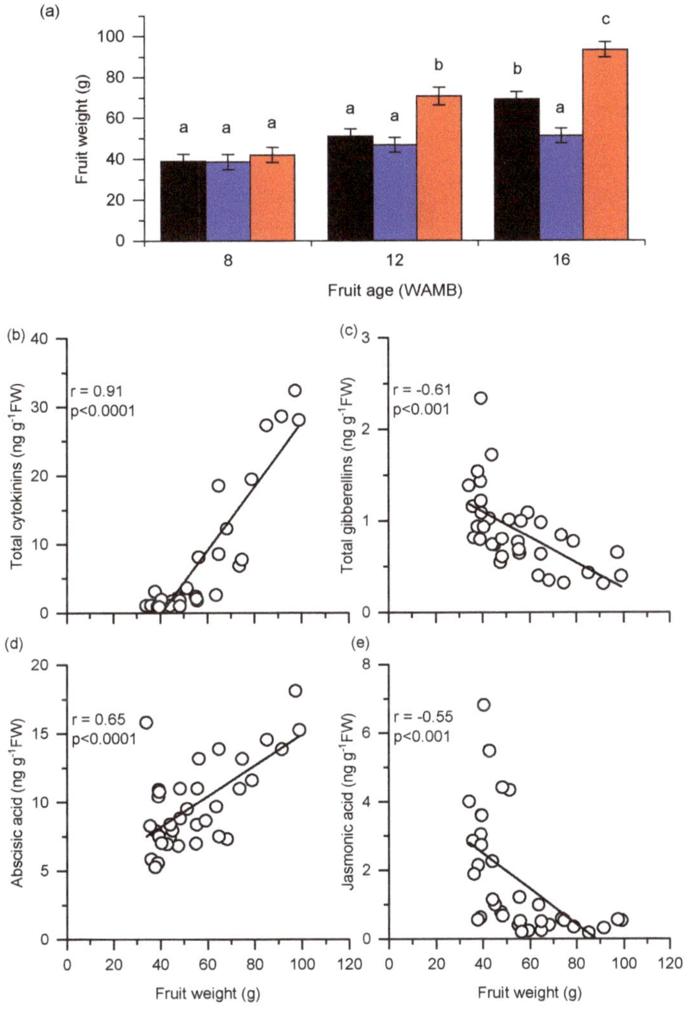

Figure 1. Fruit weight of kiwifruit (**a**) at three key stages of development: 8, 12, and 16 weeks after mid bloom (WAMB). Untreated control (black bar; vine standard), low carbohydrate supply (blue bar), and high carbohydrate supply (red bar). Values are averages ± SEM ($n = 3$ or 4). Statistical analysis by linear mixed effects model with type 3 sums of squares Kenward–Roger's method; different letters mean a statistical difference, adjusted for multiple comparison by Tukey's correction ($p < 0.05$). Pearson correlation analysis of between fruit weight and (**b**) total cytokinins, (**c**) total gibberellins, (**d**) abscisic acid, and (**e**) jasmonic acid concentration in fruit outer pericarp. $n = 33$ for cytokinins and $n = 35$ for the other phytohormones. FW, fresh weight.

2.2. Carbon Starvation Reduced Cytokinin Concentration in Fruit Outer Pericarp

The following cytokinin were quantified in kiwifruit outer pericarp tissue and were above the limit of detection (LOD): *trans*-zeatin (tZ), isopentenyl adenine (iP), *cis*-zeatin (cZ), dihydrozeatin riboside (DZR), isopentenyl adenine riboside (iPR), *trans*-zeatin riboside (tZR), *cis*-zeatin riboside (cZR), *trans*-zeatin-*O*-glucoside (tZROG), isopentenyl adenine-9-glucoside (iP9G), and *trans*-zeatin-9-glucoside (tZ9G) (Figure 2).

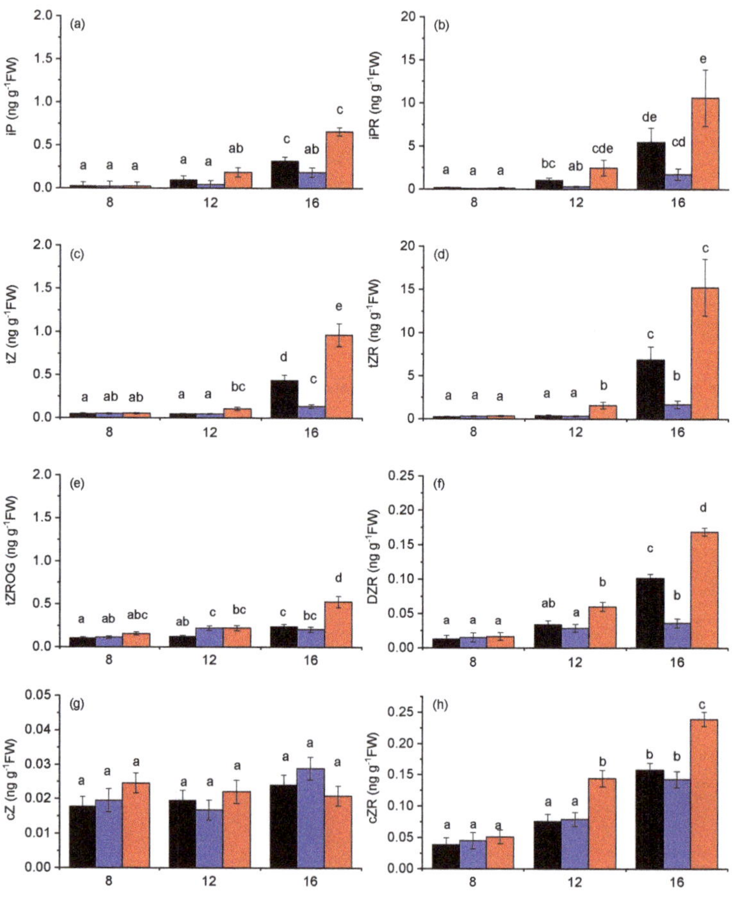

Figure 2. Cytokinin concentrations in the outer pericarp of developing kiwifruit: (**a**) iP, isopentenyl adenine; (**b**) iPR, isopentenyl adenine riboside; (**c**) tZ, *trans*-zeatin; (**d**) tZR, *trans*-zeatin riboside; (**e**) tZROG, *trans*-zeatin-*O*-glucoside; (**f**) DZR, dihydrozeatin riboside; (**g**) cZ, *cis*-zeatin; (**h**) cZR, *cis*-zeatin riboside.. Untreated control (black bar; vine standard), low carbohydrate supply (blue bar), and high carbohydrate supply (red bar). Values are averages ± SEM (n = 3 or 4). Statistical analysis by linear mixed effects model with type 3 sums of squares Kenward–Roger's method; different letters mean a statistical difference adjusted for multiple comparison by Tukey's correction ($p < 0.05$). Fruit age is in weeks after mid bloom (WAMB). FW, fresh weight.

tZ and iP had similar concentrations and represented the major active cytokinins in developing kiwifruit, followed by cZ with a concentration that was one order of magnitude lower (Figure 2a,c,g). Dihydrozeatin (DZ) concentration was below the detection level. In untreated control fruit, the proportional concentration of active cytokinins to the total cytokinin pool decreased during development from 10% at 8 WAMB to 6% at 16 WAMB (Table S1). Cytokinin ribosides iPR and tZR were the major cytokinins (Figure 2b,d). Other ribosides DZR and cZR were also detected, but their concentrations were one order of magnitude lower than those of iPR and tZR (Figure 2f,h). In untreated control fruit, the proportional concentration of cytokinin ribosides increased during development from 50% at 8 WAMB to 90% at 16 WAMB (Table S1), and this pattern was shared with the other two treatments. For the reversible *O*-glycosylated conjugates, only tZROG was above the detection level and its proportional concentration decreased in untreated control fruit from 10% at 8 WAMB to 2% at 16 WAMB (Table S1). Irreversibly *N*-glycosylated cytokinins (iP9G and tZ9G) also decreased in untreated control fruit during development, from 25% at 8 WAMB to 2% at 16 WAMB (Table S1). By 16 WAMB, carbon starvation significantly reduced the majority of the active cytokinins and their metabolic or glycosylated intermediates (Figure 2). iP and its riboside iPR were significantly lower in carbon-starved fruit (low carbohydrate supply; Figure 2a,b). Carbon starvation had a similar effect on tZ, tZR, and tZROG, and their concentrations were significantly lower in carbon-starved fruit than in high carbohydrate supply fruit (Figure 2c,d,e). DZR concentration was also significantly lowered by carbon starvation (Figure 2f). cZ was not affected by carbon starvation, but cZR concentrations were significantly higher in high carbohydrate supply fruit than in carbon-starved fruit and the untreated control (Figure 2g,h).

We further employed principal component analysis to assess the effects of the treatment on cytokinin composition. It is shown in Figure 3 that 59%, 57%, and 76% of variation can be explained by principal component 1 (PC1) at 8, 12, and 16 WAMB, respectively. An effect was first observed at 12 WAMB, with the cytokinin composition of high carbohydrate supply treatment fruit being distinct from that of the untreated control fruit and low carbohydrate supply treatment. At 16 WAMB, the data clustered into three distinct cohorts, of which high and low carbohydrate supply treated fruit showed the highest divergence in cytokinin composition. Dimension 1 was strongly and positively driven by all the measured cytokinins, with the exception of cZ, which mainly drove dimension 2 (Figure S1).

2.3. Carbon Starvation Had Lesser Effects on Other Phytohormones

Gibberellins GA1, GA19, and GA20 were present in the outer pericarp of kiwifruit in concentrations above the limit of detection (Figure 4a–c). Active gibberellin GA1 concentrations were 4-fold lower than concentrations of its precursor GA19. GA19 was the most abundant gibberellin, and its concentration decreased 2-fold during fruit development. GA20 was present at 4- to 8-fold lower concentrations than GA19, and its concentrations were stable during development. Under carbon starvation, GA19 concentrations decreased less during fruit development than in the other treatments (Figure 4b), and at 16 WAMB, it was about 3-fold higher than in high carbohydrate supply fruit ($p < 0.0001$).

Indole-3-acetic acid (IAA) was the only auxin detected by our system and its concentration slightly increased during fruit development (Figure 4d). Carbon starvation had no effect on IAA concentrations.

The concentration of the stress hormone abscisic acid (ABA) was stable during fruit development, although by 16 WAMB, its concentration in high carbohydrate supply fruit was significantly higher than in fruit from the other treatments (Figure 4e). This suggests that high carbohydrate supply fruit was the main driver of the positive correlation between fruit weight and ABA observed in Figure 1d (data not show). Jasmonic acid (JA) concentration decreased during fruit development and at 8 WAMB, its concentration was significantly higher in high carbohydrate supply fruit than in fruit from the other treatments (Figure 4f). JA bioactive isoleucine conjugate (JA-Ile) was below the detection level.

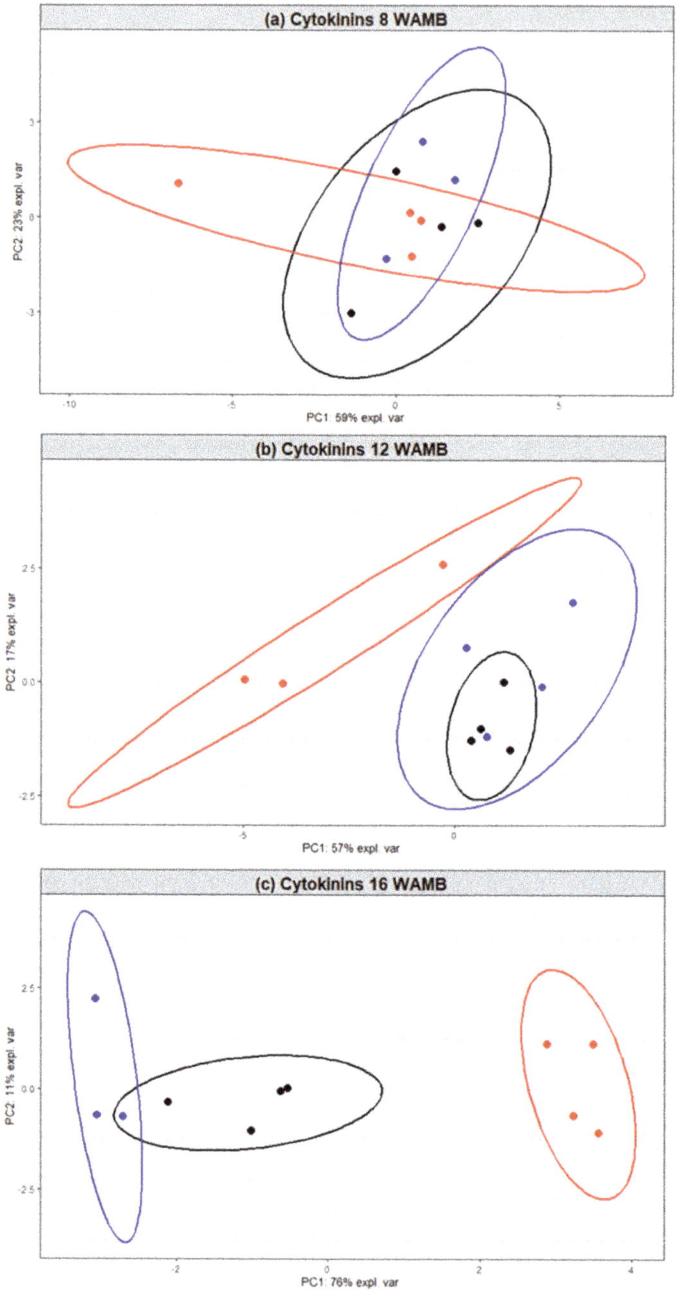

Figure 3. Principal component analysis (PCA) based on cytokinin concentrations in the outer pericarp of developing kiwifruit: (**a**) 8 weeks after mid bloom (WAMB); (**b**) 12 WAMB; (**c**) 16 WAMB. Black, untreated control (vine standard); blue, low carbohydrate supply (carbon starvation); red, high carbohydrate supply. PC, principal component. Confidence ellipses: $p < 0.05$.

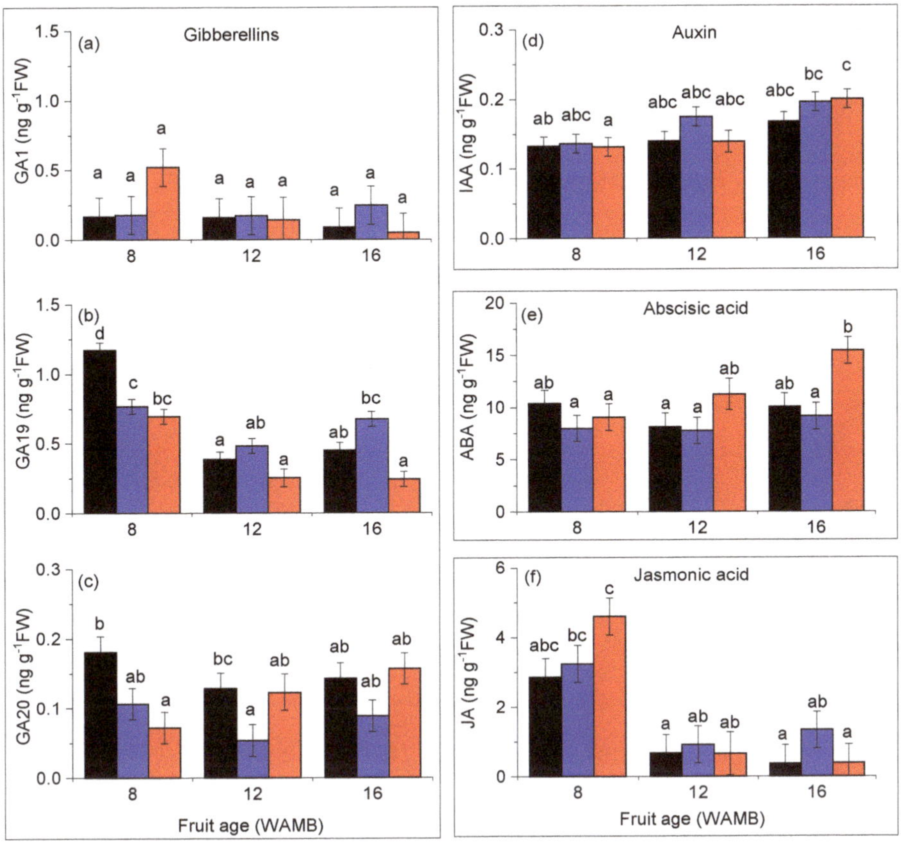

Figure 4. Gibberellins, auxin, abscisic acid, and jasmonic acid concentrations in the outer pericarp of developing kiwifruit: (**a**) GA1, gibberellin A1; (**b**) GA19, gibberellin A19; (**c**) GA20, gibberellin A20; (**d**) IAA, indole-3-acetic acid; (**e**) ABA, abscisic acid; (**f**) JA, jasmonic acid.. Untreated control (black bar; vine standard), low carbohydrate supply (carbon starvation; blue bar), and high carbohydrate supply (red bar). Values are averages ± SEM ($n = 4$). Statistical analysis by linear mixed effects model with type 3 sums of squares Kenward–Roger's method; different letters mean a statistical difference, adjusted for multiple comparison by Tukey's correction ($p < 0.05$). Fruit age is in weeks after mid bloom (WAMB). FW, fresh weight.

2.4. Carbon Starvation Downregulates Cytokinin Biosynthetic Genes

The cytokinin biosynthetic and catabolic pathway was constructed (Figure 5) using published information [28,29]. Gene models that code for the enzymes in the cytokinin pathway were identified by gene mining the kiwifruit genome [30] on the basis of previously published kiwifruit data [15] and homologues from other species [31,32], marked as blue in the chart (Figure 5). Where genes coding for a particular enzyme were unknown in Arabidopsis or were not found for kiwifruit, the enzymes were marked in grey.

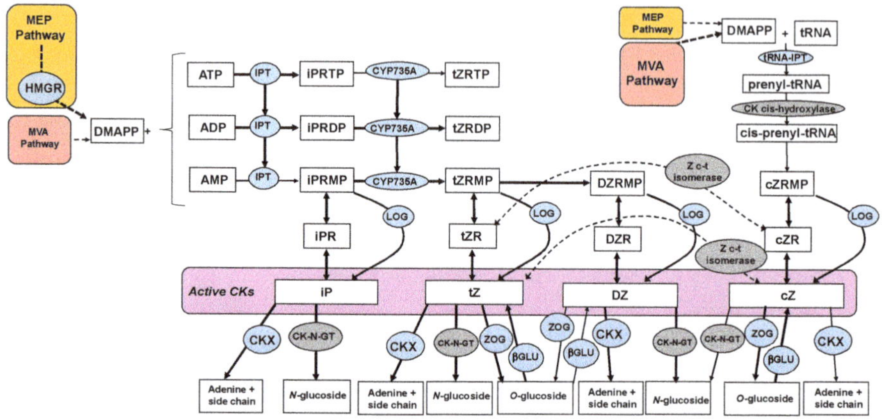

Figure 5. Cytokinin biosynthetic and catabolic pathways. Enzymes in blue are known and have been identified in kiwifruit; enzymes in grey are unknown and/or not identified in kiwifruit. Modified from [28,29]. βGLU, β-glucosidase; CK, cytokinin; CK-N-GT, cytokinin *N*-glucosyltransferase; CKX, cytokinin dehydrogenase; CYP735A, cytokinin hydroxylase; cZ, *cis*-zeatin; cZR, cZ riboside; cZRM(/D/T)P, cZ nucleotides; DMAPP, dimethylallyl pyrophosphate; DZ, dihydrozeatin; DZR, DZ, riboside; DZRM(/D/T)P, DZ nucleotides; HMGR, 3-hydroxy-3-methylglutaryl-coenzyme A reductase; iP, N^6-(Δ^2-isopentenyl)adenine; iPR, iP riboside; iPRM(/D/T)P, iP nucleotides; IPT, adenosine phosphate-isopentenyltransferase; MEP, methylerythritol phosphate pathway; LOG, LONELY GUY (cytokinin riboside 5′-monophosphate phosphoribohydrolase); MVA, mevalonate pathway; tRNA, transfer RNA; tZ, *trans*-zeatin; tZR, tZ riboside; tZRM(/D/T)P, tZ nucleotides; ZOG, zeatin *O*-glucosyltransferase.

The expression of a number of genes involved in cytokinin biosynthesis was reduced under carbon starvation. Five genes of the 3-hydroxy-3-methylglutaryl-coenzyme A reductase 1 family were expressed in kiwifruit and the *HMGR1.1* gene was significantly downregulated under carbon starvation (Figure 6). We identified nine adenylate isopentenyltransferase (*IPT*) genes: three *IPT1*, one *IPT2*, two *IPT3*, and three *IPT5*. Of these, *IPT1.2*, *IPT1.3*, and *IPT5.1* were significantly downregulated under carbon starvation from 12 WAMB. Of the four cytokinin hydroxylase 1 genes (*CYP735A1*) expressed, *CYP735A1.1* was significantly downregulated under carbon starvation from 12 WAMB.

Expression of cytokinin catabolic genes were also affected by carbon starvation, although to a lesser extent than the biosynthetic genes, with an overall upregulation of the genes involved in reversible conjugation (zeatin *O*-glucosyltransferase, ZOG; β-glucosidase, βGLU) or irreversible cleavage (cytokinin hydrogenase, CKX) (Figure 7). The main cytokinin dehydrogenase (*CKX5.3*) was only significantly upregulated at 12 WAMB.

2.5. Carbon Starvation Effects on the Genes in the Multistep Phosphorelay (MSP) Cytokinin Signalling

The genes involved in the first two steps of the multistep phosphorelay (MSP) cytokinin signalling were upregulated by carbon starvation (Figure S2). Histidine kinases *AHK2.1*, *AHK3.5*, *AHK4.2*, and *AHK4.3* were significantly upregulated at 16 WAMB (2-, 1.8-, 1.6-, and 1.6-fold, respectively). The histidine-containing phosphotransfer protein *AHP1.7* was 5-fold upregulated under carbon starvation. Four of the 12 A-type Arabidopsis Response Regulators (type-A ARR) genes identified in kiwifruit, which are negative regulators of cytokinin signalling, were downregulated under carbon starvation whilst four of the 19 type-B ARRs genes, which act as positive regulators of cytokinin signalling, were upregulated by this treatment (Figure S3). *ARR12.5* was 1.5-fold upregulated by carbon starvation from 12 WAMB.

Gene name	Gene model	Expression level (base mean)	Log2 fold-change Fruit age (WAMB)		
			8	12	16
3-hydroxy-3-methylglutaryl-coenzyme A reductase 1					
HMGR1.1	Acc01123.1	62.6	0.0	-0.3	**-1.2**
HMGR1.2	Acc23455.1	234.4	0.1	0.1	0.3
HMGR1.3	Acc04449.1	680.4	0.1	0.1	-0.3
HMGR1.4	Acc05319.1	0.4	-0.4	-1.1	-0.1
HMGR1.5	Acc10882.1	3286.9	-0.2	-0.1	0.4
Adenylate isopentenyltransferase					
IPT1.1	Acc28499.1	0.1	-0.7	-0.7	-0.1
IPT1.2	Acc31314.1	102.2	0.2	**-1.2**	**-1.3**
IPT1.3	Acc33739.1	52.0	-2.5	**-1.6**	**-1.2**
IPT2.1	Acc21528.1	150.4	-0.2	0.1	-0.1
IPT3.1	Acc19568.1	4.5	-0.1	0.9	0.7
IPT3.2	Acc31685.1	2.8	-0.2	0.8	0.0
IPT5.1	Acc27283.1	29.2	0.0	**-1.4**	**-1.4**
IPT5.2	Acc04302.1	2.4	-0.7	**-2.1**	-1.6
IPT5.3	Acc07021.1	0.7	-0.2	0.9	0.9
Cytokinin hydroxylase					
CYP735A1.1	Acc02242.1	158.3	-0.9	**-2.2**	**-1.1**
CYP735A1.2	Acc30843.1	0.0	0.0	0.2	-0.1
CYP735A1.3	Acc04002.1	0.1	0.0	0.6	-0.4
CYP735A1.4	Acc05869.1	0.0	0.0	-0.1	-0.1
LOG - Cytokinin riboside 5'-monophosphate phosphoribohydrolase					
LOG1.1	Acc30152.1	5.3	0.0	-0.7	**-2.8**
LOG1.2	Acc32210.1	1.7	-0.3	2.6	1.2
LOG1.3	Acc04561.1	0.0	-	-	-
LOG2.1	Acc11911.1	0.2	-0.4	0.0	-0.4
LOG2.2	Acc11946.1	0.1	0.4	0.7	0.6
LOG3.1	Acc15066.1	57.8	0.0	-0.1	-0.3
LOG3.2	Acc23573.1	0.1	0.0	2.1	-0.1
LOG3.3	Acc24762.1	1.9	0.2	-0.5	1.2
LOG3.4	Acc27441.1	0.9	0.4	0.0	0.7
LOG5.1	Acc11870.1	0.6	0.8	2.8	1.0
LOG5.1	Acc12721.1	0.0	0.0	-0.1	-0.1
LOG5.2	Acc07173.1	0.1	0.8	0.4	-0.1
LOG5.3	Acc33270.1	36.6	0.0	-0.2	-0.3
LOG5.4	Acc10850.1	0.5	0.4	0.5	2.1
LOG7.1	Acc13368.1	0.0	0.4	-0.1	-0.1
LOG7.2	Acc22968.1	0.0	-	-	-
LOG8.1	Acc22618.1	26.9	-0.6	0.3	**2.1**
LOG8.2	Acc28672.1	2.9	0.0	1.4	0.7
LOG8.3	Acc29501.1	135.7	0.0	0.4	**0.5**
LOG8.4	Acc09301.1	0.2	0.9	-0.1	-0.1

Log2 fold-change
-5 0 5

Figure 6. Heat-map of kiwifruit cytokinin biosynthetic genes. DESeq comparison of low carbohydrate supply (carbon starvation) versus high carbohydrate supply treatments. Differentially expressed genes were identified based on the DESeq analysis from Nardozza et al. [6]. For each gene, the expression level (base mean) and the log2 fold-change are presented. Fruit age is in weeks after mid bloom (WAMB). Bold figures mean differences are significant for adjusted $p < 0.05$ (DESeq analysis).

Gene name	Gene model	Expression level (base mean)	Log2 fold-change Fruit age (WAMB)		
			8	12	16
Zeatin O-glucosyltransferase					
ZOG1.1	Acc13178.1	0.1	0.0	-0.4	1.0
ZOG1.2	Acc02214.1	3.4	0.2	-0.2	-0.1
ZOG1.3	Acc23139.1	0.6	-0.2	0.5	-0.1
ZOG1.4	Acc03972.1	129.9	0.5	**0.8**	**0.7**
ZOG1.5	Acc10216.1	2.9	0.7	0.6	**1.4**
ZOG1.6	Acc10217.1	0.0	0.1	-0.2	-0.1
ZOG2.1	Acc15775.1	77.9	-0.4	**-1.4**	**-0.9**
ZOG2.2	Acc17636.1	0.1	-0.3	-0.1	-0.1
ZOG2.3	Acc18218.1	20.3	0.3	0.1	0.1
ZOG2.4	Acc18219.1	0.4	-0.7	-0.1	0.4
ZOG2.5	Acc21186.1	177.0	-0.3	-0.7	**-0.6**
ZOG2.6	Acc21210.1	71.7	0.2	0.4	**1.1**
ZOG2.7	Acc23933.1	9.4	0.2	2.7	**4.9**
ZOG2.8	Acc23934.1	9.2	1.1	-0.6	-0.3
ZOG2.9	Acc23935.1	31.2	-0.2	-0.2	**1.2**
ZOG2.10	Acc24009.1	0.1	0.2	-0.1	-0.1
ZOG2.11	Acc24128.1	1.7	**0.5**	**3.4**	2.2
ZOG2.12	Acc24129.1	0.4	0.5	0.2	-0.1
ZOG2.13	Acc05179.1	2.7	0.3	0.9	1.1
ZOG2.14	Acc06782.1	25.7	-0.7	**2.4**	**3.8**
ZOG2.15	Acc07982.1	1.2	0.0	-0.9	-1.7
ZOG2.16	Acc07984.1	1.4	-0.6	1.3	**2.2**
ZOG3.1	Acc20480.1	168.1	0.8	0.5	-0.1
β-glucosidase					
BGLU12.1	Acc14271.1	3.5	-0.4	0.0	-0.1
BGLU12.2	Acc17106.1	17.3	0.1	-0.3	-0.3
BGLU12.3	Acc17107.1	0.0	0.0	-0.1	-0.4
BGLU12.4	Acc17108.1	0.0	-	-	-
BGLU12.5	Acc17109.1	0.0	-	-	-
BGLU12.6	Acc17110.1	0.0	-	-	-
BGLU12.7	Acc30424.1	19.4	-0.1	0.1	0.3
BGLU12.8	Acc09332.1	0.4	0.9	-0.7	-0.1
BGLU12.9	Acc09333.1	2.6	-0.2	0.6	-0.4
BGLU12.10	Acc09334.1	0.0	0.4	-0.1	-0.1
BGLU12.11	Acc09335.1	1.2	-1.5	2.6	**3.7**
BGLU12.12	Acc09336.1	0.0	0.0	-0.1	-0.1
BGLU40.1	Acc31471.1	2.4	0.3	0.1	**3.6**
BGLU40.2	Acc06385.1	60.9	0.0	**1.3**	**2.5**
BGLU41.1	Acc21668.1	3.9	0.4	0.1	0.3
BGLU42.1	Acc24746.1	298.4	0.1	-0.2	**-0.3**
BGLU44.1	Acc18581.1	126.9	0.1	1.6	0.7
BGLU47.1.	Acc25036.1	38.2	0.4	0.4	0.4
Cytokinin dehydrogenase					
CKX1.1	Acc23129.1	40.1	-0.1	0.2	0.0
CKX3.1	Acc00936.1	0.0	-	-	-
CKX3.2	Acc12996.1	1.1	-0.4	1.7	**1.7**
CKX3.3	Acc23338.1	0.1	0.4	-0.1	0.6
CKX3.4	Acc09843.1	0.0	-	-	-
CKX5.1	Acc23617.1	2.5	0.7	**3.4**	**2.8**
CKX5.2	Acc28235.1	86.2	0.1	0.0	0.4
CKX5.3	Acc04606.1	417.4	0.3	**0.8**	0.4
CKX5.4	Acc33634.1	0.0	-	-	-
CKX6.1	Acc17044.1	130.6	0.3	0.4	0.4
CKX6.2	Acc09272.1	0.3	0.4	-1.5	1.1

Log2 fold-change

-5 0 5

Figure 7. Heat-map of kiwifruit cytokinin catabolic genes. DESeq comparison of low carbohydrate supply (carbon starvation) versus high carbohydrate supply treatments. Differentially expressed genes were identified based on the DESeq analysis from Nardozza et al. [6]. For each gene, the expression level (base mean) and the log2 fold-change are presented. Fruit age is in weeks after mid bloom (WAMB). Bold figures mean differences are significant for adjusted $p < 0.05$ (DESeq analysis).

2.6. Carbon Starvation Results in Downregulation of Expansin Genes

Expansins are a class of cell wall proteins able to drive non-enzymatic pH-dependent cell wall relaxation by partnering with H(+)-ATPAses proton pumps [16,33]. Of the expansin genes identified in kiwifruit [30], the expansin A class genes were the most affected by carbon starvation at 16 WAMB. *EXP2-7* and *EXP24* were significantly downregulated under carbon starvation (Figure 8). Highly expressed *EXP3, EXP5,* and *EXP6* were downregulated by 3.5-, 4.5-, and 9-fold, respectively. In addition, plasma membrane H(+)-ATPase genes were also downregulated (Figure S4).

Gene name	Gene model	Expression level (base mean)	Log2 fold-change Fruit age (WAMB)		
			8	12	16
EXPA - Expansin A					
EXP1	Acc08682.1	20.5	0.1	0.5	0.5
EXP2	Acc31871.1	23.3	0.5	-2.3	**-7.2**
EXP3	Acc27868.1	1483.2	0.3	-1.0	**-1.8**
EXP4	Acc14128.1	12.0	1.5	**-2.3**	**-2.3**
EXP5	Acc33002.1	647.2	0.2	**-1.7**	**-2.2**
EXP6	Acc30734.1	1243.2	0.3	0.9	**-3.2**
EXP7	Acc32338.1	31.8	0.9	-0.5	**-2.6**
EXP8	Acc01142.1	111.3	0.0	0.0	-0.3
EXP9	Acc00770.1	0.3	0.9	0.2	-0.1
EXP10	Acc10127.1	0.1	0.0	-0.7	0.3
EXP11	Acc26691.1	0.5	1.2	-0.2	0.3
EXP12	Acc21929.1	9.6	0.1	0.4	-0.4
EXP13	Acc18284.1	0.0	-	-	-
EXP14	Acc31509.1	0.0	-	-	-
EXP16	Acc06347.1	15.3	0.9	**-3.7**	**-2.8**
EXP17	Acc02787.1	14.6	0.0	-0.4	0.4
EXP18	Acc01513.1	1.7	0.8	1.0	-1.4
EXP19	Acc11893.1	3.7	0.6	-0.1	0.3
EXP20	Acc33655.1	0.0	-	-	-
EXP21	Acc03046.1	0.0	0.4	-0.1	-0.1
EXP22	Acc01870.1	0.1	-0.4	-0.1	-0.4
EXP23	Acc30273.1	4.6	0.8	-0.1	0.3
EXP24	Acc25860.1	58.6	0.4	-0.3	**-5.0**
EXP25	Acc24889.1	16.0	0.7	**2.6**	**2.1**
EXP26	Acc02877.1	1.6	0.7	2.1	1.6
EXP27	Acc01632.1	0.0	0.0	0.2	-0.1
EXP28	Acc10880.1	10.9	0.1	0.5	-0.3
EXP29	Acc33740.1	0.0	-	-	-
EXP30	Acc10482.1	2.9	-0.3	0.4	0.1
EXP31	Acc08457.1	0.1	0.0	-0.7	-0.1
EXP32	Acc18326.1	0.1	0.1	-0.4	-1.2
EXP33	Acc19956.1	0.2	0.5	1.0	1.0
EXP34	Acc15741.1	0.0	-	-	-
EXP41	Acc18873.1	0.2	0.2	-0.1	-0.1
EXPB - Expansin B					
EXPLB2.1	Acc07668.1	0.0	-	-	-
EXPLB2.2	Acc00462.1	0.1	-0.7	-0.1	-0.1
EXPLB3.1	Acc26079.1	0.8	0.9	-0.2	-0.1
EXPLB3.2	Acc19000.1	0.1	0.8	-0.1	-0.1
EXPLB3.3	Acc13823.1	1.1	0.2	-0.2	0.3
EXLA - Expansin-like A					
EXPLA1	Acc10849.1	58.5	0.4	0.6	**-1.6**
EXPLA2	Acc07172.1	14.0	0.3	**1.0**	-1.9
EXPLA3	Acc16346.1	0.3	-0.3	-0.2	-0.8
EXLB - Expansin-like B					
EXPLB1.1	Acc28101.1	0.0	-	-	-
EXPLB1.2	Acc15856.1	0.1	0.0	-1.3	-0.1
EXPLB1.3	Acc28974.1	2.5	-0.6	1.5	-0.2
EXPLB1.4	Acc16689.1	3.7	-0.2	0.6	-0.6

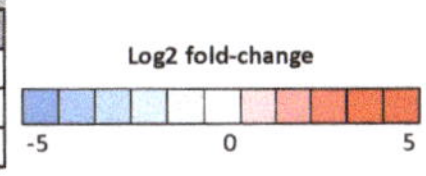

Figure 8. Heat-map of kiwifruit expansin genes. DESeq comparison of low carbohydrate supply (carbon starvation) versus high carbohydrate supply treatments. Differentially expressed genes were identified based on the DESeq analysis from Nardozza et al. [6]. For each gene, the expression level (base mean) and the log2 fold-change are presented. Fruit age is in weeks after mid bloom (WAMB). Bold figures mean differences are significant for adjusted $p < 0.05$ (DESeq analysis).

3. Discussion

In kiwifruit, fruit growth increases with an increase in carbohydrate supply [4,5]. The fruit expansion phase in kiwifruit appears to differ from those of other fleshy fruits. Auxin and/or gibberellins drive fruit expansion in strawberry [34], apple [10], and tomato [35]. Our results showed that cytokinins are key phytohormones during fruit expansion in kiwifruit. Phytohormone profiling revealed that fruit weight differences obtained under contrasting carbohydrate supplies positively correlated with cytokinin concentrations in the fruit outer pericarp. Changes in gibberellin, jasmonic acid, and abscisic acid concentrations were mostly developmental in nature and not strongly affected by carbohydrate supply. Interestingly, gibberellin precursor GA19 increased under carbon starvation, however, this effect was not maintained downstream in the biosynthetic pathway, and both GA20 and active GA1 concentrations were unaffected, suggesting that the role of gibberellin was not significant in the kiwifruit cell expansion phase. Surprisingly, unconjugated auxin concentrations were not correlated with fruit growth and were also not affected by fruit carbohydrate supply. This is in contrast to observations in fleshy fruit types [10,34,35], suggesting that free auxin is not critical during the fruit expansion phase of kiwifruit. However, we cannot rule out that conjugated auxins, which were not measured in this study, may play a role.

Similar to Arabidopsis seedlings, where de novo cytokinin synthesis is induced by sugar, our data showed that in carbon-starved kiwifruit, cytokinin synthesis is repressed. This is via downregulation of the cytokinin biosynthetic genes *IPTs* and *CYP735As*, suggesting that sugar control over cytokinin production could apply to sink tissues. Signalling of carbon starvation in kiwifruit is mediated by the sucrose non-fermenting-1 related protein kinase 1 via trehalose 6-phosphate [6]. In Arabidopsis, it has been proposed that cytokinins have a role in regulating the genes of the trehalose 6-phosphate pathway [17]. A similar scenario may occur in kiwifruit outer pericarp tissue, where the observed reduction of trehalose 6-phosphate concentration in carbon-starved fruit [6] could be triggered by cytokinin signalling.

In maturing kiwifruit, cytokinin catabolism was not critical for controlling cytokinin concentrations [15], whilst in other plant species (e.g., cabbage, maize, and wheat) it has been suggested that *IPT* and *CKX* transcription levels are coupled [31,36,37] and CKX is a key enzyme for cytokinin homeostasis. Our results show that genes coding for cytokinin catabolic enzymes were only marginally affected by carbon supply treatments and cytokinin dehydrogenase was not associated with different cytokinin concentrations (Figure 2; Figure 7). These data suggest that, in kiwifruit, carbon starvation affects de novo synthesis rather than irreversible cleavage of cytokinins during the fruit expansion phase.

Whilst transcriptional data for cytokinin biosynthetic genes supported the phytohormone data, the regulation of the multistep phosphorelay cytokinin signalling was less clear. Under carbon starvation, we observed an upregulation of the genes involved in the multistep phosphorelay pathway (histidine kinase, histidine phosphotransferase, and type-B RRs), suggesting that they might not be controlled at the transcriptional level as transcription is triggered by decreased cytokinin concentrations. Multistep phosphorelay signalling is triggered by cytokinin binding, which leads to autophosphorylation of histidine kinase. The phosphate is then transferred to histidine phosphotransferase and the final acceptor type-B RR proteins, which positively regulate transcription in the nucleus (reviewed by Kieber and Schaller [38]). In contrast, type-A RRs are negative feedback regulators in cytokinin signalling and are transcriptionally activated by type-B RRs [39]. In kiwifruit outer pericarp, type-A RRs were mostly downregulated when cytokinin concentrations were low. This is in agreement with findings from Bhargava et al. [40], where type-A response regulators were upregulated in response to exogenous cytokinin applications, suggesting an increase in endogenous cytokinins could lead to a similar response.

Expansins are proteins involved in the acidic relaxation of plant cell walls and drivers of short-term cell expansion [41]. Effects of cytokinins on expansin-mediated cell wall relaxation and cell expansion has been suggested [42]. A meta-analysis of cytokinin effects on Arabidopsis transcriptome identified expansins as targets of cytokinin signalling [43]. Expansins drive acidic cell elongation in Arabidopsis roots, where α-expansinA genes and plasma membrane H(+)-ATPase are controlled by cytokinins via ARR1 [44]. In soybean, cytokinins regulate the expression of expansin and cell wall expansion [45]. The effects of cytokinins on expansin during cell expansion and fruit expansion have not been well characterised. A study of grape berries associated cytokinin concentrations with post-veraison cell expansion and berry growth [19]. Exogenous cytokinin application induced expansin expression in white sweet clover (*Melilotus alba*) [42]. In kiwifruit, under carbon starvation, we observed significant decreases in cytokinin concentrations, cytokinin biosynthetic genes transcription, a-expansin transcription, and fruit weight, suggesting a role for expansin in driving cell expansion and fruit expansion. Cytokinin concentrations were developmentally driven as the positive correlation between total cytokinin concentration and fruit weight held true also within each treatment, in agreement with Pilkington et al. [15]. Further experimentation will be required to clarify the mechanism and if, similar to Arabidopsis roots, the signalling may be mediated by a type-B RR [44].

In conclusion, our results suggest a new role for cytokinins in kiwifruit growth, where they contribute to stimulate cell expansion via a mechanism that could involve expansin protein. This is in contrast to other fleshy fruits where cytokinins are a key phytohormone class during cell division, and further supports the responsiveness of kiwifruit to exogenous cytokinin treatments following the cell division phase. We acknowledge a limitation of this study in discussing the functional role of cytokinin in kiwifruit as developmental phases were inferred from the literature [1,2], rather than defined from direct cell number and cell size observations. Overall, these findings contribute to expand the current knowledge on fruit weight determination in kiwifruit.

4. Materials and Methods

4.1. Plant Material

Actinidia chinensis (Planch.) var. *chinensis* 'Zes006' (red fleshed) kiwifruit was sampled as per the experiments described in Nardozza et al. [6]. The outer pericarp tissue from the following treatments was used: high carbohydrate supply (girdled shoot with leaf to fruit ratio of 4), low carbohydrate supply (girdled shoot with a leaf to fruit ratio of 1), and untreated control (ungirdled shoot with a leaf to fruit ratio of 1; vine standard [5]). Individual shoots represented the experimental unit. Girdles were applied to shoots at 4 WAMB and maintained open until the end of the experiment. Only samples collected at 8, 12, and 16 WAMB were considered to match the transcriptomic data. Fresh fruit weight data are given in Nardozza et al. [6]. Outer pericarp tissues were collected to include three to four biological replicates (due to lost samples in the field or during analysis; see details in Table S2), snap frozen in liquid nitrogen, and stored at –80 °C until further analysis.

4.2. Phytohormones

4.2.1. Phytohormone Extraction and Fractionation

Frozen plant material was ground in liquid nitrogen to a fine powder using a mortar and pestle and stored at –80 °C for chemical analysis. To each sample (200 mg fresh weight), 1 mL chilled (4 °C) extraction solvent 80:20 CH_3CN:H_2O, labelled internal standard mix ([$^2H_5$] Z 10 ng, [$^2H_5$] (9R)Z 3 ng, [$^2H_5$] t-ZROG 30 ng, [$^2H_5$] (9G)Z 3 ng, [$^2H_6$] iP 0.5 ng, [$^2H_6$] (9R) iP 1 ng, [$^2H_6$] iP9G 3 ng, [$^2H_5$] t-ZOG 7 ng, [$^2H_2$] GA7 0.5 ng, [$^2H_4$] SA 1 ng, [$^2H_6$] ABA 2 ng, [$^2H_5$] JA 2 ng, [$^2H_{10}$] JA-Ile 2 ng, [$^{13}C_6$] IAA 10 ng; OlchemIm Ltd., Olomouc, Czech Republic), and 0.8 g stainless steel beads 0.9–2 mm (Next Advance Inc., Troy, NY, USA) were added. Samples were bead beaten for 5 min (Bullet Blender 24 Gold, Next Advance Inc., Troy, NY, USA) and then extracted overnight in the dark at 4 °C using an end-over-end rotator at 30 rotations/min. After centrifugation at 13,000× *g* for 5 min, the supernatant was transferred into a 96-well collection plate (Phenomenex, Torrance, CA, USA). The remaining pellet was re-extracted twice with 1 mL 80:20 CH_3CN:H_2O overnight, and combined with the first supernatant, and evaporated to dryness using a CentriVap concentrator (Labcon, Petaluma, CA, USA). Samples were reconstituted in 1 mL 1 M formic acid (aq) and placed on an orbital shaker at room temperature for 1 h at 230 rpm (IKA Labortechnik, Staufen, Germany). To remove interfering compounds, the extract was passed through a SOLA SCX 96-well plate (10 mg/2 mL, Thermo Scientific) equilibrated with 4 mL acetonitrile and conditioned with 4 mL 1 M formic acid (aq). After conditioning, the reconstituted samples were loaded, washed with 3 mL of 1 M formic acid (aq), 3 mL of water, and the acidic plant hormones were eluted with 1.5 mL acetonitrile (Fraction A). The plate was then washed with 1 mL water, followed by 1.5 mL of 0.35 M ammonium hydroxide and the cytokinins were eluted with 0.5 mL of 0.35 M ammonium hydroxide in 60% acetonitrile (Fraction B). Each fraction was evaporated to dryness using a CentriVap concentrator (Labcon, Petaluma, CA, USA).

Prior to derivatisation with bromocholine bromide (BETA), Fraction A was further purified using a modification of a method described by Kojima et al. [46] to increase the sensitivity of analysis for gibberellins. Briefly, samples were reconstituted in 1 mL 80:20 CH_3CN:H_2O + 4% trimethylamine and applied to a Hypersep 96-well plate (Hypercarb 25 mg/1 mL, Thermo Scientific) equilibrated with 1 mL acetonitrile and conditioned with 1 mL water. The sample eluate was collected into a 96-well collection plate and a further 0.5 mL CH_3CN + 4% trimethylamine was applied to the Hypersep plate and collected into the same collection plate. After evaporation, the samples were reconstituted with 160 μL CH_3CN and to each sample 40 μL of derivatisation solution (6.25 M BETA in 86% acetonitrile) and 20 μL 4% trimethylamine in CH_3CN was added. The mixed solution was incubated at 50 °C at 300 rpm for 24 h (Eppendorf, ThermoMixer C, Hamburg, Germany) and then evaporated to dryness.

4.2.2. Liquid-Chromatography Tandem Mass-Spectrometry (LC-MS/MS) Analysis

Liquid-Chromatography Tandem Mass-Spectrometry (LC-MS/MS) experiments were performed on a 5500 QTrap triple quadrupole/linear ion trap (QqLIT) mass spectrometer equipped with a TurboIon-Spray^{TM} interface (AB Sciex, Concord, ON, Canada) coupled to an Ultimate 3000 UHPLC (Dionex, Sunnyvale, CA, USA).

4.2.3. Cytokinins

For cytokinins, dried samples from Fraction B were reconstituted in 200 μL 10:90 CH_3OH:H_2O + 1% acetic acid and filtered through a conditioned (200 μL CH_3OH) 96-well 0.45 μm hydrophobic filter plate (Pall Filters, AcroPrep, Cortland, NY, USA) prior to mass spectrometric (MS) analysis. Cytokinins were separated on an Acquity UPLC BEH C18 1.7 μm 2.1 × 150 mm ID column (Waters, Wexford, Ireland) maintained at 70 °C. Solvents were (A) 15 mM ammonium formate adjusted to pH 4 with formic acid and (B) methanol with a flow rate of 350 μL min^{-1}. The initial mobile phase, 15% B was held for 6.5 min, then ramped linearly to 20% B at 9 min, then to 50% B at 12.5 min and 100% B at 14 min and holding at 100% B for 1 min before resetting to the original conditions. Injection size was 10 μL. MS data were acquired in the positive ion mode using a scheduled multiple reaction monitoring (MRM) method. The transitions monitored (Q1 and Q3) are listed in Table S3. Other operating parameters were as follows: ion spray voltage 4500 V; temperature 600 °C; curtain gas 45 psi; ion source gas 1 60 psi; ion source gas 2 60 psi; collision gas set to medium.

4.2.4. Acidic Phytohormones

For acidic phytohormones (gibberellins, auxins, jasmonates, salicylates and abscisates), Fraction A dried and derivatised samples were reconstituted in 100 μL 5% (CH_3OH:CH_3CN): 95% (5 mM ammonium formate adjusted to pH 3.7 with formic acid). An internal standard for each analyte was created by derivatising a mixed analytical standard with a deuterated analogue of bromocholine bromide ([$^2H_9$]-BETA) using a modified method to that described by Sun et al. [47], as here detailed. To each sample, 100 μL of internal standard (5 ppb [$^2H_9$]-BETA) was added. Samples were filtered through a 0.7-μm glass filter plate prior to LC-MS analysis. Acidic phytohormones were separated on an Acquity UPLC BEH C18 1.7 μm 2.1 × 150 mm ID column (Waters, Wexford, Ireland) maintained at 40 °C. Solvents were (A) 5% (CH_3OH:CH_3CN):95% (5 mM ammonium formate adjusted to pH 3.7 with formic acid) and (B) 95% (CH_3OH:CH_3CN):5% (5 mM ammonium formate adjusted to pH 3.7 with formic acid) with a flow rate of 250 μL min^{-1}. The initial mobile phase, 0% B was held for 1 min before ramping linearly to 5.3% B at 2 min, 7.5% B at 5.5 min, 40% B at 11 min, and holding for 3.5 min before ramping to 100% B at 15.5 min and holding at 100% B until 21 min before resetting to the original conditions. Injection size was 2 μL. MS data were acquired in the positive mode using a scheduled MRM method. The transitions monitored (Q1 and Q3) are listed in Table S4. Transitions for compounds other than gibberellins were detuned from optimum to reduce their sensitivity to fit within the dynamic linear range of the instrument. Other operating parameters were as follows: ion spray voltage 4500 V; temperature 600 °C; curtain gas 45 psi; ion source gas 1 60 psi; ion source gas 2 60 psi; collision gas set to medium.

4.2.5. Phytohormone Identification and Quantification

Identification and quantification of all compounds were confirmed through the comparison of the acquired spectra with spectra from the authentic standards. All data were analysed and processed using Analyst version 1.6.2 and MultiQuant version 3.0 software packages. Concentrations were calculated for each compound in equivalence to their respective stable isotope as the internal standard.

4.3. Transcriptomic Data

The Red5 *A. chinensis* var. *chinensis* genome [30] was mined for the gene models involved in the cytokinin biosynthetic, catabolic, and signalling pathways, and for plasma membrane H(+)-ATPase genes. The expansin gene models list was sourced from Pilkington et al. [30]. Gene models were then searched in the differentially expressed gene (DEG) lists generated by Nardozza et al. [6] (Bioproject ID PRJNA593615) and heat maps were created to visualise the effect of the carbohydrate supply on gene transcription, with a focus on the two girdled treatments. DESeq output is shown in Figure S5.

4.4. Statistical Analysis

The effect of the carbohydrate supply and fruit age (factors) on fruit weight and phytohormone concentrations were analysed using a linear mixed effects model (LME, type 3 sums of squares Kenward–Roger's method) in R (version 3.5.1) [48]. The biological replicates were treated as random effects. When significant effects or interactions were present, the means were separated on the basis of all pairwise comparisons of least-squares means, adjusted for multiple comparison by Tukey's correction (letters assigned; confidence level 95%). Residual plots were inspected to check for the assumptions of normality and constant variance. Where appropriate, a log-transformation was used prior to analysis with the fitted means back-transformed onto the original scale. Pearson correlation analysis was performed to identify positive or negative correlations between fruit weight and phytohormone concentration. Principal component analysis for cytokinin hormones were also performed in R using the MixOmics libraries [49]. Data were plotted using package ggplot2 [50] with 95% confidence interval ellipses.

Supplementary Materials: The following are available online at http://www.mdpi.com/2218-1989/10/1/23/s1. Figure S1: Cytokinin biplot at 16 weeks after mid bloom; Figure S2: Heat-map for multistep phosphorelay (MSP) pathway kiwifruit genes: cytokinin receptors histidine kinases and positive regulators histidine-containing phosphotransfer proteins; Figure S3: Heat-map for multistep phosphorelay (MSP) pathway kiwifruit genes: negative regulators A-type Arabidopsis Response Regulators (ARR) and positive regulators B-type ARR; Figure S4: Heat-map of kiwifruit plasma membrane H(+)-ATPase genes; Table S1: Proportion of cytokinins types on total cytokinin concentrations in different treatments; Table S2: List of collected samples and biological replicates used for the experiments; Table S3: Multiple reaction monitoring (MRM) transitions used for cytokinin analysis; Table S4: Multiple reaction monitoring (MRM) transitions used for acidic phytohormone analysis; Table S5: DESeq pairwise comparison between low carbohydrate supply (LC) and high carbohydrate supply (HC) samples at 8, 12 and 16 weeks after mid bloom.

Author Contributions: Conceptualization, S.N., H.L.B., and A.C.R.; Methodology, J.C. and S.N.; Formal analysis, K.G.H., T.T., D.J., and A.H.T.; Data curation, S.N.; Writing—original draft preparation, S.N., J.C., and A.C.A.;

Writing—review and editing, A.C.A., A.C.R., S.N., and H.L.B. All authors have read and agreed to the published version of the manuscript.

Funding: This research was supported by PFR's Premium Kiwifruit research programme, funded by the New Zealand Ministry of Business, Innovation, and Employment's Strategic Science Investment Fund.

Acknowledgments: The authors thank Catrin Guenther and Charlotte Voogd for their critical review of the manuscript, and Mark Wohlers for statistical advice.

Conflicts of Interest: The authors declare no conflicts of interest.

References

1. Gould, K.; Ferguson, I.B. Kiwifruit development: A case study. In *Plants in Action: Adaptation in Nature, Performance in Cultivation*; Atwell, B.J., Kriedeman, P.E., Turnbull, C.G.N., Eds.; 2010. Volume Chapter 11, Section 11.2.4. First published 1999 by Macmillan Education Australia, Melbourne, Australia; Available online: http://plantsinaction.science.uq.edu.au/edition1 (accessed on 20 November 2019).
2. Nardozza, S.; Boldingh, H.L.; Osorio, S.; Hohne, M.; Wohlers, M.; Gleave, A.P.; MacRae, E.A.; Richardson, A.C.; Atkinson, R.G.; Sulpice, R.; et al. Metabolic analysis of kiwifruit (*Actinidia deliciosa*) berries from extreme genotypes reveals hallmarks for fruit starch metabolism. *J. Exp. Bot.* **2013**, *64*, 5049–5063. [CrossRef]
3. Richardson, A.; Boldingh, H.; McAtee, P.; Gunaseelan, K.; Luo, Z.; Atkinson, R.; David, K.; Burdon, J.; Schaffer, R. Fruit development of the diploid kiwifruit, *Actinidia chinensis* 'Hort16A'. *BMC Plant Biol.* **2011**, *11*, 182. [CrossRef] [PubMed]
4. Lai, R.; Woolley, D.; Lawes, G.S. Effect of leaf:fruit ratio on fruit growth of kiwifruit (*Actinidia deliciosa*). *Sci. Hortic.* **1989**, *39*, 247–255. [CrossRef]
5. Minchin, P.E.H.; Snelgar, W.P.; Blattmann, P.; Hall, A.J. Competition between fruit and vegetative growth in Hayward kiwifruit. *N. Z. J. Crop Hortic. Sci.* **2010**, *38*, 101–112. [CrossRef]
6. Nardozza, S.; Boldingh, H.L.; Kashuba, P.; Feil, R.; Jones, D.; Thrimawithana, A.H.; Ireland, H.S.; Philippe, M.; Wohlers, M.W.; McGhie, T.; et al. Carbon starvation reduces carbohydrate and anthocyanin accumulation in red-fleshed fruit via trehalose 6-phosphate and MYB27. *Plant Cell Environ.* **2019**. [CrossRef] [PubMed]
7. Nardozza, S.; Boldingh, H.; Kashuba, M.; McCaughan, L.; Philippe, M.; Wohlers, M.; McGhie, T.; Currie, M.; Montefiori, M.; Richardson, A. Effects of the manipulation of carbohydrate supply on fruit dry matter and colour development in a block-red *Actinidia chinensis* var. *chinensis genotype. Acta Hortic.* **2017**, *1218*, 155–162. [CrossRef]
8. McAtee, P.; Karim, S.; Schaffer, R.; David, K. A dynamic interplay between phytohormones is required for fruit development, maturation, and ripening. *Front. Plant Sci.* **2013**, *4*, 79. [CrossRef] [PubMed]
9. Kumar, R.; Khurana, A.; Sharma, A.K. Role of plant hormones and their interplay in development and ripening of fleshy fruits. *J. Exp. Bot.* **2014**, *65*, 4561–4575. [CrossRef]
10. Devoghalaere, F.; Doucen, T.; Guitton, B.; Keeling, J.; Payne, W.; Ling, T.J.; Ross, J.J.; Hallett, I.C.; Gunaseelan, K.; Dayatilake, G.A.; et al. A genomics approach to understanding the role of auxin in apple (*Malus x domestica*) fruit size control. *BMC Plant Biol.* **2012**, *12*, 7. [CrossRef]
11. Giovannoni, J.J. Genetic regulation of fruit development and ripening. *Plant Cell* **2004**, *16* (Suppl. 1), S170–S180. [CrossRef]
12. Nardozza, S.; Boldingh, H.L.; Wohlers, M.W.; Gleave, A.P.; Luo, Z.; Costa, G.; MacRae, E.A.; Clearwater, M.J.; Richardson, A.C. Exogenous cytokinin application to *Actinidia chinensis* var. *deliciosa* 'Hayward' fruit promotes fruit expansion through water uptake. *Hortic. Res.* **2017**, *4*, 17043. [CrossRef] [PubMed]
13. Hopping, M.E. Effect of exogenous auxins, gibberellins, and cytokinins on fruit development in Chinese gooseberry (*Actinidia chinensis* Planch.). *N. Z. J. Bot.* **1976**, *14*, 69–75. [CrossRef]
14. Lewis, D.H.; Burge, G.K.; Schmierer, D.M.; Jameson, P.E. Cytokinins and fruit development in the kiwifruit (*Actinidia deliciosa*). 1. Changes during fruit development. *Physiol. Plant.* **1996**, *98*, 179–186. [CrossRef]
15. Pilkington, S.M.; Montefiori, M.; Galer, A.L.; Emery, R.J.N.; Allan, A.C.; Jameson, P.E. Endogenous cytokinin in developing kiwifruit is implicated in maintaining fruit flesh chlorophyll levels. *Ann. Bot.* **2013**, *112*, 57–68. [CrossRef]
16. Cosgrove, D.J. Growth of the plant cell wall. *Nat. Rev. Mol. Cell Biol.* **2005**, *6*, 850–861. [CrossRef]
17. Brenner, W.G.; Ramireddy, E.; Heyl, A.; Schmulling, T. Gene regulation by cytokinin in Arabidopsis. *Front. Plant Sci.* **2012**, *3*, 8. [CrossRef]

18. Ljung, K.; Nemhauser, J.L.; Perata, P. New mechanistic links between sugar and hormone signalling networks. *Curr. Opin. Plant Biol.* **2015**, *25*, 130–137. [CrossRef]
19. Bottcher, C.; Burbidge, C.A.; Boss, P.K.; Davies, C. Changes in transcription of cytokinin metabolism and signalling genes in grape (*Vitis vinifera* L.) berries are associated with the ripening-related increase in isopentenyladenine. *BMC Plant Biol.* **2015**, *15*, 223. [CrossRef]
20. Argueso, C.T.; Ferreira, F.J.; Kieber, J.J. Environmental perception avenues: The interaction of cytokinin and environmental response pathways. *Plant Cell Environ.* **2009**, *32*, 1147–1160. [CrossRef]
21. Roitsch, T.; Ehneß, R. Regulation of source/sink relations by cytokinins. *Plant Growth Regul.* **2000**, *32*, 359–367. [CrossRef]
22. Werner, T.; Holst, K.; Pors, Y.; Guivarc'h, A.; Mustroph, A.; Chriqui, D.; Grimm, B.; Schmulling, T. Cytokinin deficiency causes distinct changes of sink and source parameters in tobacco shoots and roots. *J. Exp. Bot.* **2008**, *59*, 2659–2672. [CrossRef] [PubMed]
23. Kiba, T.; Takebayashi, Y.; Kojima, M.; Sakakibara, H. Sugar-induced de novo cytokinin biosynthesis contributes to Arabidopsis growth under elevated CO_2. *Sci. Rep.* **2019**, *9*, 7765. [CrossRef] [PubMed]
24. Hwang, I.; Sheen, J. Two-component circuitry in Arabidopsis cytokinin signal transduction. *Nature* **2001**, *413*, 383–389. [CrossRef] [PubMed]
25. Ma, J.K.; Hanssen, M.; Lundgren, K.; Hernandez, L.; Delatte, T.; Ehlert, A.; Liu, C.M.; Schluepmann, H.; Droge-Laser, W.; Moritz, T.; et al. The sucrose-regulated Arabidopsis transcription factor bZIP11 reprograms metabolism and regulates trehalose metabolism. *New Phytol.* **2011**, *191*, 733–745. [CrossRef] [PubMed]
26. Nardozza, S.; Hallett, I.C.; McCartney, R.; Richardson, A.C.; MacRae, E.A.; Costa, G.; Clearwater, M.J. Is fruit anatomy involved in variation in fruit starch concentration between *Actinidia deliciosa* genotypes? *Funct. Plant Biol.* **2011**, *38*, 63–74. [CrossRef]
27. Richardson, A.; Boldingh, H.; Kashuba, P.; Knight, G.; Ellingham, D. Flowering time determines the weight and composition of *Actinidia chinensis* var. *chinensis 'Zesy002' kiwifruit. Sci. Hortic.* **2019**, *246*, 741–748. [CrossRef]
28. Hirose, N.; Takei, K.; Kuroha, T.; Kamada-Nobusada, T.; Hayashi, H.; Sakakibara, H. Regulation of cytokinin biosynthesis, compartmentalization and translocation. *J. Exp. Bot.* **2008**, *59*, 75–83. [CrossRef]
29. Sakakibara, H. Cytokinins: Activity, biosynthesis, and translocation. *Annu. Rev. Plant Biol.* **2006**, *57*, 431–449. [CrossRef]
30. Pilkington, S.M.; Crowhurst, R.; Hilario, E.; Nardozza, S.; Fraser, L.; Peng, Y.; Gunaseelan, K.; Simpson, R.; Tahir, J.; Deroles, S.C.; et al. A manually annotated *Actinidia chinensis* var. *chinensis* (kiwifruit) genome highlights the challenges associated with draft genomes and gene prediction in plants. *BMC Genom.* **2018**, *19*, 257. [CrossRef]
31. Song, J.C.; Jiang, L.J.; Jameson, P.E. Co-ordinate regulation of cytokinin gene family members during flag leaf and reproductive development in wheat. *BMC Plant Biol.* **2012**, *12*, 78. [CrossRef]
32. Kieber, J.J.; Schaller, G.E. Cytokinins. *Arab. book* **2014**, *12*, e0168. [CrossRef] [PubMed]
33. Marowa, P.; Ding, A.M.; Kong, Y.Z. Expansins: Roles in plant growth and potential applications in crop improvement. *Plant Cell Rep.* **2016**, *35*, 949–965. [CrossRef] [PubMed]
34. Csukasi, F.; Osorio, S.; Gutierrez, J.R.; Kitamura, J.; Giavalisco, P.; Nakajima, M.; Fernie, A.R.; Rathjen, J.P.; Botella, M.A.; Valpuesta, V.; et al. Gibberellin biosynthesis and signalling during development of the strawberry receptacle. *New Phytol.* **2011**, *191*, 376–390. [CrossRef] [PubMed]
35. de Jong, M.; Wolters-Arts, M.; García-Martínez, J.L.; Mariani, C.; Vriezen, W.H. The *Solanum lycopersicum* AUXIN RESPONSE FACTOR 7 (*SlARF7*) mediates cross-talk between auxin and gibberellin signalling during tomato fruit set and development. *J. Exp. Bot.* **2010**, *62*, 617–626. [CrossRef]
36. Liu, Z.N.; Lv, Y.X.; Zhang, M.; Liu, Y.P.; Kong, L.J.; Zou, M.H.; Lu, G.; Cao, J.S.; Yu, X.L. Identification, expression, and comparative genomic analysis of the IPT and CKX gene families in Chinese cabbage (*Brassica rapa* ssp *pekinensis*). *BMC Genom.* **2013**, *14*, 594. [CrossRef]
37. Brugiere, N.; Jiao, S.P.; Hantke, S.; Zinselmeier, C.; Roessler, J.A.; Niu, X.M.; Jones, R.J.; Habben, J.E. Cytokinin oxidase gene expression in maize is localized to the vasculature, and is induced by cytokinins, abscisic acid, and abiotic stress. *Plant Physiol.* **2003**, *132*, 1228–1240. [CrossRef]
38. Kieber, J.J.; Schaller, G.E. Cytokinin signaling in plant development. *Development* **2018**, *145*, dev149344. [CrossRef]

39. Taniguchi, M.; Sasaki, N.; Tsuge, T.; Aoyama, T.; Oka, A. ARR1 directly activates cytokinin response genes that encode proteins with diverse regulatory functions. *Plant Cell Physiol.* **2007**, *48*, 263–277. [CrossRef]
40. Bhargava, A.; Clabaugh, I.; To, J.P.; Maxwell, B.B.; Chiang, Y.-H.; Schaller, G.E.; Loraine, A.; Kieber, J.J. Identification of cytokinin-responsive genes using microarray meta-analysis and RNA-Seq in Arabidopsis. *Plant Physiol.* **2013**, *162*, 272–294. [CrossRef]
41. Cosgrove, D.J. Loosening of plant cell walls by expansins. *Nature* **2000**, *407*, 321–326. [CrossRef]
42. Lee, A.; Giordano, W.; Hirsch, A.M. Cytokinin induces expansin gene expression in *Melilotus alba* Desr. wild-type and the non-nodulating, non-mycorrhizal (NodMyc) mutant Masym3. *Plant Signal. Behav.* **2008**, *3*, 218–223. [CrossRef] [PubMed]
43. Brenner, W.G.; Schmulling, T. Summarizing and exploring data of a decade of cytokinin-related transcriptomics. *Front. Plant Sci.* **2015**, *6*, 29. [CrossRef] [PubMed]
44. Pacifici, E.; Di Mambro, R.; Dello Ioio, R.; Costantino, P.; Sabatini, S. Acidic cell elongation drives cell differentiation in the Arabidopsis root. *EMBO J.* **2018**, *37*, e99134. [CrossRef] [PubMed]
45. Downes, B.P.; Crowell, D.N. Cytokinin regulates the expression of a soybean beta-expansin gene by a post-transcriptional mechanism. *Plant Mol. Biol.* **1998**, *37*, 437–444. [CrossRef] [PubMed]
46. Kojima, M.; Kamada-Nobusada, T.; Komatsu, H.; Takei, K.; Kuroha, T.; Mizutani, M.; Ashikari, M.; Ueguchi-Tanaka, M.; Matsuoka, M.; Suzuki, K.; et al. Highly sensitive and high-throughput analysis of plant hormones using MS-probe modification and Liquid Chromatography—Tandem Mass Spectrometry: An application for hormone profiling in *Oryza sativa*. *Plant Cell Physiol.* **2009**, *50*, 1201–1214. [CrossRef] [PubMed]
47. Sun, X.H.; Ouyang, Y.; Chu, J.F.; Yan, J.; Yu, Y.; Li, X.Q.; Yang, J.; Yan, C.Y. An in-advance stable isotope labeling strategy for relative analysis of multiple acidic plant hormones in sub-milligram Arabidopsis thaliana seedling and a single seed. *J. Chromatogr.* **2014**, *1338*, 67–76. [CrossRef]
48. R Core Team. *R: A Language and Environment for Statistical Computing*; R Foundation for Statistical Computing: Vienna, Austria, 2018.
49. Rohart, F.; Gautier, B.; Singh, A.; Lê Cao, K.-A. mixOmics: An R package for 'omics feature selection and multiple data integration. *PLoS Comp. Biol.* **2017**, *13*, e1005752. [CrossRef]
50. Wickham, H. *Ggplot2: Elegant graphics for Data Analysis*; Springer: New York, NY, USA, 2016.

metabolites

Article

GC-MS Based Metabolite Profiling to Monitor Ripening-Specific Metabolites in Pineapple (*Ananas comosus*)

Muhammad Maulana Malikul Ikram [1], Sobir Ridwani [2], Sastia Prama Putri [1,*] and Eiichiro Fukusaki [1]

[1] Department of Biotechnology, Graduate School of Engineering, Osaka University, 2-1 Yamadaoka, Suita, Osaka 565-0871, Japan; malikul_ikram@bio.eng.osaka-u.ac.jp (M.M.M.I.); fukusaki@bio.eng.osaka-u.ac.jp (E.F.)

[2] Center for Tropical Horticulture Studies, IPB University, Jl. Baranangsiang, Bogor 16144, Indonesia; ridwanisobir@gmail.com

* Correspondence: sastia_putri@bio.eng.osaka-u.ac.jp; Tel.: +81-6-6879-7416

Received: 10 February 2020; Accepted: 25 March 2020; Published: 31 March 2020

Abstract: Pineapple is one of the most cultivated tropical, non-climacteric fruits in the world due to its high market value and production volume. Since non-climacteric fruits do not ripen after harvest, the ripening stage at the time of harvest is an important factor that determines sensory quality and shelf life. The objective of this research was to investigate metabolite changes in the pineapple ripening process by metabolite profiling approach. Pineapple (Queen variety) samples from Indonesia were subjected to GC-MS analysis. A total of 56, 47, and 54 metabolites were annotated from the crown, flesh, and peel parts, respectively. From the principal component analysis (PCA) plot, separation of samples based on ripening stages from C0–C2 (early ripening stages) and C3–C4 (late ripening stages) was observed for flesh and peel parts, whereas no clear separation was seen for the crown part. Furthermore, orthogonal projection to latent structures (OPLS) analysis suggested metabolites that were associated with the ripening stages in flesh and peel parts of pineapple. This study indicated potentially important metabolites that are correlated to the ripening of pineapple that would provide a basis for further study on pineapple ripening process.

Keywords: pineapple; metabolomics; ripening; non-climacteric

1. Introduction

Pineapple (*Ananas comosus*) market value is approximately 14.7 billion USD with a production volume of around 25 million metric tons in the world [1]. Pineapple is categorized as a non-climacteric fruit. The major difference between climacteric and non-climacteric fruit is non-climacteric fruit produces low levels of ethylene and does not show any major peak in the respiration rate during the ripening process, whereas climacteric fruit depends on ethylene bursts during ripening [2,3]. In addition, ethylene treatment does not give any effect to non-climacteric fruit with the exception of degreening (removal of chlorophyll) [4]. Another distinct characteristic of non-climacteric fruit is the fruit will not continue its ripening process after harvest, thus making it important to be harvested in the right ripening stage to ensure proper quality [3]. The ripening stage of pineapple is divided into 5 stages, C0–C4, with the green-ripe fruit at C0 and the full-ripe fruit at C4 based on United Nations Economic Commission for Europe (UNECE) Standard for pineapple (FFV-49) as seen in Figure 1 [5]. This classification is based on the peel color of pineapple, in which C0 stage contains 0% yellow color, C1 stage contain 0%–25% yellow color, C2 stage contain 25%–50% yellow color, C3 stage contain 50%–75% yellow color, and C4 stage contain 75%–100% yellow color [5]. Pineapple is usually exported in the C1 stage [6], while the fully ripe fruit (C4 stage) is mainly for domestic consumption. The fruit of pineapple consists of the fusion of individual fruits. These individual fruits are developed from a single flower and the external of these fruits were protected with a hard-polygonal shield, commonly called as pineapple peel [7]. On top of pineapple fruit, there are leaves that can be used for vegetative reproduction of pineapple commonly called the pineapple crown. This crown part is commonly harvested along with the fruit harvest [7]. Crown and peel parts develop during pineapple fruit development. Therefore, to understand the pineapple ripening process comprehensively, analysis of pineapple peel and crown is needed in addition to flesh analysis. At present, there is limited information on the differences in metabolite composition of pineapple from different ripening stages [8]. Monitoring metabolites changes using tools such as metabolomics, a comprehensive study of metabolite, is a powerful tool for further understanding pineapple ripening process.

Figure 1. Pineapple sample from all ripening stages. From left to right: C0 until C4 stages. This classification is based on the peel color of pineapple, in which C0 stage contains 0% yellow color, C1 stage contain 0%–25% yellow color, C2 stage contain 25%–50% yellow color, C3 stage contain 50%–75% yellow color, and C4 stage contain 75%–100% yellow color [5].

Recent studies about fruit ripening metabolomics mostly focused on climacteric fruits, such as banana, mango, capsicum, dates, avocado, peach, climacteric melon, and mangosteen [9–16]. On the other hand, only a few of non-climacteric ripening processed fruit had been elucidated using metabolomics approach, such as cherry, blackcurrant, blueberry, non-climacteric melon, and pineapple [15,17–21]. Previous metabolomics studies on non-climacteric fruit employed mass spectrometry-based instruments, such as gas chromatography-mass spectrometer (GC-MS) or liquid chromatography-mass spectrometer (LC-MS). Several studies performed a combination of headspace-solid phase microextraction (HS-SPME) with GC-MS to measure volatile compounds during the ripening process [15,20]. Reports on pineapple ripening have been focused in volatile and phenolic compounds using HS-SPME-GC-MS, high-performance liquid chromatography with diode array detection and electrospray ionization multiple-stage mass-spectrometry (HPLC-DAD-ESI-MS^n), and electrospray ionization mass spectrometry (ESI(-)FT-ICR MS) [21–24]. The outcome from these previous reports suggested the changes in phenolic patterns, such as coumaroyl isocitrate and *S*-*p*-coumaryl, and volatile compounds, such as methyl 3-(methylthio)propanoate and δ-octalactone, along the pineapple ripening process.

As mentioned previously, previous studies on pineapple ripening were on the targeted analysis of volatile and phenolic compounds using the flesh part as a sample. To date, there is no study that analyzed different parts of pineapple including flesh, peel, and crown parts, and incorporating broad coverage of primary metabolites such as sugar, organic acid, amino acid, sugar alcohol, sugar acid, and amine compounds. In order to suggest metabolites that are associated with ripening, there are several different multivariate analyses that can be used. The most common multivariate analysis is principal component analysis (PCA) and orthogonal projections to latent structures (OPLS) regression analysis [25,26]. In this study, a metabolite profiling approach using GC-MS in combination with PCA and OPLS was conducted to monitor the changes of primary metabolites (sugar, organic acid, amino acid, etc.) during pineapple ripening process by analyzing pineapple fruit (crown, flesh, and peel) from different ripening stage (Figure S1). OPLS model was constructed using metabolites annotated by GC-MS as an explanatory variable and ripening stages as a response variable. The constructed model from flesh and peel samples indicated several potentially important metabolites that were correlated with the pineapple ripening process. This study would be important to complement the knowledge of the pineapple ripening process and could serve as a basis for post-harvest handling strategy in pineapple industries.

2. Results

2.1. Optimization of Sample Preparation Methods in Pineapple Fruit

The first analysis was to compare two different sample preparation methods, namely food processor and freeze-drying. This analysis was conducted to optimize sample preparation of pineapple fruit in GC-MS analysis. A total of 47 metabolites belonging to various metabolite classes were annotated by GC-MS analysis in flesh samples (Table S1). These metabolites comprise of 26 metabolites belongs to sugars class, 12 metabolites belong to amino acids and amines, and 9 metabolites belong to organic acids. These annotated metabolites were subjected to PCA to clearly visualize the differences in metabolite levels in pineapple flesh prepared by food processing and freeze-drying methods.

Figure 2 shows the comparison between these two methods after analysis by GC-MS. From the score plot, flesh samples prepared by two different methods were clearly separated along the PC1 with a 95.2% variance. The loading plot showed that almost all metabolites were accumulated in the flesh samples prepared by freeze-dry method. Only sucrose was found to accumulate in samples prepared by food processor method. In the next analysis, we applied the freeze-drying method to analyze different parts of pineapple in GC-MS. A total of 54, 44, and 50 metabolites were annotated from crown, flesh, and peel, respectively (Table S2). PCA score plot from Figure 3 clustered pineapple fruit into three different parts (crown, flesh, and peel) based on the metabolite distributions. Amino acid and organic acid were found to be accumulated in crown part, while the peel and flesh part show accumulation of sugar and sugar-acid. Due to the separation for each part, this result becomes the basis to analyze three parts of pineapple separately.

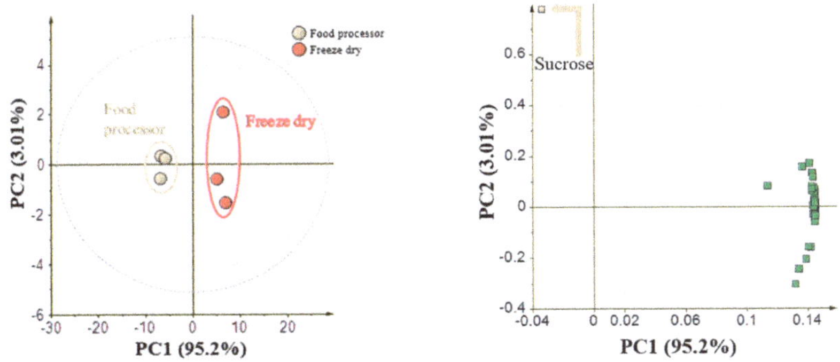

Figure 2. PCA of comparison of different sample preparation methods: Food processor method and freeze-drying method. 47 annotated metabolites from GC-MS analysis were auto-scaled prior to PCA. (Left: Score plot between food processor and freeze dry samples; Legends represent the samples and colored as follow: food processor: light brown circle, freeze dry: red circle. Right: Loading plot shown almost all metabolites (except sucrose shown by light brown arrow) showed higher accumulation by freeze-dry method).

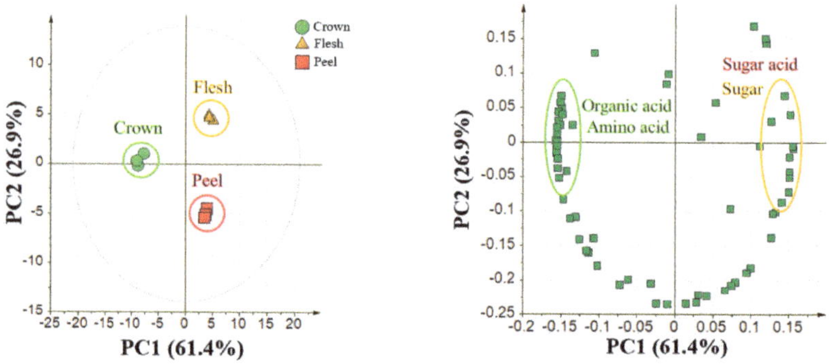

Figure 3. PCA result from different parts of pineapple. 64 tentatively identified metabolites from GC-MS analysis were auto-scaled prior to PCA. Left: Score plot between crown, flesh, and peel part. Legends represent the sample and colored as follows: crown: green circle, flesh: yellow triangle, peel: red square. Right: Loading plot shown that crown part accumulates organic and amino acid, while flesh and peel accumulate sugar and sugar acid.

2.2. GC-MS and Principal Component Analysis of Pineapple from Different Ripening Stages

Analysis of crown, flesh, and peel parts of pineapple from different ripening stages was conducted separately. Metabolite profiling approach using GC-MS instrument could detect 351 metabolite peaks in the crown part, 297 metabolite peaks in the flesh part, and 359 metabolite peaks in the peel part. Among those peaks, 85 peaks in the crown part, 74 peaks in the flesh part, and 73 peaks in the peel part were annotated using MSP Library containing RI and EI-MS from our laboratory experimental data. Metabolites from QC samples with RSD more than 20% were excluded from the analysis [27]. After exclusion of metabolites with RSD higher than 20%, the number of annotated metabolites in crown part were 56 metabolites, in the flesh part were 47 metabolites, and in the peel part were 54 metabolites. A complete list of these metabolites during ripening analysis is shown in Table S3.

Figure 4 shows the score plot from PCA for pineapple from different ripening stages (C0 to C4 stage) as observed in three different parts of pineapple. As seen in Figure 4b,c, flesh and peel part showed two distinct clusters along PC1. Less ripe samples (C0–C2) were clustered together and ripe fruit samples (C3 and C4) formed a separate cluster. This trend was explained by 63.9% and 53.3% variance in the flesh and peel part, respectively. However, this trend was not shown in the crown part of pineapple from all principal components. Loading plot in Figure 4b,c showed the metabolite accumulation in the less ripe and ripe samples for flesh and peel part, respectively. These metabolite intensities that used to create a score plot were normalized using an internal standard, ribitol. The internal standard was chosen because it is not present in pineapple samples and stable in a mixed solvent solution.

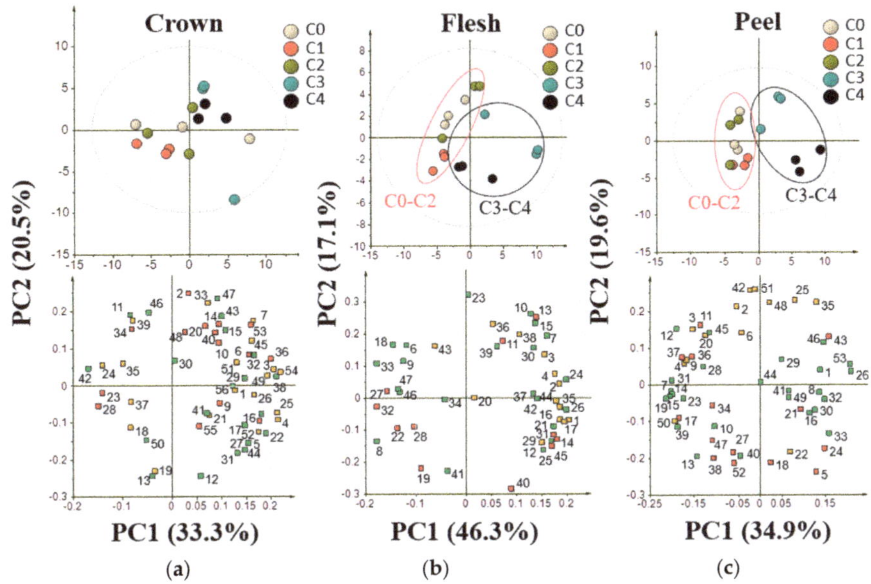

Figure 4. PCA result from flesh, crown and peel parts of pineapple from different ripening stages. Variables used for PCA were 56, 47, and 54 annotated metabolites by GC-MS from crown, flesh, and peel parts respectively. Data was auto scaled prior to PCA. (**a**) Score and loading plot from the crown part; (**b**) score and loading plot from the flesh part; (**c**) score and loading plot from the peel part. Legends represent the samples and colored as follows: brown: C0 stage, red: C1 stage, green: C2 stage, blue: C3 stage, black: C4 stage. Upper part show score plot; Bottom part show loading plot. Loading plot was colored based on metabolite classes: green: sugars; red: organic acids; yellow: amino acids and amines.

2.3. Orthogonal Projection to Latent Structures of Pineapple Ripening Process

Orthogonal projection of latent structures (OPLS) regression analysis was conducted to identify metabolites that were highly influenced by the process related to the response variable [28]. In this study, two latent variables were used to construct the model using flesh and peel parts of pineapple. Crown part was not analyzed based on the previous result in PCA which indicated that crown part was not able to show any trend in ripening process. Response variables that were used to generate the model were ripening stages from C0 as 1, C1 as 2, C2 as 3, C3 as 4, and C4 as stages 5, while the explanatory variables were metabolites annotated by GC-MS analysis. Pineapple from C0 to C4 ripening stages harvested in April 2019 were used as a training set to generate the model (Figure 5). Model validation was conducted by leave-one-out cross-validation from each replicate.

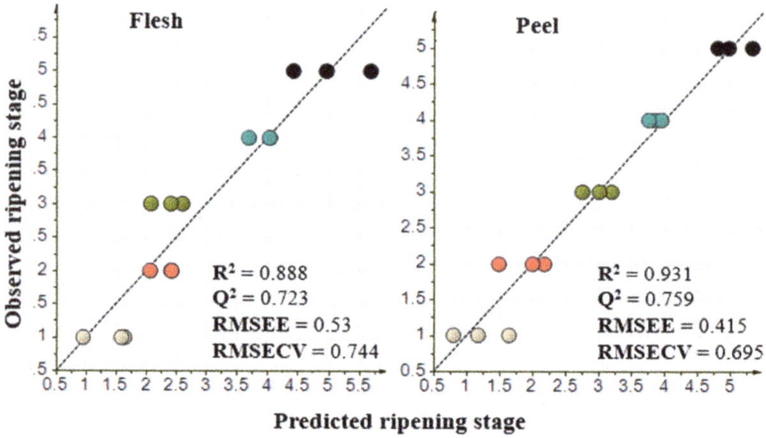

(a) (b)

Figure 5. Orthogonal projection to latent structures (OPLS) results from flesh and part of pineapple. Explanatory variables in flesh part are 47 metabolites, while in peel part are 54 metabolites. Response variable for both models is the ripening stages with numbered as follows: C0 stage as 1, C1 stage as 2, C2 stage as 3, C3 stage as 4, and C4 stage as 5. Value of R^2, Q^2, RMSEE, and RMSECV were used to evaluate the model. (**a**) OPLS model of flesh part; (**b**) OPLS model of peel part.

The constructed OPLS regression model with R^2 of 0.888 and 0.931 for flesh and peel part, respectively, are shown in Figure 5. In the OPLS regression analysis, statistically important metabolites for the models were indicated by the score of variable important in projection (VIP). Metabolites with a VIP score of more than 1 considered important for the model [28] (Table S4). Contributing metabolites were chosen based on the five highest VIP scores. The metabolites were melezitose, inositol, xylonic acid, gluconic acid, and raffinose in the flesh model, whereas inositol, mannose, galactose, sucrose, and aspartic acid were the top five highest VIP metabolites in the peel. Among these highest VIP metabolites in both flesh and peel, melezitose, xylonic acid, gluconic acid, and sucrose have a positive correlation with the ripening stages, while inositol, raffinose, mannose, galactose, and aspartic acid showed a negative correlation with the ripening process (Figure 6).

3. Discussion

Metabolite profiling is known to be useful to analyze a large group of metabolites that belong to a specific class of compounds that reflects the dynamic response to physiological change or developmental stimuli [29,30] In this study, a metabolite profiling approach using GC-MS was employed for the study of pineapple ripening process. GC-MS is suitable for metabolite profiling because it provides high sensitivity, reproducibility, and quantitation of a large number of metabolites with a single-step extraction [31,32]. Metabolite annotated in pineapple crown, peel and flesh were classified as sugars, amino acids, amines, organic acids, and other compounds. Sugars were found to be the most abundant in pineapple. This is in agreement with previous work that mentioned the high content of sugars was observed in pineapple flesh samples [21]. In this study, we conducted for the first time the analysis of peel and crown parts of pineapple in addition to flesh samples. Annotated metabolites from each part were subjected to PCA and OPLS analyses. Principal component analysis (PCA) is a multivariate data analysis that could show the variance among the samples using metabolites as the explanatory data [33]. Based on the PCA, pineapple ripening was clustered into two major phases namely C0-C2 stages (early ripening) and C3-C4 stages (late ripening). The trends above were observed only in flesh and peel samples, whereas there was no clear trend of ripening in crown part. It was previously suggested that crown photo-assimilation seems to be derived from its own photosynthesis, not from the fruit [34]. It is also well known that pineapple maturation developed from the bottom, not from the top [35]. Therefore, crown part is considered to have no correlation with fruit ripening.

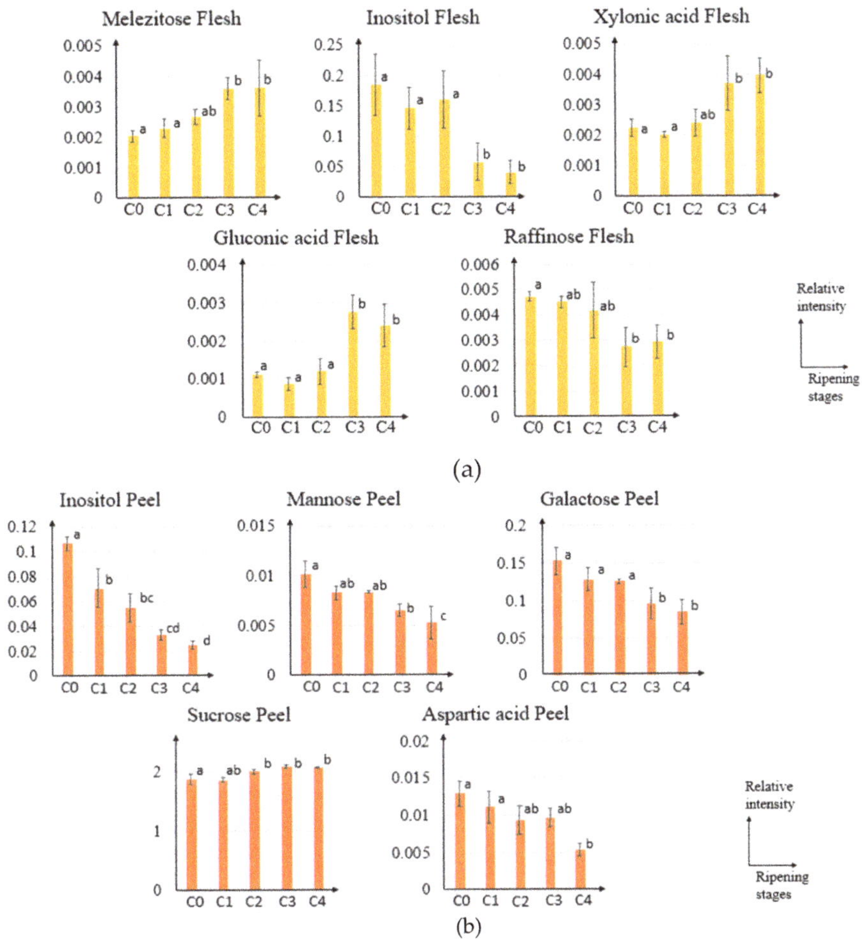

Figure 6. Bar graph of five highest variable important in projection (VIP) metabolites related to pineapple ripening process in (**a**) flesh and (**b**) peel part. The relative intensity of the five highest VIP metabolites was normalized by the internal standard. Vertical axis represents metabolites relative intensity and horizontal axis represents ripening stages. Significant differences ($p < 0.05$) are indicated with the different letters based on mean comparison Tukey's test.

Further analysis to identify potentially important metabolites that are correlated with the ripening process was conducted using OPLS regression analysis. OPLS regression analysis is known to be more powerful to explain the relationship between the response variable and explanatory variable because it is a supervised multivariate analysis [28]. The OPLS regression model shown in Figure 5 has some parameters that could be used to evaluate the quality of the model itself. These parameters are R^2, Q^2, RMSEE, and RMSECV. R^2 is defined as the square of the correlation coefficient between observed and predicted value in a regression [36]. Q^2 is known to be a reliable parameter for model predictivity [16,36]. RMSEE or root mean square error of estimation and RMSECV or root mean square error of cross-validation are the values to evaluate accuracy, prediction, and model robustness [36,37].

A good model would have an R^2 value of more than 0.6, Q^2 value of more than 0.6, and a low value of RMSEE and RMSECV [36]. We constructed 3 OPLS models from metabolites annotated in flesh, peel and crown. The constructed model of flesh and peel showed R^2 value of more than 0.6, Q^2 value of more than 0.6, and a low value of RMSEE and RMSECV, whereas crown model showed R^2 value of 0.896, Q^2 value of 0.432, RMSEE value of 0.509, and RMSECV value of 1.066. These results indicated that only peel and flesh model meet the thresholds for a valid model with a good fit. The low Q^2 value in crown model showed that the samples from crown part cannot be used to predict ripening stages in pineapple. This is in line with the results obtained from PCA.

Contributing metabolites related to ripening stages could be obtained from the variable importance in the projection (VIP) scores. Based on these scores, the five highest VIP metabolites in the flesh part are melezitose, inositol, xylonic acid, gluconic acid, and raffinose; while for peel part are inositol, mannose, galactose, sucrose, and aspartic acid. Figure 6 shows the dynamic of these VIP metabolites relative intensity (normalized with ribitol) along the ripening process of pineapple. These metabolites were shown to be increased or decreased during the pineapple ripening process. From these VIP metabolites in flesh parts, the raffinose level was in agreement with the previous report that showed a decreased level during ripening process [38]. In addition to the previously reported metabolites that correlate with the ripening process, this study also reports the dynamics of inositol, melezitose, xylonic acid, and gluconic acid in the flesh part during ripening process. Inositol or commonly known as myo-inositol was known to regulate osmotic pressure in blueberry fruit thus maintaining turgor and fruit firmness between the firm cultivar and soft cultivar [19]. In addition to that, inositol might be oxidized to D-glucuronic acid known as a major precursor of the cell wall in Arabidopsis [39]. Therefore, the presence of inositol might also relate to the cell wall in fruit. Melezitose is known to play a role in osmoregulation system [40]. However, comparing the relative intensity trend with inositol, the mechanism underlying these two metabolites might be different to regulate the osmoregulation system during the pineapple ripening process. Even though melezitose relative intensity was considered low compared to other sugar, it was reported in the previous study that it could attract ants in honeydew fruit [41]. Therefore, the accumulation of melezitose in the latter stage of ripening might reflect the attractancy of the fruit in the fully ripe stage.

Xylonic acid relative intensity was shown to be increased during the ripening process (Figure 6a). The increase of this organic acid might be related to the reactive oxygen species (ROS). During fruit ripening, oxidative stress was increased and might result in some changes in fruit, such as changes in skin color or fruit softening. Due to the presence of ROS, fruit antioxidants might act to balance the reduction–oxidation homeostasis [42]. One of the most known fruit antioxidants is ascorbic acid. The previous report stated that xylonic acid is a product of ascorbic acid degradation, thus explaining the increase of xylonic acid during the ripening process [43]. During the pineapple ripening process, raffinose was found to be decreased along with the progression of ripening (Figure 6). In agreement with this result, previous report showed that the raffinose level also decreases in Japanese plum non-climacteric cultivar during its ripening [38]. They reported that the raffinose level in non-climacteric fruit might be related to its ability to alleviate the oxidative process during fruit ripening [38]. Therefore, not only xylonic acid but the level of raffinose might also be related to the reduction–oxidation process along the ripening process. The gluconic acid concentration was shown to be increased during the late-ripening process in pineapple (Figure 6a). This increase might be triggered by the increase of carbon molecules availability during the later stage of pineapple ripening [44]. In addition, gluconic acid intensity increase might also cause by the effect of cell wall degradation, change in cuticle composition and pH of host cells that allow the transition of fungi into their aggressive colonization [44]. The presence of gluconic acid might indicate an infection that could secrets gluconic acid and acidify the pH in fruit such as in apple and mango [45,46].

Metabolite that shows the highest VIP score in peel after inositol is mannose. Mannose is known as a component of the plant cell wall, specifically hemicellulose [47]. The previous report mentioned that the concentration of mannose was decreased during fruit development, hence support our findings shown in Figure 6b [48]. Similar to mannose, decrease level of aspartic acid in the pineapple ripening process were also reported during ripening of banana [49]. Aspartic acid was known to be a source for umami taste along with glutamic acid [50]. Its level was known to be varied among the fruits. In banana and *Vasconcellea quercifolia*, the level of aspartic acid was found to be decreased along with the ripening progress, while mature or ripe tomato contained more aspartic acid that brings out the umami taste [49,51,52]. Other than the source of umami taste, the level of aspartic acid might also connect with the free auxin level in fruits that affect its ripening [53]. Aspartic acid was found to be conjugated with indole acetic acid (IAA) and lead to degradation of the IAA hormones [53].

Other than the previously reported metabolites, this study found the change of sucrose and galactose level in the peel part might correlate with the pineapple ripening process. This study showed the sucrose level in pineapple peel part was increased during ripening (Figure 6b) while previous report mentioned the increase of sucrose during ripening process in pineapple flesh [21]. Sucrose was known to become a source of sweet taste in food and commonly used as a standard solution for sweetness [54]. However, sucrose in the peel part might not directly affect the sweetness in the flesh. It is reported that sucrose in the peel is lower if compared with the flesh part [55]. Sucrose not only contributes to sweetness, but it also plays a role to regulate fruit development and ripening in strawberry fruit, a non-climacteric fruit [56]. Sucrose accumulation might induce the expression level of the key enzymes in abscisic acid (ABA) hormones pathway, hence promote ripening process in non-climacteric fruit via ABA hormones [57]. Galactose in Figure 6b was shown a decreasing trend along the ripening stages. This might be explained by the relation of galactose with the plant cell wall. The previous report showed that galactose is the major non-cellulosic sugar in the cell wall and significantly decreased during fruit ripening [58,59]. All these metabolites with high VIP score consist of sugars (melezitose, inositol, raffinose, mannose, galactose, and sucrose) and organic acids (xylonic acid, gluconic acid, and aspartic acid).

During ripening process, many biological processes occur, such as cell wall loosening, texture changes, flavor development, chlorophyll degradation, and pigment accumulation [60]. Changes in melezitose and inositol level might indicate the texture changes during pineapple ripening process. Melezitose and inositol are known to regulate fruit firmness that affects the texture or hardness of fruit [19]. In addition to that, inositol, galactose, and mannose levels might be related with cell wall loosening during ripening. Inositol were known to be precursor of D-glucuronic acid of plant cell wall, galactose is a major non-cellulosic sugar in plant cell wall, and mannose are component of hemicellulose in plant cell wall. Therefore, the decrease of these three metabolites might correlate with loosening of cell wall that usually accompanied with decrease level of firmness and increase of gluconic acid [19,44]. Reactive oxygen species also play a role in ripening process to regulate programed cell death and cell aging [61]. This report in line with our results that show decrease level of xylonic acid and increase level of raffinose during ripening process. Both metabolites were known to respond to reactive oxygen species as discussed previously. These biological processes that affected by the ripening process are product of biochemical changes that mediated by plant hormone. Abscisic acid (ABA) and auxin were known to be affecting the ripening process in non-climacteric fruit [60]. Changes in sucrose and aspartic acid during ripening process might affected abscisic acid and auxin, respectively. Therefore, it might modulate ripening process in pineapple fruit.

This study showed the significance of sample preparation to gain more metabolite coverage that is useful for further analysis. Based on metabolites data acquired from GC-MS analysis, flesh and peel data could show clustering separation between C0-C2 stages and C3-C4 stages using principal component analysis (PCA), while the crown part does not show correlation with the ripening process. Orthogonal projection to latent structures (OPLS) regression analysis reveals metabolites that have possible relations to the pineapple ripening process in flesh and peel parts. In the flesh part, melezitose, inositol, xylonic acid, gluconic acid, and raffinose were found to be the five highest important metabolites, while for the peel part are inositol, mannose, galactose, sucrose, aspartic acid. These metabolites were known to be involved during plant cell wall metabolism and osmoregulation system thus affecting the firmness and shelf life of pineapple, in addition to the redox defense system and non-climacteric ripening hormones. For future applications, these VIP metabolites could be added exogenously to regulate specific effects, for example, the addition of polyamine and ascorbic acid to regulate the shelf life of fruit, as shown in the previous reports [62,63]. In addition, influencing the level of the metabolites through post-harvest treatment was also feasible to be conducted, such as regulating inositol, galactose, and raffinose by cold or heat treatment [64–66]. It must be noted that this study only limited to the "Queen" cultivar. Future study using other widely known cultivars, such as "Smooth Cayenne", is still needed to enrich the information regarding pineapple ripening process. Regardless, this study might become a basis for resolving the post-harvest issue in the pineapple industry by controlling important metabolites influenced in the ripening process.

4. Materials and Methods

4.1. Plant Materials

Pineapple (*Ananas comosus*) fruit from Indonesia corresponding to 5 different ripening stages were used in this study (Figure 1). To set the same harvest time at the end of April 2019, ethephon treatment was used to induce fruit development around November to December 2018. Cultivars of pineapple used in this study to represent the important cultivar from the pineapple Industry is cv. Mahkota Bogor 'Queen'. Three samples (biological replicates) from different plants were collected from each ripening stage for Queen cultivars from the cultivation period of November 2018—April 2019 in Center for Tropical Horticulture Studies, Bogor Agricultural University (CENTROHS, Bogor Agriculture University), Bogor, Indonesia (minimum temperature 21 °C and maximum temperature 35 °C). Ripening stage determination was conducted with the help of a trained panelist according to peel color changes [5].

4.2. Optimization of Sample Preparation

4.2.1. Comparison between Food Processor and Freeze-Drying Methods

Pineapple fruit was cut into half then the flesh was diced using a stainless-steel knife. Diced flesh with 1 × 1 cm size from one half of fruit was subjected to freeze dry-extraction method using VD-800F Freeze dryer (Taitec, Saitama, Japan). This flesh was put into a Pyrex tube and closed with holed parafilm before being quenched with liquid nitrogen and lyophilized. Diced flesh with 1 × 1 cm size from the other half of the fruit was homogenized using hand immersion-blender WSB-33XJ (Waring Commercial, Pennsylvania, United States of America) before extraction. Ten milligrams of both samples were subjected to extraction following the method described previously prior to GC-MS analysis [66].

4.2.2. Comparison of Crown, Flesh, and Peel of Pineapple Fruit

The pineapple was cut into three different parts, crown, flesh, and peel. Crown part was analyzed by cutting the leaves into a 1x1 cm size before quenching by liquid nitrogen and freeze-dried. Peel part was analyzed by scraping the peel into a 1x1 cm size using a stainless-steel knife before quenching and freeze-drying. Flesh part was analyzed according to above description. Lyophilized sample was homogenized using Multi-beads shocker (Yasui Kikai, Osaka, Japan). Ten milligrams of the homogenized sample of each part of pineapple was subjected to extraction and GC-MS analysis [66].

4.3. Sample Preparation and Extraction of Pineapple from Different Ripening Stages

Pineapple fruit of different ripening stages collected in Indonesia was divided into three parts: Crown, flesh, and peel. Each part was cut into small pieces and placed into a Pyrex tube covered with holed parafilm. Samples were quenched by immersing the Pyrex tubes in liquid nitrogen prior to lyophilization using the VD-800R Freeze dryer (Taitec, Saitama, Japan). Freeze-dried pineapple samples were transported from Indonesia to Japan within a day. The samples were homogenized and ground into a fine powder using Multi-beads shocker. This extraction method was conducted based on the method described in our previous study [66]. Pineapple samples (10 mg), blank samples and quality control (QC) samples were extracted together and lyophilized in a single day. QC samples were prepared by collecting small aliquots of each sample obtained in this study.

All samples were extracted using the mixture of methanol (Wako Chemical, Osaka, Japan), chloroform (Kishida Chemical Co. Ltd, Osaka, Japan), ultrapure water (Wako Chemical, Osaka, Japan); in the ratio of 2.5/1/1 (*v*/*v*/*v*) containing 100 μg/mL ribitol as an internal standard. The mixture was incubated at 37 °C, 1200 rpm for 30 min followed by centrifugation for 3 min at 40 °C. Six hundred microliters of supernatant was transferred to a new 1.5 mL microtube and 300 μL water was added into the mixture. The sample mixture was centrifuged for 3 min at 40 °C and 400 μL of supernatant was transferred to a new microtube and closed with a holed cap. The solvent from the sample mixture was evaporated for 1 h at room temperature followed by lyophilization overnight. All samples were analyzed in triplicates (n = 3). One hundred microliters of methoxyamine hydrochloride (20 mg/mL in pyridine) was added into lyophilized samples and incubated in thermomixer for 90 min at 30 °C. Subsequently, 50 μL N-Methyl-N-trimethylsilyl-trifluoroacetamide (MSTFA) (GL Sciences) was added to the samples and incubated for 30 min at 37 °C.

4.4. GC-MS Analysis

GC-Q/MS analysis was performed on a GC-MS QP2010 Ultra (Shimadzu, Kyoto, Japan) equipped with an InertCap 5 MS/NP column (GL Sciences). Tuning and calibration of the mass spectrometer were done prior to analysis. One microliter of the derivatized sample was injected in split mode, 25:1 (*v*/*v*), with an injection temperature of 230 °C. The carrier gas (He) flow was 1.12 mL/min with a linear velocity of 39 cm/s. The column temperature was held at 80 °C for 2 min, increased by 15 °C/min to 330 °C, and then held for 6 min. The transfer line and ion source temperatures were 250 and 200 °C, respectively. Ions were generated by electron ionization (EI) at 0.94 kV. Spectra were recorded at 10,000 u/s (check value) over the mass range m/z 85–500. A standard alkane mixture (C8–C40) was injected prior to analysis for peak identification.

4.5. GC-MS Data Analysis

The raw data obtained from the analysis was converted to the AIA file using GCMS solution software package (Shimadzu, Kyoto, Japan). Peak alignment, peak filtering and annotation was conducted by MS-DIAL ver. 4.00 using GCMS-5MP Library (Riken, Kanagawa, Japan). Peak confirmation of important metabolites, namely inositol, mannose, galactose, melezitose were conducted by co-injection with authentic standard (Wako Pure Chemical Industries Ltd., Osaka, Japan; Sigma-Aldrich Japan Ltd., Tokyo, Japan; Alfa Aesar Ltd., Heysham, UK).

4.6. Statistical Analysis

Annotated metabolites from GC-MS analysis were pre-treated by normalizing each metabolite peak height to internal standard (ribitol). Normalized data were scaled by autoscaling and without transformation subjected to PCA (Principal Component Analysis) using SIMCA-P+ version 13 (Umetrics, Umea, Sweden). Principal component analysis is an unsupervised analysis that is useful as a dimension-reduction tool in order to easily observe trends, clusters and outliers [25]. Other than PCA, a projections to latent structures (PLS) regression model that is constructed from the maximal correlation of explanatory variable (x-variable) with response variable (y-variable) offers ranking of metabolites correlation with a certain quantitative phenotype [67,68]. In particular, orthogonal projections to latent structures (OPLS) regression model is very useful for reducing many variables to limited latent variables [26]. Parts that show ripening trends in PCA were subjected to OPLS (Orthogonal Projections to Latent Structures) analyses using SIMCA-P+ version 13. From OPLS analyses, variable importance in projection (VIP) were calculated for each metabolite. The top five highest VIP score's metabolites in each pineapple part were statistically analyzed by analysis of variance (ANOVA) with Tukey's post hoc test performed using JASP Version 0.11.1 (JASP Team, Amsterdam, Netherlands). The statistical analysis was conducted to evaluate the differences among mean values of VIP metabolites obtained from all ripening stages. Differences were considered significant if $p < 0.05$.

Supplementary Materials: The following are available online at http://www.mdpi.com/2218-1989/10/4/134/s1, Table S1: Complete list of annotated metabolites with RSD < 20% for sample preparation analysis, Table S2: Complete list of annotated metabolites with RSD < 20% for different part analysis, Table S3: Complete list of annotated metabolites with RSD < 20% for ripening process analysis, Table S4: metabolites with VIP score (more than 1) and its coefficient, Figure S1: Visual experimental design on pineapple ripening study, Figure S2: Venn diagram of annotated metabolites in crown, flesh, and peel part on different parts analysis, Figure S3: Venn diagram of annotated metabolites in crown, flesh, and peel part on pineapple ripening analysis.

Author Contributions: M.M.M.I. and S.P.P. designed the experiments; M.M.M.I. performed the experiments, analyzed the data, and wrote the manuscript; S.R. advised sample treatment design and aided in data interpretation; S.P.P. analyzed the data and wrote the manuscript. E.F. conceived the study and participated in its design and coordination. All authors have read and agreed to the published version of the manuscript.

Funding: This research received no external funding.

Acknowledgments: The authors thank Baesuni and Rizal from the Center for Tropical Horticultural Studies for their help during sample collection. The authors also thank Fenny Martha Dwivany and Intan Taufik from School of Life Sciences and Technology, Institut Teknologi Bandung for assistance in sample preparation and freeze-drying. This study represents a portion of a dissertation submitted by MM. Malikul Ikram to Osaka University in partial fulfillment of requirements for his Ph.D.

Conflicts of Interest: The authors declare no conflict of interest.

References

1. Mulderij, R. Overview Global Pineapple Market. Available online: https://www.freshplaza.com/article/2194086/overview-global-pineapple-market/ (accessed on 28 June 2019).
2. Hui, Y.H.; Chen, F.; Nollet, L.M.L.; Guiné, R.P.F.; Martín-Belloso, O.; Mínguez-Mosquera, M.I.; Paliyath, G.; Pessoa, F.L.P.; Le Quéré, J.-L.; Sidhu, J.S. *Handbook of Fruit and Vegetable Flavors*; Wiley Online Library: Hoboken, NJ, USA, 2010; Volume 64.
3. Kader, A.A. Fruit Maturity, Ripening, and Quality Relationship. In *Acta Horticulturae*; International Society for Horticultural Science: Leuven, Belgium, 1999; pp. 203–208.
4. Symons, G.M.; Chua, Y.-J.; Ross, J.J.; Quittenden, L.J.; Davies, N.W.; Reid, J.B. Hormonal Changes during Non-Climacteric Ripening in Strawberry. *J. Exp. Bot.* **2012**, *63*, 4741–4750. [CrossRef] [PubMed]
5. Agricultural Standards Unit United Nations Economic Commission for Europe. *Standard on the Marketing and Commercial Quality Control of Pineapples—Explanatory Brochure*; UNECE: Geneva, Switzerland, 2013.
6. Ahmad, M.S.; Siddiqui, M.W. *Postharvest Quality Assurance of Fruits*; Springer: Cham, Switzerland, 2015.
7. Medina, J.D.L.C.; García, H.S. *Post-Harvest Operations*; Food and Agriculture Organization of the United Nations: Roma, Italy, 2005.
8. Kamol, S.; Howlader, J.; Dhar, G.S.; Aklimuzzaman, M. Effect of Different Stages of Maturity and Postharvest Treatments on Quality and Storability of Pineapple. *J. Bangladesh Agric. Univ.* **2016**, *12*, 251–260. [CrossRef]
9. Yun, Z.; Li, T.; Gao, H.; Zhu, H.; Gupta, V.K.; Jiang, Y.; Duan, X. Integrated Transcriptomic, Proteomic, and Metabolomics Analysis Reveals Peel Ripening of Harvested Banana under Natural Condition. *Biomolecules* **2019**, *9*, 167. [CrossRef] [PubMed]
10. White, I.R.; Blake, R.S.; Taylor, A.J.; Monks, P.S. Metabolite Profiling of the Ripening of Mangoes Mangifera Indica, L. Cv. 'Tommy Atkins' by Real-Time Measurement of Volatile Organic Compounds. *Metabolomics* **2016**, *12*, 1–11. [CrossRef]
11. Aizat, W.M.; Dias, D.A.; Stangoulis, J.C.R.; Able, J.A.; Roessner, U.; Able, A.J. Metabolomics of Capsicum Ripening Reveals Modification of the Ethylene Related-Pathway and Carbon Metabolism. *Postharvest Biol. Technol.* **2014**, *89*, 19–31. [CrossRef]
12. Diboun, I.; Mathew, S.; Al-Rayyashi, M.; Elrayess, M.; Torres, M.; Halama, A.; Méret, M.; Mohney, R.P.; Karoly, E.D.; Malek, J.; et al. Metabolomics of Dates (Phoenix Dactylifera) Reveals a Highly Dynamic Ripening Process Accounting for Major Variation in Fruit Composition. *BMC Plant Biol.* **2015**, *15*, 291. [CrossRef]

13. Pedreschi, R.; Munoz, P.; Robledo, P.; Becerra, C.; Defilippi, B.G.; van Eekelen, H.D.L.M.; Mumm, R.; Westra, E.H.; de Vos, R.C.H. Metabolomics Analysis of Postharvest Ripening Heterogeneity of 'Hass' Avocadoes. *Postharvest Biol. Technol.* **2014**, *92*, 172–179. [CrossRef]
14. Monti, L.L.; Bustamante, C.A.; Osorio, S.; Gabilondo, J.; Borsani, J.; Lauxmann, M.A.; Maulion, E.; Valentini, G.; Budde, C.O.; Fernie, A.R.; et al. Metabolic Profiling of a Range of Peach Fruit Varieties Reveals High Metabolic Diversity and Commonalities and Differences during Ripening. *Food Chem.* **2016**, *190*, 879–888. [CrossRef]
15. Allwood, J.W.; Cheung, W.; Xu, Y.; Mumm, R.; De Vos, R.C.H.; Deborde, C.; Biais, B.; Maucourt, M.; Berger, Y.; Schaffer, A.A.; et al. Metabolomics in Melon: A New Opportunity for Aroma Analysis. *Phytochemistry* **2014**, *99*, 61–72. [CrossRef]
16. Parijadi, A.A.R.; Putri, S.P.; Ridwani, S.; Dwivany, F.M.; Fukusaki, E. Metabolic Profiling of Garcinia Mangostana (Mangosteen) Based on Ripening Stages. *J. Biosci. Bioeng.* **2018**, *125*, 238–244. [CrossRef]
17. Karagiannis, E.; Michailidis, M.; Karamanoli, K.; Lazaridou, A.; Minas, I.S.; Molassiotis, A. Postharvest Responses of Sweet Cherry Fruit and Stem Tissues Revealed by Metabolomic Profiling. *Plant Physiol. Biochem.* **2018**, *127*, 478–484. [CrossRef] [PubMed]
18. Jarret, D.A.; Morris, J.; Cullen, D.W.; Gordon, S.L.; Verrall, S.R.; Milne, L.; Hedley, P.E.; Allwood, J.W.; Brennan, R.M.; Hancock, R.D. A Transcript and Metabolite Atlas of Blackcurrant Fruit Development Highlights Hormonal Regulation and Reveals the Role of Key Transcription Factors. *Front. Plant Sci.* **2018**, *9*, 1–22. [CrossRef] [PubMed]
19. Montecchiarini, M.L.; Margarit, E.; Morales, L.; Rivadeneira, M.F.; Bello, F.; Gollán, A.; Vázquez, D.; Podestá, F.E.; Tripodi, K.E.J. Proteomic and Metabolomic Approaches Unveil Relevant Biochemical Changes in Carbohydrate and Cell Wall Metabolisms of Two Blueberry (Vaccinium Corymbosum) Varieties with Different Quality Attributes. *Plant Physiol. Biochem.* **2019**, 230–244. [CrossRef] [PubMed]
20. Steingass, C.B.; Dell, C.; Lieb, V.; Mayer-Ullmann, B.; Czerny, M.; Carle, R. Assignment of Distinctive Volatiles, Descriptive Sensory Analysis and Consumer Preference of Differently Ripened and Post-Harvest Handled Pineapple (Ananas Comosus [L.] Merr.) Fruits. *Eur. Food Res. Technol.* **2016**, *242*, 33–43. [CrossRef]
21. Ogawa, E.M.; Costa, H.B.; Ventura, J.A.; Caetano, L.C.S.; Pinto, F.E.; Oliveira, B.G.; Barroso, M.E.S.; Scherer, R.; Endringer, D.C.; Romão, W. Chemical Profile of Pineapple Cv. Vitória in Different Maturation Stages Using Electrospray Ionization Mass Spectrometry. *J. Sci. Food Agric.* **2018**, *98*, 1105–1116. [CrossRef] [PubMed]
22. Steingass, C.B.; Grauwet, T.; Carle, R. Influence of Harvest Maturity and Fruit Logistics on Pineapple (Ananas Comosus [L.] Merr.) Volatiles Assessed by Headspace Solid Phase Microextraction and Gas Chromatography-Mass Spectrometry (HS-SPME-GC/MS). *Food Chem.* **2014**, *150*, 382–391. [CrossRef]
23. Steingass, C.B.; Jutzi, M.; Müller, J.; Carle, R.; Schmarr, H.G. Ripening-Dependent Metabolic Changes in the Volatiles of Pineapple (Ananas Comosus (L.) Merr.) Fruit: II. Multivariate Statistical Profiling of Pineapple Aroma Compounds Based on Comprehensive Two-Dimensional Gas Chromatography-Mass Spectrometry. *Anal. Bioanal. Chem.* **2015**, *407*, 2609–2624. [CrossRef]
24. Steingass, C.B.; Glock, M.P.; Schweiggert, R.M.; Carle, R. Studies into the Phenolic Patterns of Different Tissues of Pineapple (Ananas Comosus [L.] Merr.) Infructescence by HPLC-DAD-ESI-MS (n) and GC-MS Analysis. *Anal. Bioanal. Chem.* **2015**, *407*, 6463–6479. [CrossRef]
25. Werth, M.T.; Halouska, S.; Shortridge, M.D.; Zhang, B.; Powers, R. Analysis of Metabolomic PCA Data Using Tree Diagrams. *Anal. Biochem.* **2010**, *399*, 58–63. [CrossRef]
26. Teoh, S.T.; Putri, S.; Mukai, Y.; Bamba, T.; Fukusaki, E. A Metabolomics-Based Strategy for Identification of Gene Targets for Phenotype Improvement and Its Application to 1-Butanol Tolerance in Saccharomyces Cerevisiae. *Biotechnol. Biofuels* **2015**, *8*, 1–14. [CrossRef]
27. Zhao, X.; Chen, M.; Zhao, Y.; Zha, L.; Yang, H.; Wu, Y. GC–MS-Based Nontargeted and Targeted Metabolic Profiling Identifies Changes in the Lentinula Edodes Mycelial Metabolome under High-Temperature Stress. *Int. J. Mol. Sci.* **2019**, *20*, 2330. [CrossRef]
28. Aebersold, R.; Mann, M. Mass spectrometry-based proteomics. *Nature* **2003**, *422*, 198–207. [CrossRef] [PubMed]
29. Wolfender, J.L.; Marti, G.; Thomas, A.; Bertrand, S. Current Approaches and Challenges for the Metabolite Profiling of Complex Natural Extracts. *J. Chromatogr. A* **2015**, *1382*, 136–164. [CrossRef] [PubMed]
30. Clarke, C.J.; Haselden, J.N. Metabolic Profiling as a Tool for Understanding Mechanisms of Toxicity. *Toxicol. Pathol.* **2008**, *36*, 140–147. [CrossRef] [PubMed]

31. Hanifah, A.; Maharijaya, A.; Putri, S.P.; Laviña, W.A.; Sobir. Untargeted Metabolomics Analysis of Eggplant (Solanum Melongena, L.) Fruit and Its Correlation to Fruit Morphologies. *Metabolites* **2018**, *8*, 49. [CrossRef] [PubMed]
32. Papadimitropoulos, M.-E.P.; Vasilopoulou, C.G.; Maga-Nteve, C.; Klapa, M.I. Untargeted GC-MS Metabolomics. *Methods Mol. Biol.* **2018**, *1738*, 133–147. [CrossRef] [PubMed]
33. Jolliffe, I.T.; Cadima, J.; Cadima, J. Principal Component Analysis: A Review and Recent Developments Subject Areas. *Phil. Trans. R. Soc. A* **2016**, *374*, 1–16. [CrossRef]
34. Sanewski, G.M.; Bartholomew, D.P.; Paull, R.E. *The Pineapple: Botany, Production and Uses*, 2nd ed.; CABI: Wallingford, UK, 2018.
35. Joy, P.P.; Rejuva TA, R. *Harvesting and Post-Harvest Handling of Pineapple*; Pineapple Research Station: Kerala, India, 2016.
36. Alexander, D.L.J.; Tropsha, A.; Winkler, D.A. Beware of R(2): Simple, Unambiguous Assessment of the Prediction Accuracy of QSAR and QSPR Models. *J. Chem. Inf. Model.* **2015**, *55*, 1316–1322. [CrossRef]
37. Fausto Rivero-Cruz, J.; Rodríguez de San Miguel, E.; Robles-Obregón, S.; Hernández-Espino, C.C.; Rivero-Cruz, B.E.; Pedraza-Chaverri, J.; Esturau-Escofet, N. Prediction of Antimicrobial and Antioxidant Activities of Mexican Propolis by 1H-NMR Spectroscopy and Chemometrics Data Analysis. *Molecules* **2017**, *22*, 1184. [CrossRef]
38. Farcuh, M.; Li, B.; Rivero, R.M.; Shlizerman, L.; Sadka, A.; Blumwald, E. Sugar Metabolism Reprogramming in a Non-Climacteric Bud Mutant of a Climacteric Plum Fruit during Development on the Tree. *J. Exp. Bot.* **2017**, *68*, 5813–5828. [CrossRef]
39. Siddique, S.; Endres, S.; Sobczak, M.; Radakovic, Z.S.; Fragner, L.; Grundler, F.M.W.; Weckwerth, W.; Tenhaken, R.; Bohlmann, H. Myo-Inositol Oxygenase Is Important for the Removal of Excess Myo-Inositol from Syncytia Induced by Heterodera Schachtii in Arabidopsis Roots. *New Phytol.* **2014**, *201*, 476–485. [CrossRef] [PubMed]
40. Panizzi, A.R.; Parra, J.R.P. *Insect Bioecology and Nutrition for Integrated Pest Management*; CRC Press: Boca Raton, FL, USA, 2012.
41. Lin, Y.-C.; Liu, Y.; Nakamura, Y. The Choline/Ethanolamine Kinase Family in Arabidopsis: Essential Role of CEK4 in Phospholipid Biosynthesis and Embryo Development. *Plant Cell* **2015**, *27*, 1497–1511. [CrossRef] [PubMed]
42. Decros, G.; Baldet, P.; Beauvoit, B.; Stevens, R.; Flandin, A.; Colombié, S.; Gibon, Y.; Pétriacq, P. Get the Balance Right: ROS Homeostasis and Redox Signalling in Fruit. *Front. Plant Sci.* **2019**, *10*. [CrossRef] [PubMed]
43. Musara, A. Determination of Ascrobic Acid in Some Citrus Fruits by Using Pencil Graphite Electrode. Ph.D. Thesis, Near East University, Nicosia, Turky, 2019.
44. Paliyath, G.; Subramanian, J.; Lim, L.-T.; Subramanian, K.S.; Handa, A.K.; Matto, A.K. *Postharvest Biology and Nanotechnology*; New York Academy of Sciences: New York, NY, USA, 2019.
45. Alkan, N.; Fortes, A.M. Insights into Molecular and Metabolic Events Associated with Fruit Response to Post-Harvest Fungal Pathogens. *Front. Plant Sci.* **2015**, *6*, 1–14. [CrossRef]
46. Davidzon, M.; Alkan, N.; Kobiler, I.; Prusky, D. Acidification by Gluconic Acid of Mango Fruit Tissue during Colonization via Stem End Infection by Phomopsis Mangiferae. *Postharvest Biol. Technol.* **2010**, *55*, 71–77. [CrossRef]
47. Moirangthem, K.; Tucker, G. How Do Fruits Ripen? *Front. Young Minds* **2018**, *6*, 1–7. [CrossRef]
48. Dheilly, E.; Gall, S.L.; Guillou, M.-C.; Renou, J.-P.; Bonnin, E.; Orsel, M.; Lahaye, M. Cell Wall Dynamics during Apple Development and Storage Involves Hemicellulose Modifications and Related Expressed Genes. *BMC Plant Biol.* **2016**, *16*, 201. [CrossRef]
49. Wu, Q.; Li, T.; Chen, X.; Wen, L.; Yun, Z.; Jiang, Y. Sodium Dichloroisocyanurate Delays Ripening and Senescence of Banana Fruit during Storage. *Chem. Cent. J.* **2018**, *12*, 1–11. [CrossRef]
50. Zgola-Grześkowiak, A.; Grześkowiak, T. Determination of Glutamic Acid and Aspartic Acid in Tomato Juice by Capillary Isotachophoresis. *Int. J. Food Prop.* **2012**, *15*, 628–637. [CrossRef]
51. Agius, C.; von Tucher, S.; Poppenberger, B.; Rozhon, W. Quantification of Glutamate and Aspartate by Ultra-High Performance Liquid Chromatography. *Molecules* **2018**, *23*, 1389. [CrossRef]

52. Folharini, Z.F.; Orlandi, C.R.; Martini, M.C.; Bruxel, F.; Altmayer, T.; Brietzke, D.T.; Hoehne, L. Nutritional Characterization of Vasconcellea Quercifolia, A.St-Hil.: Potential for the Development of Functional Food. *Food Sci. Technol.* **2019**, *2061*, 432–438. [CrossRef]
53. Bernales, M.; Monsalve, L.; Ayala-Raso, A.; Valdenegro, M.; Martínez, J.P.; Travisany, D.; Defilippi, B.; González-Agüero, M.; Cherian, S.; Fuentes, L. Expression of Two Indole-3-Acetic Acid (IAA)-Amido Synthetase (GH3) Genes during Fruit Development of Raspberry (Rubus Idaeus Heritage). *Sci. Hortic.* **2019**, *246*, 168–175. [CrossRef]
54. Iatridi, V.; Hayes, J.E.; Yeomans, M.R. Quantifying Sweet Taste Liker Phenotypes: Time for Some Consistency in the Classification Criteria. *Nutrients* **2019**, *11*, 129. [CrossRef] [PubMed]
55. Siti Roha, A.M.; Zainal, S.; Noriham, A.; Nadzirah, K.Z. Determination of Sugar Content in Pineapple Waste Variety N36. *Int. Food Res. J.* **2013**, *20*, 1941–1943.
56. Jia, H.; Wang, Y.; Sun, M.; Li, B.; Han, Y.; Zhao, Y.; Li, X.; Ding, N.; Li, C.; Ji, W.; et al. Sucrose Functions as a Signal Involved in the Regulation of Strawberry Fruit Development and Ripening. *New Phytol.* **2013**, *198*, 453–465. [CrossRef] [PubMed]
57. Fuentes, L.; Figueroa, C.R.; Valdenegro, M. Recent Advances in Hormonal Regulation and Cross-Talk during Non-Climacteric Fruit Development and Ripening. *Horticulturae* **2019**, *5*. [CrossRef]
58. O'Donoghue, E.M.; Somerfield, S.D.; Watson, L.M.; Brummell, D.A.; Hunter, D.A. Galactose Metabolism in Cell Walls of Opening and Senescing Petunia Petals. *Planta* **2009**, *229*, 709–721. [CrossRef]
59. Ogasawara, S.; Abe, K.; Nakajima, T. Pepper Beta-Galactosidase 1 (PBG1) Plays a Significant Role in Fruit Ripening in Bell Pepper (Capsicum Annuum). *Biosci. Biotechnol. Biochem.* **2007**, *71*, 309–322. [CrossRef]
60. Osorio, S.; Scossa, F.; Fernie, A.R. Molecular Regulation of Fruit Ripening. *Front. Plant Sci.* **2013**, *4*, 1–8. [CrossRef]
61. Kumar, V.; Irfan, M.; Ghosh, S.; Chakraborty, N.; Chakraborty, S.; Datta, A. Fruit Ripening Mutants Reveal Cell Metabolism and Redox State during Ripening. *Protoplasma* **2016**, *253*, 581–594. [CrossRef]
62. Sharma, S.; Pareek, S.; Sagar, N.A.; Valero, D.; Serrano, M. Modulatory Effects of Exogenously Applied Polyamines on Postharvest Physiology, Antioxidant System and Shelf Life of Fruits: A Review. *Int. J. Mol. Sci.* **2017**, *18*. [CrossRef] [PubMed]
63. Golly, M.K.; Ma, H.; Sarpong, F.; Dotse, B.P.; Oteng-Darko, P.; Dong, Y. Shelf-Life Extension of Grape (Pinot Noir) by Xanthan Gum Enriched with Ascorbic and Citric Acid during Cold Temperature Storage. *J. Food Sci. Technol.* **2019**, *56*, 4867–4878. [CrossRef] [PubMed]
64. Yun, Z.; Gao, H.; Liu, P.; Liu, S.; Luo, T.; Jin, S.; Xu, Q.; Xu, J.; Cheng, Y.; Deng, X. Comparative Proteomic and Metabolomic Profiling of Citrus Fruit with Enhancement of Disease Resistance by Postharvest Heat Treatment. *BMC Plant Biol.* **2013**, *13*. [CrossRef] [PubMed]
65. Bustamante, C.A.; Monti, L.L.; Gabilondo, J.; Scossa, F.; Valentini, G.; Budde, C.O.; Lara, M.V.; Fernie, A.R.; Drincovich, M.F. Differential Metabolic Rearrangements after Cold Storage Are Correlated with Chilling Injury Resistance of Peach Fruits. *Front. Plant Sci.* **2016**, *7*, 1–15. [CrossRef] [PubMed]
66. Anjaritha, A.A.; Ridwani, S.; Dwivany, F.M.; Putri, S.P.; Fukusaki, E. A Metabolomics-Based Approach for the Evaluation of off-Tree Ripening Conditions and Different Postharvest Treatments in Mangosteen (Garcinia Mangostana). *Metabolomics* **2019**, *15*, 1–16. [CrossRef]
67. Geladi, P.; Kowalski, B.R. Partial Least-Squares Regression: A Tutorial. *Anal. Chim. Acta* **1986**, *185*, 1–17. [CrossRef]
68. Nitta, K.; Laviña, W.A.; Pontrelli, S.; Liao, J.C.; Putri, S.P.; Fukusaki, E. Orthogonal Partial Least Squares / Projections to Latent Structures Regression-Based Metabolomics Approach for Identi Fi Cation of Gene Targets for Improvement of 1-Butanol Production in Escherichia Coli. *J. Biosci. Bioeng.* **2017**, *124*, 498–505. [CrossRef]

metabolites

MDPI

Article

Cross-Species Comparison of Fruit-Metabolomics to Elucidate Metabolic Regulation of Fruit Polyphenolics Among Solanaceous Crops

Carla Lenore F. Calumpang, Tomoki Saigo, Mutsumi Watanabe and Takayuki Tohge *

Graduate School of Science and Technology, Nara Institute of Science and Technology (NAIST), Ikoma, Nara 630-0192, Japan; calumpang.carla_lenore.bw2@bs.naist.jp (C.L.F.C.); saigo.tomoki.sn6@bs.naist.jp (T.S.); mutsumi@bs.naist.jp (M.W.)

* Correspondence: tohge@bs.naist.jp; Tel.: +81-743-72-5480

Received: 20 March 2020; Accepted: 14 May 2020; Published: 19 May 2020

Abstract: Many solanaceous crops are an important part of the human daily diet. Fruit polyphenolics are plant specialized metabolites that are recognized for their human health benefits and their defensive role against plant abiotic and biotic stressors. Flavonoids and chlorogenates are the major polyphenolic compounds found in solanaceous fruits that vary in quantity, physiological function, and structural diversity among and within plant species. Despite their biological significance, the elucidation of metabolic shifts of polyphenols during fruit ripening in different fruit tissues, has not yet been well-characterized in solanaceous crops, especially at a cross-species and cross-cultivar level. Here, we performed a cross-species comparison of fruit-metabolomics to elucidate the metabolic regulation of fruit polyphenolics from three representative crops of Solanaceae (tomato, eggplant, and pepper), and a cross-cultivar comparison among different pepper cultivars (*Capsicum annuum* cv.) using liquid chromatography-mass spectrometry (LC-MS). We observed a metabolic trade-off between hydroxycinnamates and flavonoids in pungent pepper and anthocyanin-type pepper cultivars and identified metabolic signatures of fruit polyphenolics in each species from each different tissue-type and fruit ripening stage. Our results provide additional information for metabolomics-assisted crop improvement of solanaceous fruits towards their improved nutritive properties and enhanced stress tolerance.

Keywords: polyphenolics; solanaceous crops; *capsicum annuum*; pepper; tomato; eggplant; fruit ripening; metabolomics; tissue-specificity; flavonoid

1. Introduction

Solanaceae (nightshade family) is an agronomically- and botanically-diverse plant taxonomic group, with members ranging from vegetable crops through medicinal plants to ornamentals. Few representative crops of economic importance include tomato (*Solanum lycopersicum*), eggplant (*Solanum melongena*), and pepper (*Capsicum annuum*). With available genome sequences [1,2] and genetic resources from different tomato varieties and natural mutants [3], tomato has become the first model crop for fleshy fruit ripening, fruit pigmentation, specialized (secondary) metabolism, and plant defense [4–9]. Subsequently, metabolomic data of specialized metabolites from tomato fruits [4,10–12] and a tomato metabolite database [5,13–15] have been published and developed via a mass-spectrometry (MS)-based metabolomic analysis for fruit-omics approach. The metabolomics approach, focusing on specialized metabolism, is also being used on other solanaceous plants, including tobacco (*Nicotiana* spp.) [16,17], potato (*Solanum tuberosum*) [18–20], petunia (*Petunia* spp.) [21], and *Atropa* spp. [22]. However, developing similar resources for related crops is still a goal of the scientific community [23].

In the point of fruit-omics, pepper (*Capsicum* spp.) is increasingly gaining recognition as an excellent model plant for solanaceous fruit-omics [8,12,24,25]. Pepper (*Capsicum* spp.) has a world yield of 184,742 hg/ha and a world production of almost 40 million tons in 2018 [26]. Consumed either raw, cooked, as a spice, or food colorant [27], pepper have known human health benefits, such as weight reduction [28], pain relief [29], and cancer prevention [30]. Peppers can also be used as sprays for crowd control and personal defense since some cultivars cause skin irritation [31,32]. Capsaicinoids (e.g., capsaicin, 8-methyl-*N*-vanillyl-6-nonenamide) have been the main focus of metabolic diversity studies in pepper due to their dramatic bioactivities; however, pepper polyphenols are less highlighted, since fruit metabolomics focusing on polyphenols have already been well-investigated in tomato species. Furthermore, tissue-specific patterns of polyphenol distribution in pepper fruit have not been well-characterized during fruit ripening, specifically in the peel and pericarp. There is still a need to investigate metabolic shifts of polyphenols among different fruit tissues during different stages of ripening in pepper through fruit-omics analysis, because of the fact that significant metabolic shift of polyphenolic compounds during fruit ripening is observed in different fruit tissues, such as in the case of tomato species [10,33–35]. Furthermore, capsaicinoids share the same biosynthetic precursors, e.g., phenylalanine and *p*-coumarate, which can cause differences in metabolic shifts of polyphenols in pepper versus tomato. The pungent properties in pepper are due to the bioactivity of capsaicinoids, an exclusive trait amongst capsicum species and is not exhibited by other solanaceous crops. Pepper also has a variety of fruit colors among cultivars during fruit ripening, which mainly indicate variation in fruit carotenoids and chlorophylls, but also in fruit polyphenolics, in the case of purple pepper cultivars. While total amount of flavonoid aglycones have been compared between pungent and non-pungent pepper cultivars [36], further metabolomic analysis of fruit polyphenolics in terms of their tissue specificity and metabolic shift during fruit ripening should be highlighted.

On the other hand, fruit-metabolomics on eggplant polyphenolics is still in progress with only a few studies having been performed. The metabolomic analysis of different *Solanum* species, including five eggplant (*S. melongena*) accessions, three accessions from an eggplant wild ancestor (*Solanum insanum*), and two from scarlet eggplant (*Solanum aethiopicum*), through a LC-MS based strategy revealed metabolic diversity of anthocyanins, chlorogenic acid derivatives, flavonoids, triterpenoid alkaloids and triterpenoids, and novel biosynthetic frameworks [37]. In the untargeted metabolomics of the fruit of twenty-one eggplant (*S. melongena*) accessions using GC-MS and LC-MS, some accession-specific specialized metabolites were putatively identified [38]. Moreover, the metabolic quantitative trait locus (mQTL) from eggplant fruit was investigated for identification of a genomic region of productivity for chlorogenic acid and two anthocyanin pigments [39].

Plant polyphenols are a large group of plant specialized metabolites, which can be subdivided into several large sub-groups of major phyto-antioxidants, for example flavonoids and hydroxycinnamates [40,41]. In solanaceous plants, flavonoids involved in the stress resistance against abiotic and biotic stressors, such as pathogen infection [42,43], ultraviolet light (UV) [44,45], nitrogen deficiency [46], and cold temperature [47]. Flavonoids are also involved in plant reproduction, such as attracting insect pollinators and seed dispersers [48]. Pigment flavonoids, such as anthocyanins, upon absorbing visible (VIS) light contribute to the red and purple pigmentation of solanaceous fruits [8]. Intake of flavonoids and hydroxycinnamates also reduce the risk of human disease due to their anti-cancer and antioxidant activities based on in vitro assays [49–51]. Chlorogenates have recognized human health benefits, such as antioxidant, antiviral, hepatoprotective, and anti-hypoglycemic properties [52]. With their beneficial roles in human health, tissue-specific accumulation patterns of polyphenols in pepper fruit during ripening can be focused. Since polyphenols contribute a role in plant stress response, there is a propensity for such compounds to accumulate more in the fruit peel than in the pericarp, which has already been observed in tomato fruit [10,53–56].

To develop a baseline on the polyphenolic compounds already detected in our solanaceous crops of study, a phytochemical survey of polyphenolics reported in the fruits of tomato, eggplant, and pepper was investigated via a phytochemical database KNApSAcK (http://www.knapsackfamily.

com/KNApSAcK_Family) and literature search of fruit metabolomic analyses conducted on these species [4,10,25,57–59] (Figure 1A, Table 1). In tomato species, major polyphenolic compounds include derivatives of hydroxycinnamates, flavonols, and anthocyanins [5,10,13–15,57]. For eggplant, its fruit peel is rich in anthocyanins, mainly delphinidin derivatives [60], while the pericarp consists mainly of chlorogenates (CGAs) [61]. Flavonol glycosides have also been detected in the pericarp of eggplant fruit [62]. Polyphenolic compounds detected from pepper are much more diverse, with flavonoids being the main compound group and flavones and flavonols the major flavonoid subfamilies detected. Chlorogenates have also been identified from pepper fruit [59]. Based on the data for all three solanaceous crops of study, common polyphenolic compounds present in their fruit include hydroxycinnamates (specifically chlorogenates and their positional-isomers), chalconoids/stilbenoids, flavanones, and flavonol derivatives, such as rutin (quercetin-3-*O*-rutinoside) and kaempferol-3-*O*-glucosides (Figure 1A, Table 1). A survey of representative polyphenolic compounds have been reported in the metabolic analysis of the solanaceous fruits, a metabolic framework of major polyphenolics among three solanaceous species is illustrated (Figure 1B).

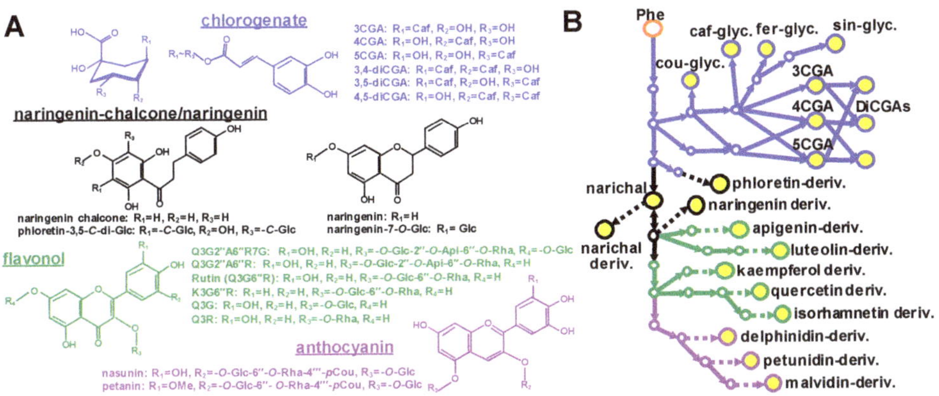

Figure 1. Major polyphenolic compounds reported in mature fruit of solanaceous crops. (**A**) chemical structure of major polyphenolic specialized metabolites found in solanaceous crops. Chlorogenates (CGAs), naringenins, and flavonols are shown. (**B**) the biosynthetic framework of polyphenolic compounds in solanaceous crops. Biosynthetic pathways were constructed by an online database search and literature review of major polyphenolic compounds in solanaceous crops. Color: Blue, hydroxycinnamates; black, stilbenoids/flavanones and chalcones; purple, anthocyanins; and green, flavonols/flavones; orange, amino acid. Yellow color inside of circle indicates major accumulation form in the plant. Abbreviations: Phe, phenylalanine; cou, coumaroyl; caf, caffeoyl; fer, feruloyl; sin, sinapoyl; glyc, glycoside; deriv, derivative. K, kaempferol; Q, quercetin; Narichal, naringenin chalcone; CGA, chlorogenate; DiCGA, dicaffeoyl-chlorogenate; Glc, glucose; Rha, rhamnose; Api, apiose; and hex, hexose.

Peppers consist of approximately 35 species [63] with five of them domesticated independently (*C. annuum*, *C. chinense*, *C. frutescens*, *C. baccatum*, and *C. pubescens*). *C. annuum* is the largest and most widely-cultivated species having both spicy (chili) and sweet varieties with different kinds of pigmentation in their fruits [8,64]. However, due to hybridization, pepper currently has around 50,000 varieties, providing a wide range of chemical variability within the same species exhibiting similar physical characteristics [12]. Despite such physical similarity within species, such genetic diversity still provides a large pool of chemical variability within species in terms of tissue-specificity, developmental ripening stage, and cultivar. Furthermore, the process of fruit ripening involves a tight metabolic regulation in conjunction with developmental stage [65], involving biochemical reactions resulting in changes in fruit flavor, texture, aroma, hardness, nutrient composition, and color [25,66].

Since some polyphenolic compounds are involved with flavor and fruit color, changes in polyphenolic content can occur during fruit ripening [15,67]. Accumulation patterns of major flavonoid glycosides in terms of subspecies and cultivar–specificity were evaluated from ripe fruits across thirty-two pepper accessions, including *C. annuum*, *C. chinense*, *C. frutescens*, and *C. baccatum* [25]. As one of the results given by this research, main pepper flavonoid decorations were observed as a metabolic polymorphism within these pepper species. Previously, primary metabolites (sugars, amino acids, and organic acids) were compared between tomato (*S. lycopersicum*) and pepper (*Capsicum chilense*) during fruit ripening [66,68] but not specialized metabolites, specifically polyphenols. Since some metabolomic studies of primary metabolites have been conducted for pepper fruit, such fruit-metabolomics studies should be expanded to include tissue-specificity of polyphenol accumulation in pepper peel versus pericarp during different stages of fruit ripening among various pepper cultivars. Furthermore, after metabolic polymorphisms of polyphenolic profiling were characterized through metabolic profiling with acid hydrolysis [49], a complete representation of the polyphenolic profile for this pepper variety can additionally be focused with metabolic profiling of the individual derivative forms. Meanwhile in hot pepper (*Capsicum annuum* "CM334") fruit, only the pericarp was analyzed for polyphenolic accumulation patterns during ripening [69]. Additionally, the identification of polyphenolic compounds specific for pepper against other solanaceous crops has not yet been performed, nor have the patterns of accumulation in relation with fruit tissue type been reported in some pepper species. Therefore, polyphenolic fruit-omics data developed in tomato research can possibly be used for the extension to other solanaceous crops, such as pepper species and cultivars through other fruit-omics approaches.

Table 1. Major polyphenolic compound groups reported in the solanaceous fruits.

Compound Family	**Compound Group**	**Tomato (*S. Lycopersicum*)**	**Eggplant (*S. Melongena*)**	**Pepper (*C. Annuum*)**
Hydroxycinnamates				
	CGAs	√	√	√
	di-CGAs	√	√	
	tri-CGA	√		
	p-coumaroyl-glycosides	√		√
	caffeoyl-glycosides	√	√	√
	feruloyl-glycosides	√	√	√
	sinapoyl-glycosides	√	√	√
Chalconoids/stilbenoids				
	phloretin deriv.	√		√
	naringenin chalcone deriv.	√	√	
	naringenin deriv.	√	√	
Flavonoids				
flavonols	kaempferol deriv.	√	√	√
	quercetin deriv.	√	√	√
	isorhamnetin deriv.	√		√
	myricetin deriv.	√	√	
anthocyanins	delphinidin deriv.	√	√	√
	petunidin deriv.	√	√	√
	malvidin deriv.	√		
flavones	apigenin deriv.			√
	luteolin deriv.			√

Here, we performed a cross-species comparison of representative phenolic compounds among three solanaceous crops (tomato, eggplant, and pepper) during fruit ripening, additionally including different pepper cultivars to address metabolic regulation of fruit polyphenolic metabolism among pepper varieties exhibiting differences in pigmentation and pungency. We focused on differences in the metabolic accumulation pattern between tissue types (peel and pericarp) and three fruit ripening stages (immature, green/purple mature, and final mature) to develop a better understanding of metabolic trade-off in fruit polyphenolics. Such information can allow breeders to optimize the biosynthesis of health-related polyphenolic compounds in pepper during their developmental stage of harvest. Metabolic signatures provided here, provide significant information for future functional genomics approach as well as for the metabolomics-assisted crop breeding of solanaceous fruits towards their nutritive improvement and stress tolerance enhancements.

2. Results

2.1. Metabolite Profiling of Major Polyphenolic Compounds in Fruit Tissues Among Solanaceous Crops

Metabolomic profiling of representative polyphenolic compounds from the fruit extracts of two tissue types (peel and pericarp) was conducted on three major solanaceous crops through liquid-chromatography-mass spectrometry (LC-MS) (Supplemental Table S1). Tomato, eggplant, and pepper were cultivated in the field and harvested for fruit-metabolomics analysis. Jalapeño pepper, a variety of chili pepper producing capsaicinoid compounds [70], was chosen for our cross-species analysis in order to understand tissue-specific metabolic trade-offs during metabolic shifts in fruit polyphenolics. Due to available studies on tissue-specific accumulation patterns of polyphenolic compounds in tomato fruit, tomato was used for the reference extracts for the peak annotation and analysis of metabolic change during fruit ripening. In our analysis, thirty-nine polyphenolic compounds were detected and annotated (Figure 2A; $\log_2$FC (mature/immature), up, > 2.0; down, < 0.5), including eggplant and pepper-specific flavonoid-derivatives. An upregulation of 19 polyphenolic compounds in the peel (4 chalcones, 9 flavonols, and 6 hydroxycinnamates) and 14 polyphenols in the pericarp (4 chalcones, 5 flavonols, and 5 hydroxycinnamates), were observed (Figure 2B) in tomato. Additionally, 8 polyphenolic compounds had higher upregulation in the peel than in the pericarp for tomato, which include one hydroxycinnamate (cou_hex_II, coumaroyl-hexoside II) and 7 flavonols (Q3G2′′A6′′R7G, quercetin-3-*O*-(2′′-*O*-apiosyl-6′′-*O*-rhamnosyl)glucoside-7-*O*-glucoside; Q3G2′′A6′′R, quercetin-3-*O*-(2′′-*O*-apiosyl-6′′-*O*-rhamnosyl)glucoside; K3G2′′A6′′R, kaempferol-3-*O*-(2′′-*O*-apiosyl-6′′-*O*-rhamnosyl) glucoside; rutin; K3G6′′R, kaempferol-3-*O*-(6′′-*O*-rhamnosyl)glucoside; Q3G, quercetin-3-*O*-glucoside; Q3R, quercetin-3-*O*-rhamnoside; and K3R, kaempferol-3-*O*-rhamnoside), as reported by Mintz-Oron et al. 2008 [10] in metabolomic analysis of tomato cv. Ailsa Craig. As reported in the other reports [53–56], the most abundant flavonol-glycoside, rutin, had more elevated levels in the peel than the pericarp (Figure 2A).

A cross-species comparison of the accumulation patterns of representative polyphenolic compounds between two tissue types from the fruits of three solanaceous crops was conducted. Two hydroxycinnamates (cou_hex_II; caf_hex_III, caffeoyl-hexoside III) were upregulated in the fruit peel among all three crops, while none were commonly upregulated in the pericarp. Between tomato cv. M82 and Jalapeño pepper, six polyphenols were commonly upregulated among their fruit peel while none were upregulated in the pericarp. Out of the six polyphenols upregulated in the fruit peel of both tomato cv. M82 and jalapeño pepper, two were hydroxycinnamates (caff_hex_I and caf_hex_II), two were flavonols (Q3G6′′R7G; K3G6′′R), and two were chalcones (narichal, naringenin chalcone; narichal_hex_I, naringenin chalcone-hexoside I). Meanwhile, a comparison between jalapeño pepper and eggplant indicated an upregulation of three hydroxycinnamates (cou_hex_III; feru_hex_I, feruloyl-hexoside I; feru_hex_II) in both crops for their fruit peel, while only feru_hex_I was upregulated in the pericarp for both. Species-specific polyphenols were also upregulated in jalapeño pepper, wherein coumaroyl-hexoside I (cou_hex_I) was specific for pepper in both the fruit peel and pericarp. Two other hydroxycinnamates were specifically upregulated only in pepper, with 5-chlorogenate (5CGA) upregulated in the peel and feru_hex_II in the pericarp. Jalapeño pepper showed a greater number of downregulated polyphenols in the peel (14) and pericarp (26), with an especially significant reduction of flavonol-derivatives in both peel and pericarp at fruit mature stage. Our cross-species comparison of fruit-omics with tissue specificity, indicated that pepper polyphenolic metabolism clearly has an opposite metabolic shift between hydroxycinnamate and flavonoid biosynthesis during fruit ripening in both peel and pericarp (Figure 2).

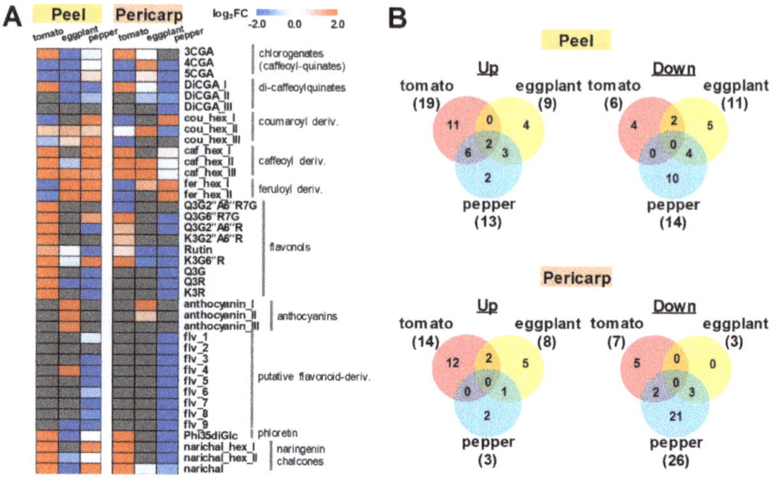

Figure 2. Metabolite profiling of major polyphenolic compounds between fruit tissues among three major solanaceous crops. (**A**) metabolic shift during fruit ripening ($\log_2$FC(mature/immature)) in peel and pericarp of tomato, eggplant and pepper fruits is presented. (**B**) venn's diagrams show conserved metabolic changes within and between fruit tissues among three solanaceous crops. Metabolites which are commonly up- (>2.0) or down-regulated (<0.5) are shown. MeV software (http://www.mev.tm4.org/) was used for data visualization. Abbreviations: cou, coumaroyl; caf, caffeoyl; fer, feruloyl; CGA, chlorogenate; DiCGA, dicaffeoyl-chlorogenate; K, kaempferol; Q, quercetin; Glc/G, glucose; Rha/R, rhamnose; Api/A, apiose; hex, hexose; Phi, phloretin; narichal, naringenin-chalcone; flv, putative flavonol-derivatives.

2.2. Metabolic Shifts of Polyphenolics Different Pepper Cultivars During Fruit Ripening

The metabolic analysis of polyphenolic compounds in pepper fruits was extended from the pungent jalapeño pepper to include other sweet pepper cultivars. Due to the high variability among pepper cultivars in terms of color, shape, and pungency, six pepper cultivars of different visible phenotypes caused by differences in pigment composition (elphinidinnone, carotenoid-type, and anthocyanin-type) were selected. Six different cultivars of pepper (*C. annuum*) including four sweet cultivars (green paprika, green pepper, yellow paprika, and red paprika), a pungent cultivar (jalapeño pepper), and an anthocyanin-type pepper (purple pepper) were selected for comparison during three fruit ripening stages (immature, green mature, and red or final mature) (Figure 3). In terms of color, five cultivars had red-colored final mature fruits while yellow paprika was yellow at final maturity. During the immature and green mature stages of ripening, purple pepper was purple-colored while the other five cultivars were colored green. In terms of pungency, jalapeño pepper was the only pungent cultivar and the other cultivars were non-pungent or sweet. In terms of fruit shape, purple and jalapeño peppers were pointed while the other cultivars were bell-shaped. Purple pepper differs from other sweet pepper cultivars since most sweet cultivars are bell-shaped [71].

Upon comparing the polyphenol distribution patterns between fruit peel and pericarp, no clear common pattern was observed among the six pepper cultivars, however they are slightly separated in each tissues and cultivars (Figure 4A). All anthocyanins which are involved in pigmentation, specifically purple to black pigmentation in pepper tissues [72], were only detected in purple pepper and not in other pepper cultivars, since purple pepper is the only cultivar exhibiting a developmental stage with purple color fruit (Figure 4B and Supplemental Table S1).

Figure 3. Pepper cultivars used in this study. Fruits from jalapeño pepper cv. "Jalapeño", red paprika pepper cv. "Flupy Red EX", purple pepper cv. "Nara Murasaki", green pepper cv. "Miogi", yellow paprika cv. "Sonia Gold", and green paprika pepper cv. "New Ace", were cultivated for our analysis.

In the pungent cultivar (jalapeño pepper), most polyphenols were upregulated in the peel during both green mature (14 polyphenols) and final mature (14 polyphenols) stages. Meanwhile, most polyphenols were upregulated in the pericarp during its immature stage (24 polyphenols). More hydroxycinnamates were upregulated in the peel (9) than in the pericarp (4) during its red mature stage. An elevation of eight hydroxycinnamates, namely coumaroyl-hexosides (cou_hex I, II, III), caffeoyl-hexosides (caff_hex I, II, III), and feruloyl-hexosides (feru_hex I and II), as well as a reduction of almost all flavonoids except rutin and putative flavonoid one (flv_1), were suggested as candidate fruit ripening markers in the peel in pungent jalapeño pepper. However, these metabolic shifts were not observed in the other pepper cultivars except purple pepper. For the non-pungent cultivar (green pepper), more hydroxycinnamates were upregulated in the pericarp (11) than in the peel (1) at its red mature stage. Reduction of chlorogenates (3CGA, 4CGA, and 5CGA), were observed in the peel, but CGAs were elevated in the pericarp. Di-caffeoyl-chlorogenate_I (DiCGA_I) is the product of both 3CGA and 4CGA while DiCGA_II is the product of both 3CGA and 5CGA. The upregulation of both DiCGA_I and DiCGA_II suggested to explain the absence of 3CGA from the pericarp. In the sweet green paprika pepper, flavonoids were upregulated in the pericarp during its red mature stage. Furthermore, more hydroxycinnamates were upregulated in the pericarp (11) than in the peel (1) at its red mature stage. Eleven candidate ripening markers were identified for green paprika pepper, which include: Three chlorogenates (3CGA, 4CGA, and 5CGA), three di-chlorogenates (DiCGA I, II, III), two coumaroyl-hexosides (cou_hex II, III), one caffeoyl-hexoside (caff_hex_III), and two feruloyl-hexosides (feru_hex I, II). In red paprika pepper, most polyphenols were upregulated in both peel (21 polyphenols) and pericarp (22 polyphenols) during its immature stage. Only two hydroxycinnamates were upregulated its red mature stage in the peel (cou_hex_I and cou_hex_III) while one hydroxycinnamate accumulated more at its red mature stage in the pericarp (cou_hex_I). In yellow paprika pepper, only two hydroxycinnamates (3CGA and DiCGA_III) were upregulated in the peel, while three hydroxycinnamates (DiCGA_III, fer_hex_I and fer_hex_II) were upregulated in the pericarp at its red mature stage. Flavonoid mono- and di-glycosides were upregulated during the course of ripening for both tissues, specifically K3G6″R in the peel and K3R in the pericarp. In purple pepper, more hydroxycinnamates were upregulated in the peel (7) than in the pericarp (3) in its red mature stage. Finally, one chlorogenate (3CGA), two di-chlorogenates (DiCGA I, III), one coumaroyl-hexoside (cou_hex_I), two caffeoyl-hexosides (caff_hex I, II), and one feruloyl-hexoside (fer_hex_II), were selected as ripening marker for polyphenolics of purple pepper (Figure 4B). K3G6''R, Q3R, rutin, and Q3G2''A6''R were upregulated in both tissues during immature stage in purple pepper. Finally, the metabolic shifts observed in the cross-species comparison was the opposing direction of metabolic shift between hydroxycinnamate and flavonoid biosynthesis during fruit ripening in both the pungent cultivar (jalapeño pepper) and anthocyanin-producing purple pepper, but not in other carotenoid-type pepper cultivars.

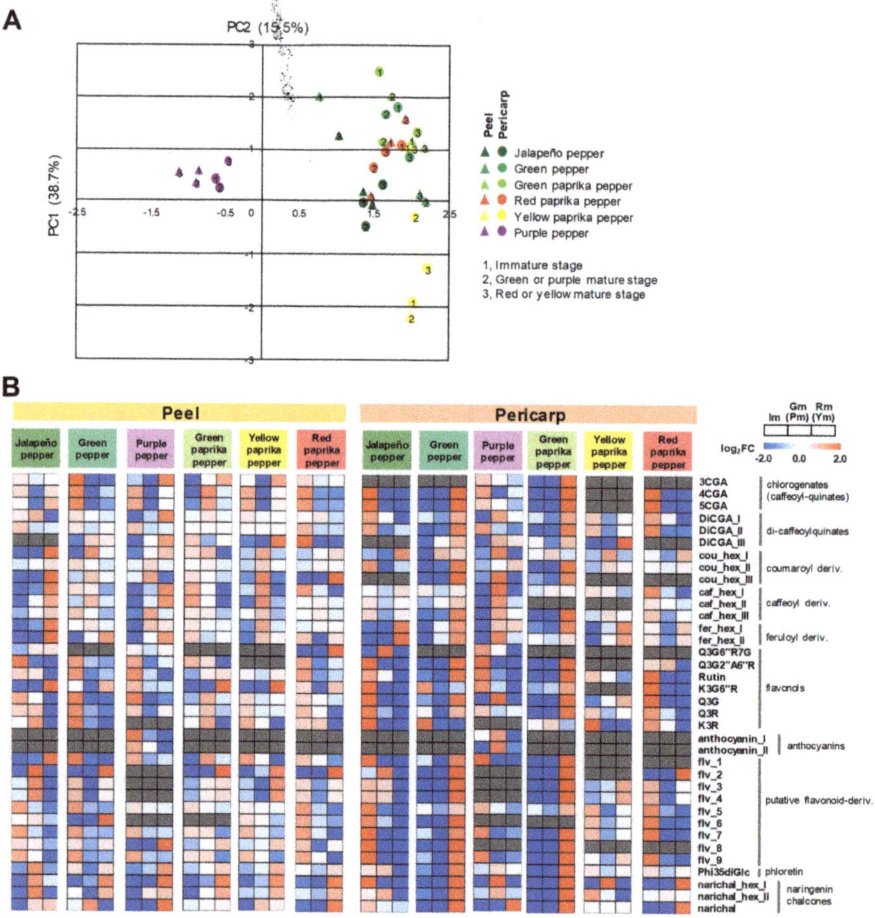

Figure 4. Metabolic shifts of polyphenolics in different fruit tissue among six pepper cultivars during fruit ripening. (**A**) principal component analysis (PCA) of polyphenolics in different pepper cultivars during fruit ripening. The plots were applied for the 39 metabolites with the average values from 3 biological replications. PCA was conducted by the MultiExperiment Viewer [73]. Principal component (PC) triangles and circles indicate peel and pericarp, respectively. Coefficient correlation was estimated by person correlation method using MeV software. (**B**) heatmap visualization of metabolite data is normalized and scaled by $\log_2$FC (mean/average_mean) for each metabolite. Abbreviations: cou, coumaroyl; caf, caffeoyl; fer, feruloyl; CGA, chlorogenate; DiCGA, dicaffeoyl-chlorogenate; Glc/G, glucose; Rha/R, rhamnose; Api/A, apiose; hex, hexose; Phi, phloretin; flv, putative flavonol-derivatives; Narichal, naringenin chalcone; Im, immature stage; Gm(Pm), green mature (or purple mature stage); and Rm(Ym), red mature (or yellow mature stage).

3. Discussion

Polyphenols are primarily involved in physiological response against abiotic and biotic stressors [42,44,46,47] and occasionally in plant reproduction [48]. Due to the phytochemical functions of these compounds, polyphenolic which are accumulating on the surface of the fruit, have been focused. Tomato polyphenols including flavonoids and hydroxycinnamates indeed were identified at higher levels in the fruit peel than in the pericarp [10]. Because of the recognized human health benefits of these polyphenolic compounds, their accumulation patterns during different stages of ripening in economically-important crops would be important to both farmers and consumers. Comparison of representative polyphenolic compounds that accumulate in the peel and pericarp during three different stages of fruit ripening, particularly polyphenols that are specifically up-regulated only in jalapeño pepper. These polyphenolic compounds are also upregulated in the other sweet pepper cultivars, but at different stages of fruit ripening (Figure 4).

Comparison of individual polyphenolic compounds suggest possible metabolic markers that are specifically upregulated during fruit ripening. In the pungent pepper cultivar (jalapeño pepper), we observed an elevation of three coumaroyl-hexosides, three caffeoyl-hexosides, and two feruloyl-hexosides, as well as reduction of almost all flavonoids in the peel at the final mature stage (Figure 4B). However, these metabolic shifts were not observed in the other pepper cultivars except purple pepper, which happens to be non-pungent but is the only cultivar in our study that has purple pigmentation before reaching its final maturity stage. In the previous study of hot and semi-hot pepper cultivars (*C. annuum* cvs. Cyklon, Bronowicka Ostra, Tajfun, and Tornado), individual hydroxycinnamate content increased in the pericarp from green mature to red mature stage of development in all cultivars [74]. Additionally, in the analysis of a sweet bell pepper cultivar (*C. annuum* L. cv. Vergasa) exhibited a decrease in total hydroxycinnamate content in the pericarp from their immature green to green mature stage and then slightly increasing to immature red and red mature stages [75]. Taking into both results and our result in this study, results from the previous studies are consistent with our results such that hydroxycinnamate content increased in pungent cultivars from green to red stages of fruit ripening while hydroxycinnamate content decreased in sweet cultivars from immature to mature stages. Interestingly, purple pepper which is an anthocyanin-producing type of pepper cultivar showed similar metabolic changes with the pungent pepper cultivar, wherein the accumulation of hydroxycinnamates is inversely related with that of flavonoids. In both cultivars, hydroxycinnamates were generally upregulated while flavonoids were downregulated in the peel during their red mature stage. The pungent nature of jalapeño pepper due to its production of capsaicinoids could explain its downregulation of flavonoids during its red mature stage, given that both compounds groups share the same biosynthetic precursor.

Previously, in the metabolic analysis of fruit pericarp of the pungent hot pepper (*Capsicum annuum* "CM334") during six stages of fruit ripening, the flavonoid, Q3R, had high levels during earlier stages (16 and 25 DPA) and then gradually decreased until their last stage (48 DPA) [69]. In our study, Q3R in the pericarp of the hot cultivar (jalapeño pepper) was upregulated during its immature stage and then downregulated at its green and red mature stages, which are consistent with the decrease in Q3R observed from the previous study. Total flavonoid content, total *O*-glycosylflavone content and total *C*-glycosylflavone content in the pericarp of the sweet cultivar (*C. annuum* L. cv. Vergasa) decreased during four stages of ripening. Q3R decreased in the pericarp during the four stages of ripening [75], while in our study, Q3R was upregulated in the pericarp during immature stage and then downregulated in the middle stage (green or purple mature) and final mature (red or yellow mature) stages in the sweet cultivars (red paprika, yellow paprika), a non-pungent cultivar (purple pepper), and a pungent cultivar (jalapeño pepper). However, in another sweet cultivar (green paprika), Q3R was downregulated in immature and green mature stages and upregulated during red mature stage. In another non-pungent cultivar (green pepper), Q3R was upregulated during immature and red mature stages and downregulated in green mature stage of ripening. Variability in Q3R regulation among the six pepper cultivars is consistent with the subspecies or genotype-specific accumulation pattern of flavonoid glycosides among 32 pepper accessions from a previous study, wherein specific accessions contained higher levels of flavones, flavanones, and flavonol glycosides [25].

In our results, flavonol mono-glycosides (Q3G, Q3R, and K3R) were generally decreased during ripening in the peel for almost all pepper cultivars. This metabolic shift pattern was observed inversely in the level of naringenin chalcones which increased in the peels during fruit ripening. In spite of the metabolic changes of flavonol mono-glycosides in peel, these flavonols were increased in green paprika pericarp during fruit ripening which also the same for the naringenin chalcones, including Phi35diGlc. Rutin and flavonol-tri-glycosides showed a slight shift in decrease at late stages among all pepper cultivars for both peel and pericarp during ripening. In green pepper, flavonol di- and tri-glycosides decreased in the peel during ripening, were increased in the pericarp during fruit development. The flavonol-tetra-glycosides, Q3G2‴A6″R7G which was detected in tomato fruits, was not detected in all pepper species in both tissue types. In red paprika pepper, flavonol di- and tri-glycosides decreased in the peel during fruit ripening, while in the pericarp, flavonol di-glycosides also decreased. There were nine putative flavonoids identified during analysis with most of them decreasing in the peel and pericarp of three cultivars (red paprika, jalapeño, and purple peppers). For green pepper, most of the putative flavonoids decreased in the fruit peel during ripening but increased in the pericarp. No obvious pattern among the nine flavonoids were observed for green and yellow paprika peppers. Flavonol mono-glycosides (Q3G, Q3R, and K3R) decreased in the peel during fruit development for most pepper cultivars, except in green paprika pepper which exhibited highest accumulation during its green mature stage. Meanwhile, in pepper pericarp, flavonol mono-glycosides decreased during maturity only for both red paprika and jalapeño peppers, while no concrete pattern was present for the other pepper cultivars.

Anthocyanin derivatives (anthocyanin_I, delphinidin-3-*O*-(-feruloyl) rutinoside; anthocyanin_II, delphinidin-3-*O*-(-*p*-coumaroyl) rutinoside-5-*O*-glucoside) were upregulated in purple pepper during either the purple immature or purple mature stages in both fruit tissues (Figure 4). Anthocyanin II (delphinidin-3-*O*-(-*p*-coumaroyl)rutinoside-5-*O*-glucoside) was also previously detected in immature purple pepper whole fruit [76] and peel [77]. Anthocyanin I being one of the major anthocyanins identified from eggplant fruit peel [78]. Anthocyanins I, II, and III were also upregulated in eggplant peel in our study (Figure 2). Anthocyanins are involved in pigmentation, specifically purple to black pigmentation in pepper fruit tissues [72] and in the purple pigmentation in the peel of eggplant fruit [78]. Metabolic changes of these anthocyanins are slightly different in terms of ripening stage in peel and pericarp. Anthocyanins I and II were upregulated in the peel during immature stage in purple pepper while they were upregulated in the pericarp during purple mature stage. Purple pepper is the only cultivar that was unable to detect more than one of the putative flavonoid compounds and the only cultivar that detected anthocyanins, indicating a possible inverse relationship between the two compound groups.

Most naringenin chalcones and Phi35diGlc (phloretin-3′,5′-di-*C*-glucoside) were upregulated in the fruit peel in sweet and non-pungent cultivars during their final mature stage (red or yellow mature). Most naringenin chalcones and Phi35diGlc were upregulated in the peel of pungent jalapeño pepper during green mature stage. Due to the pungent nature of jalapeño pepper, differences in metabolic regulation with sweet cultivars could account for the down-regulation of most naringenin chalcones during red mature stage in jalapeño pepper. The pattern of accumulation is different in the fruit pericarp, with all three naringenin chalcones being upregulated at final maturity only for green pepper and green paprika peppers. Only naringenin-chalcone-hexose and phloretin dihexoside have been previously detected in non-pungent pepper (*C. annuum)* cultivars [79]. Accumulation of naringenin chalcones in the fruit peel is related with their functions of attracting seed dispersers, moderating damage against UV-light [55], and providing a structural role in the cuticle by controlling water movement [80]. Concentration of naringenin chalcone and its derivatives might have important roles after fruit maturation, particularly in the non-pungent and sweet cultivars where naringenin chalcones are upregulated in the peel during final mature stage of ripening.

Environmental stress, such as high temperature and UV-light, can increase reactive oxygen species (ROS) concentration resulting in fruit oxidative damage. Specialized metabolites can act as antioxidants protecting the fruit against photooxidative damage [81,82]. Polyphenols, such as flavonoids and hydroxycinnamates, act as antioxidants [40,41]. In solanaceous crop tomato (*S. lycopersicum*), total antioxidant activity in ripe fruit was in agreement with their total polyphenol content which corroborates that polyphenols can act as antioxidants [83]. In the perennial shrub *Ribes stenocarpsum* Maxim, polyphenol content was significantly higher in immature than in mature fruits, with antioxidant activities consequently being higher in immature fruits [84]. In Alphonso mango (*Mangifera indica*) fruit, a positive correlation between phenolic and antioxidant content in healthy tissues was present during the course of fruit ripening [85]. Ripening in sweet pepper (*C. annuum*) is associated with oxidative stress due to an increase in lipid peroxidation [86,87] and decrease in antioxidant enzymes during its red ripe stage [86]. In our study, polyphenols were generally upregulated during red mature stage in the pericarp of sweet green paprika (*C. annuum*) and non-pungent green pepper *(C. annuum)*, which is in agreement that fruit ripening in sweet pepper is related with oxidative stress since polyphenols can act as antioxidants. In tomato, total phenolics and free radical scavenging activity increased during ripening for all cultivars under study [88]. Decrease in antioxidant enzyme activity during fruit ripening involved the enzymes catalase [89] and ascorbate peroxidase [86] which are not relevant with polyphenol biosynthesis.

Comparing the number of polyphenolic compounds that accumulated more in the peel against the pericarp showed difference between fruit tissue type for the six pepper cultivars. However, not in all cases were the polyphenolic compounds more abundant during the same stage of development between tissue types. In peel, the compounds upregulated were 5CGA and fer_hex_II. 5CGA only increased during fruit ripening for the pungent cultivar and decreased for the sweet cultivars. Fer_hex_I and fer_hex_II increased remarkably in the peel of the pungent cultivar (jalapeño pepper) although some increase was observed in purple pepper as well. In the fruit pericarp, fer_hex_I increased noticeably in the pungent jalapeño pepper; however, there was also some increase in yellow and green paprika peppers. To determine whether such accumulation patterns are specific to pungent cultivars or whether this accumulation is only cultivar-specific, accumulation patterns for these polyphenolic compounds need to be conducted among pungent pepper cultivars during fruit ripening. We also observed that similar metabolic shift of pepper specific flavonoid derivatives between green pepper and green paprika, but these metabolic changes were negatively correlated to the metabolic shift in the jalapeño pepper which is the capsaicinoids producing-type of cultivar. This result provides a metabolic trade-off of fruit polyphenolics metabolism in capsaicinoids-producing cultivars, since polyphenolics and capsaicinoids share the biosynthetic precursors. In contrast to this point, jalapeño pepper showed clear elevation of hydroxycinnamates, such as caffeoyl-hexosides, in both peel and pericarp. However, these metabolic changes were also observed in the peel of green pepper and green paprika, but not in the other red, yellow, and purple peppers. Taking into account both metabolic changes of polyphenolic subgroups, changes of metabolic flux in polyphenolics are specific metabolic regulation in each types of peppers with different tissue specific manners. Finally, our results may have been convoluted by the absence of any pattern among polyphenols in terms of maturity stage and cultivar type, however integration of metabolomics data with previous studies [24] will provide other novel insights to understand this convoluted metabolic regulations. Importantly, naringenin chalcone which is one of the major ripening marker metabolites in tomato fruits, was conserved among almost all pepper species. Metabolomics analysis presented here suggested metabolic shift including convoluted metabolic trade-off in the solanaceous crops and provided hints for metabolomics-assisted crop improvement for the polyphenolic metabolism in the solanaceous fruits for the improvement of nutritive properties and enhanced stress tolerance.

4. Materials and Methods

4.1. Plant Material and Sampling

Fruits from tomato (*S. lycopersicum* cv. "M82") (TGRC, Tomato Genetics Resource Center, Davis, CA, USA), eggplant cv. "Ryoma" (*S. melongena*) (Takii, Japan), and different pepper (*C. annuum*) cultivars (green paprika cv. "New Ace" (Takii, Japan), red paprika cv. "Flupy Red EX" (e-taneya, Japan), purple pepper cv. "Nara Murasaki" (Tsurushin, Japan), green pepper cv. "Miogi" (e-taneya, Japan), yellow paprika cv. "Sonia Gold" (Sakata-no-Tane, Japan), and jalapeño pepper cv. "Jalapeño" (Marche, Japan)) were grown from May–December 2019 in standard soil under open field (longitude, 34.734433; latitude 135.736754) conditions. Three developmental stages were defined for analysis from fruit peel and flesh. Immature fruit, mid-stage, and final mature stages were selected and defined based on the height of each fruit and the color of peel and seeds. Immature fruit are half the height of mature fruit and green in color. In the case of purple pepper and eggplant, the immature fruit color was purple. Green mature fruit had approximately the same height as the final mature stage but still green in color. Purple pepper was also purple during its green mature stage. The final mature stage is when peel color has completely changed from green to its ripe fruit color (red, yellow). Mature eggplant fruit were distinguished through their dark brown seed color. Three independent biological replicates per plant tissue during each ripening stage were harvested and used for metabolomic analysis. After tissue separation, fruit peel and pericarp were grounded into powder using liquid nitrogen and stored at −80 °C for further analysis.

4.2. Sample Extraction and LC-MS Analysis

Metabolite extraction was performed as described previously [90,91]. Fifty milligrams of powdered frozen tissue sample were aliquoted and weighed in a 2.0 mL centrifuge tube. Two hundred fifty microliters of extraction solvent (80% methanol in LC-MS grade water with 5 µg/mL isovitexin standard) were added per tissue sample in liquid nitrogen and a Zirconia bead. All frozen tissue samples were ground into powder using Mixer Mill TissueLyser II (Qiagen, Hilden, Germany) for 3 min at 25l/s at room temperature, and centrifuged at 15,366× *g* at 4 °C for 10 min. After centrifugation, to take an additional cleaning step of extracts to exclude plant tissues, two hundred microliters of supernatant per sample were transferred into a 1.5 mL centrifuge tube then all 1.5 mL tubes were centrifuged at 15,366× *g* at 4 °C for 10 min. One hundred microliters supernatant per sample were transferred into LC-MS vials and stored at 4 °C in the dark until analysis. For detection of polyphenolic metabolites, LC-electrospray ionization (ESI)-MS was used. Nanoflow-HPLC "Paradigm MS4 system" (Michrom BioResources, Inc., Auburn, CA, USA), equipped with a Luna C18 column (150 by 2.00 mm i.d. 3 micron particle size, Phenomenex, Torrance, CA, USA) operated at a temperature of 25 °C was used. The mobile phases consisted of 0.1% formic acid in water (Solvent A) and 0.1% formic acid in acetonitrile (Solvent B). The flow rate of the mobile phase was 200 µL/min, and 10 µL sample were loaded per injection. The following gradient profile was applied: The concentration of mobile phase A was 100% at 0 min, 93% at 1 min, decreased to 80% at 8 min, 60% at 17 min, 15% at 21 min, and 0% at 25 min and 28 min for column wash, then increased to 100% at 28.01 min and 31 min for the equilibration of the column in the gradient description. The LC was connected to an MS TSQ Vantage (Thermo Fisher Scientific, San Jose, CA, USA). The spectra were recorded using full scan mode, covering a mass range from *m/z* 200–1500 by both positive and negative ion detection. The transfer capillary temperature was set to 350 °C. The spray voltage was fixed at 3.00 kV. Peak identification of major polyphenolic compounds (rutin, quercetin-3-*O*-rutinoside; Q3G, quercetin-3-*O*-glucoside; and 3CGA, chlorogenate) was performed using standard compounds. Peak annotation of major polyphenolic compounds in Solanaceae species was performed via combination approach of co-elution profile of common tomato fruit extracts and phytochemical database (KomicMarket [14] and MotoDB [13,92]) as well as metabolites table in the literatures [5,15].

4.3. Data Analysis

Molecular masses, retention time, and associated peak intensities were extracted from the raw files using the Xcalibur software version 4.1.31.9 (Thermo Fisher Scientific, San Jose, CA, USA). Compounds were identified and putatively identified by comparing with corresponding retention time (minutes) and molecular weight with those provided by tomato cv. M82 and eggplant. Previous information on polyphenolic compounds identified from different pepper species and varieties from published journal articles were also used for cross-referencing. Peak picking in the Xcalibur software was performed with the parameter of RT tolerance window (20 s), base window 80, area noise factor 5.0, peak noise factor 10, and "nearest RT". MeV software version 4.9.0 (http://www.mev.tm4.org/, Dana Farber Cancer Institute, Boston, MA, USA) was used for data visualization and PCA analysis. The plots were applied for the 39 metabolites with the average values from 3 biological replications. Heatmap visualization of metabolite data is normalized and scaled by $\log_2$FC (mean/average_mean) for each metabolite. Coefficient correlation was estimated by person correlation method using MeV software.

5. Conclusions

With several fruit-metabolomics studies on the specialized metabolism of tomato recently available, such an approach could, therefore, be extended to generate information on other members from Solanaceae. Comparing the number of polyphenolic compounds that accumulated more in the peel against the pericarp showed differences between fruit tissue type in the six pepper cultivars. However, not in all cases were the polyphenolic compounds more abundant during the same stage of development between tissue types. We also observed that a similar metabolic shift of pepper-specific flavonoid derivatives between green pepper and green paprika cultivars, but these metabolic changes were negatively correlated to the metabolic shift in jalapeño pepper, which biosynthesizes capsaicinoids. This result exhibits a metabolic trade-off in fruit polyphenolics metabolism in the capsaicinoid-producing cultivar, since polyphenolics and capsaicinoids share the biosynthetic precursors. In support of this point, hydroxycinnamates in the pungent jalapeño and anthocyanin-producing purple pepper cultivars were clearly elevated in both peel and pericarp during red mature stage. However, flavonoids from both cultivars were downregulated during red mature stage suggesting a metabolic trade-off between both compound groups during fruit development. Taking into account both metabolic changes of polyphenolic subgroups, changes of metabolic flux in polyphenolics are specifically regulated in each pepper cultivar with different tissue specific manners. Finally, our results may have been convoluted by the absence of any accumulation patterns of polyphenol in terms of ripening stage and cultivar type, however integration of metabolomics data with previous studies [24] will provide other novel insights to understand these convoluted metabolic regulations. Importantly, naringenin chalcone which is one of the major ripening marker metabolites in tomato fruits, was conserved among almost all pepper cultivars. Metabolomic analysis presented here suggested metabolic shift including convoluted metabolic trade-off in three solanaceous crops and provided hints for metabolomics-assisted crop improvement of polyphenolic metabolism in three solanaceous crops towards their improved nutritive properties and enhanced stress tolerance.

Supplementary Materials: The following are available online at http://www.mdpi.com/2218-1989/10/5/209/s1, Figure S1: LC-MS TIC (total ion chromatograms) of fruit extracts of tomato, eggplant and pepper; Supplemental Table S1: Metabolite table of peaks detected in this study using LC-MS.

Author Contributions: Conceptualization, M.W. and T.T.; software, C.L.F.C. and T.T; validation, C.L.F.C., T.S. and T.T.; formal analysis, C.L.F.C.; project administration, M.W. and T.T.; supervision, M.W. and T.T.; investigation, C.L.F.C.; writing—original draft preparation, C.L.F.C. and T.T. All authors have read and agreed to the published version of the manuscript.

Funding: This research was funded by Japan Society for the Promotion of Science (JSPS) Scientific Research B (19H03249) and C (19K06723).

Acknowledgments: C.L.F.C.: T.S., M.W. and T.T gratefully acknowledge partial support by NAIST. We are grateful Jay Camisora Delfin who helped for growing tomato plants. Research activity of M.W and T.T. were additionally supported by JSPS KAKENHI Grant-in-Aid for Scientific Research B (19H03249) and C (19K06723).

Conflicts of Interest: The authors declare no conflict of interest.

References

1. Bolger, M.; Scossa, F.; Bolger, M.; Lanz, C.; Maumus, F.; Tohge, T.; Quesneville, H.; Alseekh, S.; Sørensen, I.; Lichtenstein, G.; et al. The genome of the stress-tolerant wild tomato species Solanum pennellii. *Nat. Genet.* **2014**, *46*, 1034–1038. [CrossRef]
2. The Tomato Genome Consortium. The tomato genome sequence provides insights into fleshy fruit evolution. *Nature* **2012**, *485*, 635–641. [CrossRef] [PubMed]
3. Klee, H.J.; Giovannoni, J.J. Genetics and Control of Tomato Fruit Ripening and Quality Attributes. *Annu. Rev. Genet.* **2011**, *45*, 41–59. [CrossRef] [PubMed]
4. Alseekh, S.; Tohge, T.; Wendenberg, R.; Scossa, F.; Omranian, N.; Li, J.; Kleessen, S.; Giavalisco, P.; Pleban, T.; Mueller-Roeber, B.; et al. Identification and Mode of Inheritance of Quantitative Trait Loci for Secondary Metabolite Abundance in Tomato[OPEN]. *Plant Cell* **2015**, *27*, 485–512. [CrossRef] [PubMed]
5. Tohge, T.; Fernie, A.R. Metabolomics-Inspired Insight into Developmental, Environmental and Genetic Aspects of Tomato Fruit Chemical Composition and Quality: Fig. 1. *Plant Cell Physiol.* **2015**, *56*, 1681–1696. [CrossRef]
6. Alseekh, S.; Tong, H.; Scossa, F.; Brotman, Y.; Vigroux, F.; Tohge, T.; Ofner, I.; Zamir, D.; Nikoloski, Z.; Fernie, A.R. Canalization of Tomato Fruit Metabolism. *Plant Cell* **2017**, *29*, 2753–2765. [CrossRef] [PubMed]
7. Zhu, G.; Wang, S.; Huang, Z.; Zhang, S.; Liao, Q.; Zhang, C.-Z.; Lin, T.; Qin, M.; Peng, M.; Yang, C.; et al. Rewiring of the Fruit Metabolome in Tomato Breeding. *Cell* **2018**, *172*, 249–261. [CrossRef]
8. Scossa, F.; Roda, F.; Tohge, T.; Georgiev, M.I.; Fernie, A.R. The Hot and the Colorful: Understanding the Metabolism, Genetics and Evolution of Consumer Preferred Metabolic Traits in Pepper and Related Species. *Crit. Rev. Plant Sci.* **2019**, *38*, 339–381. [CrossRef]
9. Fernie, A.R.; Aharoni, A. Pan-Genomic Illumination of Tomato Identifies Novel Gene-Trait Interactions. *Trends Plant Sci.* **2019**, *24*, 882–884. [CrossRef]
10. Mintz-Oron, S.; Mandel, T.; Rogachev, I.; Feldberg, L.; Lotan, O.; Yativ, M.; Wang, Z.; Jetter, R.; Venger, I.; Adato, A.; et al. Gene Expression and Metabolism in Tomato Fruit Surface Tissues1[C][W]. *Plant Physiol.* **2008**, *147*, 823–851. [CrossRef]
11. Rohrmann, J.; McQuinn, R.P.; Giovannoni, J.J.; Fernie, A.R.; Tohge, T. Tissue specificity and differential expression of transcription factors in tomato provide hints of unique regulatory networks during fruit ripening. *Plant Signal. Behav.* **2012**, *7*, 1639–1647. [CrossRef] [PubMed]
12. Antonio, A.S.; Wiedemann, L.S.M.; Veiga-Junior, V.F. The genus *Capsicum*: A phytochemical review of bioactive secondary metabolites. *RSC Adv.* **2018**, *8*, 25767–25784. [CrossRef]
13. Moco, S.; Bino, R.J.; Vorst, O.; Verhoeven, H.A.; De Groot, J.; Van Beek, T.A.; Vervoort, J.; De Vos, C.R. A Liquid Chromatography-Mass Spectrometry-Based Metabolome Database for Tomato1. *Plant Physiol.* **2006**, *141*, 1205–1218. [CrossRef] [PubMed]
14. Iijima, Y.; Nakamura, Y.; Ogata, Y.; Tanaka, K.; Sakurai, N.; Suda, K.; Suzuki, T.; Suzuki, H.; Okazaki, K.; Kitayama, M.; et al. Metabolite annotations based on the integration of mass spectral information. *Plant J.* **2008**, *54*, 949–962. [CrossRef] [PubMed]
15. Tohge, T.; Alseekh, S.; Fernie, A.R. On the regulation and function of secondary metabolism during fruit development and ripening. *J. Exp. Bot.* **2013**, *65*, 4599–4611. [CrossRef]
16. Ruprecht, C.; Tohge, T.; Fernie, A.; Mortimer, C.; Kozlo, A.; Fraser, P.D.; Funke, N.; Cesarino, I.; Vanholme, R.; A Boerjan, W.; et al. Transcript and metabolite profiling for the evaluation of tobacco tree and poplar as feedstock for the bio-based industry. *J. Vis. Exp.* **2014**, *87*, e51393. [CrossRef]
17. Ruprecht, C.; Mendrinna, A.; Tohge, T.; Sampathkumar, A.; Klie, S.; Fernie, A.R.; Nikoloski, Z.; Persson, S.; Mutwil, M. FamNet: A Framework to Identify Multiplied Modules Driving Pathway Expansion in Plants1. *Plant Physiol.* **2016**, *170*, 1878–1894. [CrossRef]

18. Price, E.J.; Drapal, M.; Perez-Fons, L.; Amah, D.; Bhattacharjee, R.; Heider, B.; Rouard, M.; Swennen, R.; Lopez-Lavalle, L.A.B.; Fraser, P.D. Metabolite database for root, tuber, and banana crops to facilitate modern breeding in understudied crops. *Plant J.* **2020**, *101*, 1258–1268. [CrossRef]
19. Iwaki, T.; Guo, L.; Ryals, J.A.; Yasuda, S.; Shimazaki, T.; Kikuchi, A.; Watanabe, K.N.; Kasuga, M.; Yamaguchi-Shinozaki, K.; Ogawa, T.; et al. Metabolic Profiling of Transgenic Potato Tubers Expressing Arabidopsis Dehydration Response Element-Binding Protein 1A (DREB1A). *J. Agric. Food Chem.* **2013**, *61*, 893–900. [CrossRef]
20. Oertel, A.; Matros, A.; Hartmann, A.; Arapitsas, P.; Dehmer, K.J.; Martens, S.; Mock, H. Metabolite profiling of red and blue potatoes revealed cultivar and tissue specific patterns for anthocyanins and other polyphenols. *Planta* **2017**, *246*, 281–297. [CrossRef]
21. Sasse, J.; Schlegel, M.; Borghi, L.; Ullrich, F.; Lee, M.; Liu, G.; Giner, J.; Kayser, O.; Bigler, L.; Martinoia, E.; et al. Petunia hybridaPDR2 is involved in herbivore defense by controlling steroidal contents in trichomes. *Plant Cell Environ.* **2016**, *39*, 2725–2739. [CrossRef] [PubMed]
22. Qiu, F.; Zeng, J.; Wang, J.; Huang, J.; Zhou, W.; Yang, C.; Lan, X.; Chen, M.; Huang, S.-X.; Kai, G.; et al. Functional genomics analysis reveals two novel genes required for littorine biosynthesis. *New Phytol.* **2019**, *225*, 1906–1914. [CrossRef] [PubMed]
23. Berry, H.M.; Rickett, D.V.; Baxter, C.J.; Enfissi, E.M.A.; Fraser, P.D. Carotenoid biosynthesis and sequestration in red chilli pepper fruit and its impact on colour intensity traits. *J. Exp. Bot.* **2019**, *70*, 2637–2650. [CrossRef] [PubMed]
24. Wahyuni, Y.; Ballester, A.R.; Sudarmonowati, E.; Bino, R.J.; Bovy, A.G. Metabolite biodiversity in pepper (*Capsicum*) fruits of thirty-two diverse accessions: Variation in health-related compounds and implications for breeding. *Phytochemistry* **2011**, *72*, 1358–1370. [CrossRef]
25. Wahyuni, Y.; Ballester, A.R.; Tikunov, Y.; De Vos, R.C.H.; Pelgrom, K.T.B.; Maharijaya, A.; Sudarmonowati, E.; Bino, R.J.; Bovy, A.G. Metabolomics and molecular marker analysis to explore pepper (*Capsicum* sp.) biodiversity. *Metabolomics* **2012**, *9*, 130–144. [CrossRef]
26. Food and Agriculture Organization of the United Nations. Available online: www.fao.org/faostat/en/#data/QC (accessed on 18 March 2020).
27. Arimboor, R.; Natarajan, R.B.; Menon, K.R.; Chandrasekhar, L.P.; Moorkoth, V. Red pepper (*Capsicum annuum*) carotenoids as a source of natural food colors: Analysis and stability—A review. *J. Food Sci. Technol.* **2014**, *52*, 1258–1271. [CrossRef]
28. Leung, F.W. Capsaicin as an Anti-Obesity Drug. In *Capsaicin as a Therapeutic Molecule*; Springer Science and Business Media LLC: Berlin, Germany, 2014; Volume 68, pp. 171–179.
29. Rains, C.; Bryson, H.M. Topical Capsaicin. *Drugs Aging* **1995**, *7*, 317–328. [CrossRef]
30. Luo, X.-J.; Peng, J.; Li, Y.-J. Recent advances in the study on capsaicinoids and capsinoids. *Eur. J. Pharmacol.* **2011**, *650*, 1–7. [CrossRef]
31. Smith, J.; Greaves, I. The use of chemical incapacitant sprays: A review. *J. Trauma Inj. Infect. Crit. Care* **2002**, *52*, 595–600. [CrossRef]
32. Haar, R.J.; Iacopino, V.; Ranadive, N.; Weiser, S.D.; Dandu, M. Health impacts of chemical irritants used for crowd control: A systematic review of the injuries and deaths caused by tear gas and pepper spray. *BMC Public Health* **2017**, *17*, 831. [CrossRef]
33. Rohrmann, J.; Tohge, T.; Alba, R.; Osorio, S.; Caldana, C.; McQuinn, R.P.; Arvidsson, S.; Van Der Merwe, M.J.; Riaño-Pachón, D.M.; Mueller-Roeber, B.; et al. Combined transcription factor profiling, microarray analysis and metabolite profiling reveals the transcriptional control of metabolic shifts occurring during tomato fruit development. *Plant J.* **2011**, *68*, 999–1013. [CrossRef] [PubMed]
34. Matas, A.J.; Yeats, T.H.; Buda, G.J.; Zheng, Y.; Chatterjee, S.; Tohge, T.; Ponnala, L.; Adato, A.; Aharoni, A.; Stark, R.; et al. Tissue- and Cell-Type Specific Transcriptome Profiling of Expanding Tomato Fruit Provides Insights into Metabolic and Regulatory Specialization and Cuticle Formation[W][OA]. *Plant Cell* **2011**, *23*, 3893–3910. [CrossRef]
35. Tohge, T.; Scossa, F.; Wendenburg, R.; Frasse, P.; Balbo, I.; Watanabe, M.; Alseekh, S.; Jadhav, S.S.; Delfin, J.C.; Lohse, M.; et al. Exploiting the natural variation in tomato to define pathway structure and metabolic regulation of fruit polyphenolics in the lycopersicum complex. *Mol. Plant* **2020**. [CrossRef] [PubMed]

36. Bae, H.; Jayaprakasha, G.K.; Crosby, K.; Yoo, K.S.; Leskovar, D.I.; Jifon, J.L.; Patil, B.S. Ascorbic acid, capsaicinoid, and flavonoid aglycone concentrations as a function of fruit maturity stage in greenhouse-grown peppers. *J. Food Compos. Anal.* **2014**, *33*, 195–202. [CrossRef]
37. Wu, S.-B.; Meyer, R.S.; Whitaker, B.D.; Litt, A.; Kennelly, E.J. A new liquid chromatography–mass spectrometry-based strategy to integrate chemistry, morphology, and evolution of eggplant (*Solanum*) species. *J. Chromatogr. A* **2013**, *1314*, 154–172. [CrossRef] [PubMed]
38. Hanifah, A.; Maharijaya, A.; Putri, S.P.; Laviña, W.A.; Ridwani, S. Untargeted Metabolomics Analysis of Eggplant (*Solanum melongena* L.) Fruit and Its Correlation to Fruit Morphologies. *Metabolites* **2018**, *8*, 49. [CrossRef] [PubMed]
39. Toppino, L.; Barchi, L.; Scalzo, R.L.; Palazzolo, E.; Francese, G.; Fibiani, M.; D'Alessandro, A.; Papa, V.; Laudicina, V.A.; Sabatino, L.; et al. Mapping Quantitative Trait Loci Affecting Biochemical and Morphological Fruit Properties in Eggplant (*Solanum melongena* L.). *Front. Plant Sci.* **2016**, *7*, 3017. [CrossRef]
40. Tohge, T.; Watanabe, M.; Hoefgen, R.; Fernie, A.R. Shikimate and Phenylalanine Biosynthesis in the Green Lineage. *Front. Plant Sci.* **2013**, *4*, 62. [CrossRef]
41. Tohge, T.; Watanabe, M.; Hoefgen, R.; Fernie, A.R. The evolution of phenylpropanoid metabolism in the green lineage. *Crit. Rev. Biochem. Mol. Boil.* **2013**, *48*, 123–152. [CrossRef]
42. Treutter, D. Significance of Flavonoids in Plant Resistance and Enhancement of Their Biosynthesis. *Plant Boil.* **2005**, *7*, 581–591. [CrossRef]
43. Landi, M.; Tattini, M.; Gould, K.S. Multiple functional roles of anthocyanins in plant-environment interactions. *Environ. Exp. Bot.* **2015**, *119*, 4–17. [CrossRef]
44. Tohge, T.; Wendenburg, R.; Ishihara, H.; Nakabayashi, R.; Watanabe, M.; Suplice, R.; Hoefgen, R.; Takayama, H.; Saito, K.; Stitt, M.; et al. Characterization of a recently evolved flavonol-phenylacyltransferase gene provides signatures of natural light selection in *Brassicaceae*. *Nat. Commun.* **2016**, *7*, 12399. [CrossRef] [PubMed]
45. Fernie, A.R.; Tohge, T. The Genetics of Plant Metabolism. *Annu. Rev. Genet.* **2017**, *51*, 287–310. [CrossRef] [PubMed]
46. Fritz, C.; Feil, R.; Stitt, M.; Palacios-Rojas, N. Regulation of secondary metabolism by the carbon-nitrogen status in tobacco: Nitrate inhibits large sectors of phenylpropanoid metabolism. *Plant J.* **2006**, *46*, 533–548. [CrossRef]
47. Schulz, E.; Tohge, T.; Zuther, E.; Fernie, A.R.; Hincha, D.K. Flavonoids are determinants of freezing tolerance and cold acclimation in *Arabidopsis thaliana*. *Sci. Rep.* **2016**, *6*, 34027. [CrossRef]
48. Pott, D.M.; Osorio, S.; Vallarino, J. From Central to Specialized Metabolism: An Overview of Some Secondary Compounds Derived From the Primary Metabolism for Their Role in Conferring Nutritional and Organoleptic Characteristics to Fruit. *Front. Plant Sci.* **2019**, *10*, 835. [CrossRef]
49. Di Sotto, A.; Vecchiato, M.; Abete, L.; Toniolo, C.; Giusti, A.M.; Mannina, L.; Locatelli, M.; Nicoletti, M.; Di Giacomo, S. *Capsicum annuum* L. var. Cornetto di Pontecorvo PDO: Polyphenolic profile and in vitro biological activities. *J. Funct. Foods* **2018**, *40*, 679–691. [CrossRef]
50. Sun, T.; Xu, Z.; Wu, C.-T.; Janes, M.; Prinyawiwatkul, W.; No, H.K. Antioxidant Activities of Different Colored Sweet Bell Pepper (*Capsicum annum* L.). *J. Food Sci.* **2007**, *72*, S98–S102. [CrossRef]
51. Jeong, W.Y.; Jin, J.S.; Cho, Y.A.; Lee, J.H.; Park, S.; Jeong, S.W.; Kim, Y.-H.; Lim, C.-S.; El-Aty, A.M.A.; Kim, G.-S.; et al. Determination of polyphenols in three *Capsicum annuum* L. (bell pepper) varieties using high-performance liquid chromatography-tandem mass spectrometry: Their contribution to overall antioxidant and anticancer activity. *J. Sep. Sci.* **2011**, *34*, 2967–2974. [CrossRef]
52. Farah, A.; Donangelo, C.M. Phenolic compounds in coffee. *Braz. J. Plant Physiol.* **2006**, *18*, 23–36. [CrossRef]
53. Stewart, A.J.; Bozonnet, S.; Mullen, W.; Jenkins, G.I.; Lean, M.E.J.; Crozier, A. Occurrence of flavonols in tomatoes and tomato-based products. *J. Agric. Food Chem.* **2000**, *48*, 2663–2669. [CrossRef] [PubMed]
54. Muir, S.R.; Collins, G.J.; Robinson, S.; Hughes, S.; Bovy, A.; De Vos, C.R.; Van Tunen, A.J.; Verhoeyen, M.E. Overexpression of petunia chalcone isomerase in tomato results in fruit containing increased levels of flavonols. *Nat. Biotechnol.* **2001**, *19*, 470–474. [CrossRef] [PubMed]
55. Bovy, A.; De Vos, R.; Kemper, M.; Schijlen, E.; Pertejo, M.A.; Muir, S.; Collins, G.; Robinson, S.; Verhoeyen, M.; Hughes, S.; et al. High-Flavonol Tomatoes Resulting from the Heterologous Expression of the Maize Transcription Factor Genes LC and C1. *Plant Cell* **2002**, *14*, 2509–2526. [CrossRef] [PubMed]

56. Schijlen, E.; De Vos, C.R.; Martens, S.; Jonker, H.H.; Rosin, F.M.; Molthoff, J.W.; Tikunov, Y.M.; Angenent, G.C.; Van Tunen, A.J.; Bovy, A.G. RNA Interference Silencing of Chalcone Synthase, the First Step in the Flavonoid Biosynthesis Pathway, Leads to Parthenocarpic Tomato Fruits[C]. *Plant Physiol.* **2007**, *144*, 1520–1530. [CrossRef]
57. Slimestad, R.; Verheul, M. Review of flavonoids and other phenolics from fruits of different tomato (*Lycopersicon esculentum* Mill.) cultivars. *J. Sci. Food Agric.* **2009**, *89*, 1255–1270. [CrossRef]
58. Niño-Medina, G.; Urías-Orona, V.; Muy-Rangel, M.D.; Heredia, J. Structure and content of phenolics in eggplant (*Solanum melongena*)—A review. *S. Afr. J. Bot.* **2017**, *111*, 161–169. [CrossRef]
59. Lemos, C.; Reimer, J.J.; Wormit, A. Color for Life: Biosynthesis and Distribution of Phenolic Compounds in Pepper (*Capsicum annuum*). *Agriculture* **2019**, *9*, 81. [CrossRef]
60. Wu, X.; Beecher, G.R.; Holden, J.M.; Haytowitz, D.B.; Gebhardt, S.E.; Prior, R.L. Concentrations of Anthocyanins in Common Foods in the United States and Estimation of Normal Consumption. *J. Agric. Food Chem.* **2006**, *54*, 4069–4075. [CrossRef]
61. Whitaker, B.D.; Stommel, J.R. Distribution of Hydroxycinnamic Acid Conjugates in Fruit of Commercial Eggplant (*Solanum melongena* L.) Cultivars. *J. Agric. Food Chem.* **2003**, *51*, 3448–3454. [CrossRef]
62. Singh, A.P.; Luthria, D.L.; Wilson, T.; Vorsa, N.; Singh, V.; Bañuelos, G.S.; Pasakdee, S. Polyphenols content and antioxidant capacity of eggplant pulp. *Food Chem.* **2009**, *114*, 955–961. [CrossRef]
63. Carrizo García, C.; Sterpetti, M.; Volpi, P.; Ummarino, M.; Saccardo, F. Wild capsicums: Identification and in Situ Analysis of Brazilian Species. In *Breakthroughs in the Genetics and Breeding of Capsicum and Eggplant*; Lanteri, S., Rotino, G.L., Eds.; Eucarpia: Turin, Italy, 2013; pp. 205–213.
64. Eshbaugh, W. History and Exploitation of a Serendipitous New Crop Discovery. In *New Crops*; Janick, J., Simon, J., Eds.; Wiley: New York, NY, USA, 1993; pp. 132–139.
65. Carrari, F.; Baxter, C.; Usadel, B.; Urbanczyk-Wochniak, E.; Zanor, M.-I.; Nunes-Nesi, A.; Nikiforova, V.; Centeno, D.C.; Ratzka, A.; Pauly, M.; et al. Integrated Analysis of Metabolite and Transcript Levels Reveals the Metabolic Shifts That Underlie Tomato Fruit Development and Highlight Regulatory Aspects of Metabolic Network Behavior1[W]. *Plant Physiol.* **2006**, *142*, 1380–1396. [CrossRef] [PubMed]
66. Osorio, S.; Alba, R.; Nikoloski, Z.; Kochevenko, A.; Fernie, A.R.; Giovannoni, J.J. Integrative comparative analyses of transcript and metabolite profiles from pepper and tomato ripening and development stages uncovers species-specific patterns of network regulatory behavior. *Plant Physiol.* **2012**, *159*, 1713–1729. [CrossRef] [PubMed]
67. Cin, V.D.; Tieman, D.M.; Tohge, T.; McQuinn, R.P.; De Vos, R.C.H.; Osorio, S.; Schmelz, E.A.; Taylor, M.G.; Smits-Kroon, M.T.; Schuurink, R.C.; et al. Identification of Genes in the Phenylalanine Metabolic Pathway by Ectopic Expression of a MYB Transcription Factor in Tomato Fruit[W]. *Plant Cell* **2011**, *23*, 2738–2753. [CrossRef]
68. Klie, S.; Osorio, S.; Tohge, T.; Drincovich, M.F.; Fait, A.; Giovannoni, J.J.; Fernie, A.R.; Nikoloski, Z. Conserved Changes in the Dynamics of Metabolic Processes during Fruit Development and Ripening across Species1[C][W]. *Plant Physiol.* **2013**, *164*, 55–68. [CrossRef]
69. Jang, Y.K.; Jung, E.S.; Lee, H.-A.; Choi, D.; Lee, C.H. Metabolomic Characterization of Hot Pepper (*Capsicum annuum* "CM334") during Fruit Development. *J. Agric. Food Chem.* **2015**, *63*, 9452–9460. [CrossRef]
70. Alvarez-Parrilla, E.; De La Rosa, L.A.; Amarowicz, R.; Shahidi, F. Antioxidant Activity of Fresh and Processed Jalapeño and Serrano Peppers. *J. Agric. Food Chem.* **2011**, *59*, 163–173. [CrossRef]
71. Kirii, E.; Goto, T.; Yoshida, Y.; Yasuba, K.-I.; Tanaka, Y. Non-pungency in a Japanese Chili Pepper Landrace (*Capsicum annuum*) is Caused by a Novel Loss-of-function Pun1 Allele. *Hortic. J.* **2017**, *86*, 61–69. [CrossRef]
72. Peterson, P.A. Linkage of Fruit Shape and Color Genes in *Capsicum*. *Genetics* **1959**, *44*, 407–419.
73. Saeed, A.; Sharov, V.; White, J.; Li, J.; Liang, W.; Bhagabati, N.; Braisted, J.; Klapa, M.; Currier, T.; Thiagarajan, M.; et al. TM4: A Free, Open-Source System for Microarray Data Management and Analysis. *Biotechniques* **2003**, *34*, 374–378. [CrossRef]
74. Materska, M.; Perucka, I. Antioxidant Activity of the Main Phenolic Compounds Isolated from Hot Pepper Fruit (*Capsicum annuum* L.). *J. Agric. Food Chem.* **2005**, *53*, 1750–1756. [CrossRef]
75. Marín, A.; Ferreres, F.; Tomás-Barberán, F.A.; Gil, M.I. Characterization and Quantitation of Antioxidant Constituents of Sweet Pepper (*Capsicum annuum* L.). *J. Agric. Food Chem.* **2004**, *52*, 3861–3869. [CrossRef] [PubMed]

76. Lightbourn, G.J.; Griesbach, R.J.; Novotny, J.A.; A Clevidence, B.; Rao, D.D.; Stommel, J.R. Effects of Anthocyanin and Carotenoid Combinations on Foliage and Immature Fruit Color of *Capsicum annuum* L. *J. Hered.* **2008**, *99*, 105–111. [CrossRef] [PubMed]
77. Sadilova, E.; Stintzing, F.C.; Carle, R. Anthocyanins, colour and antioxidant properties of eggplant (*Solanum melongena* L.) and violet pepper (*Capsicum annuum* L.) peel extracts. *Z. Naturforsch. C* **2006**, *61*, 527–535. [CrossRef] [PubMed]
78. Azuma, K.; Ohyama, A.; Ippoushi, K.; Ichiyanagi, T.; Takeuchi, A.; Saito, T.; Fukuoka, H. Structures and Antioxidant Activity of Anthocyanins in Many Accessions of Eggplant and Its Related Species. *J. Agric. Food Chem.* **2008**, *56*, 10154–10159. [CrossRef]
79. Morales-Soto, A.; Maqueda, M.; García-Salas, P.; Segura-Carretero, A.; Gutierrez, A.F. High-performance liquid chromatography coupled to diode array and electrospray time-of-flight mass spectrometry detectors for a comprehensive characterization of phenolic and other polar compounds in three pepper (*Capsicum annuum* L.) samples. *Food Res. Int.* **2013**, *51*, 977–984. [CrossRef]
80. Luque, P.; Bruque, S.; Heredia, A. Water Permeability of Isolated Cuticular Membranes: A Structural Analysis. *Arch. Biochem. Biophys.* **1995**, *317*, 417–422. [CrossRef]
81. Torres, C.; Andrews, P. Developmental changes in antioxidant metabolites, enzymes, and pigments in fruit exocarp of four tomato (*Lycopersicon esculentum* Mill.) genotypes: β-carotene, high pigment-1, ripening inhibitor, and 'Rutgers'. *Plant Physiol. Biochem.* **2006**, *44*, 806–818. [CrossRef]
82. Torres, C.; Andrews, P.K.; Davies, N.M. Physiological and biochemical responses of fruit exocarp of tomato (*Lycopersicon esculentum* Mill.) mutants to natural photo-oxidative conditions. *J. Exp. Bot.* **2006**, *57*, 1933–1947. [CrossRef]
83. Minoggio, M.; Bramati, L.; Simonetti, P.; Gardana, C.; Iemoli, L.; Santangelo, E.; Mauri, P.; Spigno, P.; Soressi, G.; Pietta, P. Polyphenol pattern and antioxidant activity of different tomato lines and cultivars. *Ann. Nutr. Metab.* **2003**, *47*, 64–69. [CrossRef]
84. Wang, Y.; Qi, D.; Wang, S.; Cao, X.; Ye, Y.; Suo, Y. Comparison of Phenols Content and Antioxidant Activity of Fruits from Different Maturity Stages of *Ribes stenocarpum Maxim*. *Molecules* **2018**, *23*, 3148. [CrossRef]
85. Oak, P.; Deshpande, A.; Giri, A.P.; Gupta, V. Metabolomic Dynamics Reveals Oxidative Stress in Spongy Tissue Disorder During Ripening of *Mangifera indica* L. Fruit. *Metabolites* **2019**, *9*, 255. [CrossRef] [PubMed]
86. González-Gordo, S.; Bautista, R.; Claros, M.G.; Cañas, A.; Palma, J.M.; Corpas, F.J. Nitric oxide-dependent regulation of sweet pepper fruit ripening. *J. Exp. Bot.* **2019**, *70*, 4557–4570. [CrossRef] [PubMed]
87. Chu-Puga, Á.; Gonzalez-Gordo, S.; Rodríguez-Ruiz, M.; Palma, J.M.; Corpas, F.J. NADPH Oxidase (Rboh) Activity is Up Regulated during Sweet Pepper (*Capsicum annuum* L.) Fruit Ripening. *Antioxidants* **2019**, *8*, 9. [CrossRef] [PubMed]
88. Anton, D.; Bender, I.; Kaart, T.; Roasto, M.; Heinonen, M.; Luik, A.; Püssa, T. Changes in Polyphenols Contents and Antioxidant Capacities of Organically and Conventionally Cultivated Tomato (*Solanum lycopersicum* L.) Fruits during Ripening. *Int. J. Anal. Chem.* **2017**, *2017*, 1–10. [CrossRef]
89. Rodríguez-Ruiz, M.; González-Gordo, S.; Cañas, A.; Campos, M.J.; Paradela, A.; Corpas, F.J.; Palma, J.M.; Ruiz, R.-; Gordo, G.-. Sweet Pepper (*Capsicum annuum* L.) Fruits Contain an Atypical Peroxisomal Catalase That Is Modulated by Reactive Oxygen and Nitrogen Species. *Antioxidants* **2019**, *8*, 374. [CrossRef]
90. Tohge, T.; Fernie, A.R. Combining genetic diversity, informatics and metabolomics to facilitate annotation of plant gene function. *Nat. Protoc.* **2010**, *5*, 1210–1227. [CrossRef]
91. Tohge, T.; Mettler, T.; Arrivault, S.; Carroll, A.J.; Stitt, M.; Fernie, A.R. From Models to Crop Species: Caveats and Solutions for Translational Metabolomics. *Front. Plant Sci.* **2011**, *2*, 2. [CrossRef]
92. Tohge, T.; Fernie, A.R. Web-based resources for mass-spectrometry-based metabolomics: A user's guide. *Phytochemistry* **2009**, *70*, 450–456. [CrossRef]

metabolites

MDPI

Review

Metabolite Changes during Postharvest Storage: Effects on Fruit Quality Traits

Delphine M. Pott, José G. Vallarino *,† and Sonia Osorio *

Departamento de Biología Molecular y Bioquímica, Instituto de Hortofruticultura Subtropical y Mediterránea "La Mayora", Universidad de Málaga-Consejo Superior de Investigaciones Científicas, Campus de Teatinos, 29071 Málaga, Spain; dpott@uma.es

* Correspondence: vallarino@uma.es (J.G.V.); sosorio@uma.es (S.O.); Tel.: +34-952134271 (J.G.V. & S.O.)

† Current address: Max-Planck-Institut für Molekulare Pflanzenphysiologie, 14476 Potsdam-Golm, Germany.

Received: 14 April 2020; Accepted: 6 May 2020; Published: 8 May 2020

Abstract: Metabolic changes occurring in ripe or senescent fruits during postharvest storage lead to a general deterioration in quality attributes, including decreased flavor and 'off-aroma' compound generation. As a consequence, measures to reduce economic losses have to be taken by the fruit industry and have mostly consisted of storage at cold temperatures and the use of controlled atmospheres or ripening inhibitors. However, the biochemical pathways and molecular mechanisms underlying fruit senescence in commercial storage conditions are still poorly understood. In this sense, metabolomic platforms, enabling the profiling of key metabolites responsible for organoleptic and health-promoting traits, such as volatiles, sugars, acids, polyphenols and carotenoids, can be a powerful tool for further understanding the biochemical basis of postharvest physiology and have the potential to play a critical role in the identification of the pathways affected by fruit senescence. Here, we provide an overview of the metabolic changes during postharvest storage, with special attention to key metabolites related to fruit quality. The potential use of metabolomic approaches to yield metabolic markers useful for chemical phenotyping or even storage and marketing decisions is highlighted.

Keywords: fruit; postharvest; metabolomics; quality traits; stress; biomarkers

1. Introduction

Fruit growth, ripening and senescence are complex processes, controlled by multiple developmental and environmental signals, and their molecular mechanisms remain unclear [1]. Fruits undergo important metabolic changes during ripening, including chlorophyll breakdown, anthocyanin or carotenoid pigment accumulation, cell wall degradation and the synthesis of low-weight metabolites (such as sugars, acids and volatiles), which function to increase their attractiveness to seed dispersers [2]. Once fruits are removed from the plant and until they reach consumers on the market, a period known as postharvest ripening or senescence occurs—the duration of which is variable (from days to weeks) and the effects of which mainly depend on fruit metabolism and ripening status at harvest. Indeed, climacteric fruits, such as tomatoes, kiwi or avocados, which exhibit a concomitant peak of ethylene production and a sudden rise in respiration at the onset of ripening [3], can ripen after harvest. In this sense, the control of ethylene production is fundamental to optimize the postharvest storage of these types of fruits [4]. On the other hand, non-climacteric fruits, e.g., strawberries and grapes, do not exhibit respiration and ethylene production peaks, and have to be harvested (almost) fully ripe. Postharvest storage initiates fruit senescence—the effects of which on biological processes are unavoidable and largely negative. Senescence leads to protein, lipid and nucleic acid degradation and cell dysfunction, disintegration and death [5]. Several factors influence and accelerate fruit senescence, with the most relevant being respiration, providing energy for maintaining metabolism, dehydration and fungal activity [6]. Consequently, the degradative processes associated with postharvest senescence impact fruit quality traits, i.e., aspect, texture, taste, aroma and nutritional characteristics, leading to consumer rejection and important economic losses for the fruit industry.

Currently, depending on fruit crops, different postharvest strategies are commercially practiced in order to adapt ripening to industry needs, delay senescence, maintain fruit quality attributes and, thus, prolong shelf-life. In general, fruits are highly perishable at ambient temperature. Thus, refrigerated storage is the most common method used to delay ripening, fruit respiration, enzymatic activities, and the development of pathogen infections, and, therefore, extend fruit shelf-life [7]. However, cold storage can provoke the development of a physiological disorder called chilling injury (CI). Although CI symptoms are species dependent, CI includes internal and external browning, mealiness, flesh bleeding, pitting or the inability to soften. These physiological disorders tend to appear once the fruits are acquired by consumers, having a negative impact on palatability and acceptance [8,9]. To reduce CI symptoms and depending on the type of fruit, the industry combines low-temperature storage with some complementary strategies. For example, prior heat treatment to cold storage is widely used in several crops, including *Citrus* and loquat [10], while controlled atmosphere (CA, increased CO_2 and decreased O_2 levels) is commonly applied to apple, strawberry, peach and pear, among others [11]. Additionally, delayed cooling has been successfully applied in apple to reduce soft scald, a chilling-dependent physiological disorder [12]. While CA reduces fruit respiration, heat treatment has a protective effect by acting on membrane integrity and heat shock protein accumulation and by promoting antioxidant and sugar metabolism [13]. Further, it is known that heat treatment induces defense mechanisms and induces physiological changes that allow *Citrus* fruit to withstand stressful conditions during storage. For example, GC–MS analysis in heat-treated oranges during storage showed a higher concentration of sugars while no changes were observed in organic acid levels [14].

In the case of climacteric fruits, such as tomato or banana, the application of ethylene antagonist 1-methylcyclopropene (1-MCP) is commonly used to increase shelf-life [15]. However, the aforementioned strategies have different degrees of effectiveness at reducing CI and prolonging shelf-life, depending on fruit species and varieties. In addition, it must be highlighted that these postharvest techniques constitute abiotic stresses for the fruits, which have to adapt their metabolism to maintain homeostasis [16]. In particular, stress situations induce the synthesis of compounds involved in plant protection, and trigger the accumulation of compatible metabolites, reactive oxygen species (ROS)-scavenging enzymes and changes in carbon metabolism [17,18]. In this sense, metabolomic platforms, allowing the simultaneous detection and quantification of hundreds of metabolites, offer the possibility to improve our knowledge about the molecular mechanisms underlying fruit senescence under commercial storage conditions.

2. Metabolomic Platforms in Postharvest Studies

The plant metabolome comprises a wide range of small molecules, with a large variety in physico-chemical properties and extremely variable concentrations. Metabolomics is defined as the field of the research that generates a profile of small molecules in a biological system. Thus, it can directly reflect the outcome of complex networks of biochemical reactions and, therefore, provides essential information about the underlying biological status on the system in question.

For these reasons, successful analysis of the complex network of fruit metabolites requires highly sensitive and selective analytical techniques, with each displaying both advantages and limitations and showing differential coverage depending on the nature of the metabolite. In particular, mass spectrometry (MS) coupled with gas chromatography (GC), liquid chromatography (LC) and, to a lesser extent, capillary chromatography (CE) and nuclear magnetic resonance spectroscopy (NMR) have been the most extensively applied methodologies to study the plant metabolome, including its reconfiguration during fruit postharvest senescence [17,19,20].

GC–MS is the technique of choice for measuring small polar metabolites, which are thermally stable and can be made volatile through a derivatization approach [21]. The main advantages of GC–MS are its robustness and reproducibility, which have allowed the establishment of libraries and databases facilitating the identification of metabolites. As a result of its characteristics, GC–MS is mainly used in plant metabolomic studies to investigate central primary metabolism, which includes sugars, sugar alcohols, amino acids, organic acids and polyamines [17,19]. In addition, GC–MS can be coupled with headspace solid-phase microextraction (HS-SPME), which allows the detection of specific volatiles present in a sample [22]. Both volatile and primary metabolite changes occurring during fruit postharvest storage have been extensively studied, as they are key compounds of fruit taste and aroma.

To overcome the limitations of GC–MS, which is restricted to volatile and thermally stable molecules, LC–MS is broadly used to detect a wider range of metabolites. In particular, the enormous diversity of plant secondary metabolites, which includes tens of thousands of different compounds [23], is mainly studied using LC–MS due to its versatility. However, and as a consequence of LC–MS flexibility, metabolite identification remains difficult, as no universal mass spectral library has been created [24].

Another technique used to study plant metabolomics, although rather uncommon, is capillary electrophoresis (CE)–MS. This technique allows the detection of a wide range of highly polar or charged metabolites by separating them based on their mass-to-charge ratio [25]. In this sense, this method has been proposed as a valuable complementary approach for samples that cannot be readily resolved by the more established GC– and LC–MS platforms [26].

Even if NMR presents a low sensitivity compared to that of MS approaches, it provides a series of advantages over the previously mentioned approaches by providing structural information, involving non-destructive sample preparation and providing rapid metabolite screening [19,27]. Integrated NMR platforms, allowing the monitoring of changes in both primary and secondary metabolites, have been developed and can be useful to study metabolic shifts in senescent fruits during postharvest [28].

3. Primary Metabolic Pathways Affected by Postharvest Storage: Effects on Fruit Texture and Taste

Fruit organoleptic quality is a complex trait that is influenced by taste, aroma, color and texture. In particular, fruit acceptance by consumers is directly influenced by sugar and acid content and the ratios of both groups of primary metabolites [29,30]. Fruit respiration during postharvest storage directly affects primary metabolic pathways, such as glycolysis, starch metabolism, and the tricarboxylic acid cycle (TCA), which account for changes in sugar, amino and organic acid levels. Indeed, carbohydrates, organic acids, proteins and fats are the main respiratory substrates during fruit storage. Furthermore, they are involved in gluconeogenesis, a process which has been described to be upregulated during the postharvest in orange and apple fruits [31–33]. Thus, it contributes to fruit depletion and also to important changes in primary metabolite composition. In the next paragraphs, we described alterations in sugars, organic and amino acids as a consequence of postharvest storage.

3.1. Sugars and Sugar Derivatives

Sugar content, which is commonly estimated by the soluble solid content (SSC), shows differential behavior during postharvest, depending mainly on the species and storage conditions. The SSC trend normally coincides with changes in the main sugar profiles present in ripe fruits, i.e., glucose, fructose and sucrose.

While main carbohydrates tend to decrease in some species, as profiled by GC–MS and NMR analysis in tomatoes kept at room temperature [34] or blackberries stored at 4 °C [35], in other fruits, such as bananas and kiwis, their level increases as a consequence of starch hydrolysis, which takes place during postharvest storage [36,37]. In turn, sucrose can be hydrolyzed, leading to a concomitant increase in hexoses, as monitored by GC–MS in Powell oranges stored at room temperature [33]. Interestingly, growing evidence seems to point to sugars playing a regulatory role in senescence processes [38,39]. During fruit ripening and senescence, cross-talk between sugars and hormones involved in ripening and senescence processes, such as abscisic acid, ethylene and auxin, has been described [40–43], and sucrose degradation during postharvest storage can be crucial for inducing senescence [33,44]. Furthermore, sugar uptake during fruit ripening may affect postharvest water loss by interfering with cuticle development. Indeed, stable silencing of the cell wall invertase *LIN5*, a key determinant of SSC content, led to a diminished water loss rate and wrinkling in transgenic tomato fruits kept at room temperature for 12 days. Even though the complete molecular mechanism has not been described, it was clearly established that sugar entry during fruit development impacts the cell wall and cuticle structure, resulting in a radical effect on tomato senescence [45].

Apart from the most abundant sugars, i.e., sucrose, glucose and fructose, fruits also contain minor sugar and alcohol derivatives, such as sorbitol, galactinol, raffinose, *myo*-inositol and trehalose [46]. Even if those compounds may be at low concentrations, they seem to be crucial for fruit behavior during storage, as they can alleviate the negative effects of the abiotic stresses underlying postharvest conditions. Indeed, soluble sugars are important metabolites in ROS metabolism, being the primary carbon and energy source and contributing to the generation of reducing power generation via the oxidative pentose phosphate pathway [41,47,48]. Furthermore, they play key roles in osmoprotection and cell membrane stabilization [49–51]. As an example, important increases in raffinose and galactinol levels were measured by GC–MS in peaches after heat treatment (three days at 39 °C) followed by storage at 0 °C for two days and may confer improved tolerance to CI [52]. Moreover, comparing the levels of galactinol (detected by LC–MS/MS), raffinose, trehalose and *myo*-inositol (identified by NMR) in climacteric and non-climacteric plum varieties during postharvest storage at 20 °C and in presence of 1-MCP, propylene (ethylene analogue) or control air, Farcuh et al. [46] noticed that the levels were more enhanced in the latter variety. These data could explain the capacity of the non-climacteric variety to cope better with postharvest stress conditions, and the identified sugars could be used as biomarkers to evaluate fruit physiological status during storage (Table 1).

Softening during postharvest storage is a key physiological process leading to ripe fruit firmness; however, excessive loss of firmness as a consequence of overripening can prompt physical damage and pathogen attack, and consequently lead to an important decrease in fruit quality. Softening is the result of several factors, including cell wall disassembling metabolism. Metabolites originated from cell wall disassembly, mainly monosaccharides, can be monitored by primary metabolite profiling. Indeed, in pitaya fruit, the content of several monosaccharides, including xylose, galactose, arabinose, and mannonic acid and glucuronic acid, which originate from cell wall disassembly, was measured by GC–MS [53]. Interestingly, these metabolites were decreased after blue light treatment (2 h at 25 °C under blue light emitting diode) compared to control fruits kept in the dark, suggesting that this treatment has a significant effect in delaying cell wall degradation and postharvest decay of pitaya fruit [53]. Another study, using LC coupled with tandem MS (LC–MS/MS), detected an increase in glucuronic acid, a component of pectin, among the major elements of plant cell wall, in pears stored 18 days at room temperature [54]. Pectin de-polymerization and de-esterification were also evidenced by the detection of galacturonic acid by two-dimensional GC–MS (GC x GC–MS) in overripe kiwi fruits stored at 20 °C and treated for 24 h with 200 ppm ethylene. On the contrary, no increase in galactose was observed using GC–MS measurements, suggesting that this sugar is directly metabolized after its release from cell wall, or that it is liberated as different form [55]. Among the main symptoms of CI in peach is mealiness, which is the result of a cell wall metabolism disorder. Xylose, the central constituent of hemicellulose, among the key components of plant cell wall, was increased during cold storage in peach chilling-susceptible genotypes, but not in the varieties resistant to CI, confirming a link between cell wall disassembly and mealiness in sensitive cultivars [56].

Table 1. Primary metabolites (sugars, organic and amino acids) identified as putative biomarkers by metabolomic profiling studies to assess fruit quality changes during postharvest storage. d: day; RT: room temperature; HPLC: high-performance liquid chromatography; UHPLC–MS/MS: ultra-high-pressure liquid chromatography–tandem mass spectrometry; ^{1}H-NMR: proton-NMR; CI: chilling injury; ROS: reactive oxygen species.

Metabolite	Effect on Fruit	Postharvest Treatment	Behavior during Postharvest	Fruit Species	Metabolomic Platform	Reference
raffinose, galactinol	Tolerance to CI	39 °C, 3 d + 0 °C, 2 d	Increase	peach	GC–MS	[52]
raffinose, galactinol, myo-inositol, trehalose	Enhanced capacity to cope with postharvest stress conditions	20 °C, 14 d	Increase	non-climacteric plum	NMR, UHPLC–MS/MS	[46]
malate	Decrease in water loss and in susceptibility to opportunistic fungal infections	RT, 20 d	Decrease	tomato	GC–MS	[57]
malate	Correlation with fruit firmness and shelf-life	25 °C until first symptoms of deterioration	Decrease	tomato	GC–MS, ^{1}H-NMR	[58]
mannose, citramalate, gluconate, keto-gulonate		18 °C, 10 d	Increase	tomato	GC–MS	[36,59]
dopamine	Conversion to salsolinol at late postharvest stages, decrease in fruit quality	25 °C until senescence	Decrease	banana	^{1}H-NMR	[36]
proline	Osmoprotection and ROS-scavenging functions	0 °C, 20 kPa CO_2/20 kPa O_2/60 kPa N_2, 3 d + 0 °C, air, 30 d + 20 °C, 2 d	Increase	grape	LC–MS	[60]
proline	Osmoprotection and ROS-scavenging functions	1 mM GABA treatment, 20 min + 4 °C, 18 d (dark)	Increase	zucchini	HPLC	[61]

3.2. Organic Acids

Several organic acids are related to fruit postharvest metabolism. Surprisingly, particularly in tomato, the levels of malate, among the most abundant organic acids, impact fruit shelf-life (Table 1). The malate content, measured by GC–MS, decreases during the ripening and postharvest storage at room temperature of several tomato genotypes, including *delayed fruit deterioration*, *non-ripening* and *ripening inhibitor* mutants as well as genotypes that are commercially used because of their delayed maturation and senescence. Interestingly, it was also shown that malate levels were lower in mature tomato fruits that ripened on the vine than off the vine [34]. However, when the malate concentration in tomato fruits was manipulated by reducing the expression of two TCA cycle enzymes (*fumarase* and *malate dehydrogenase* (*MDH*)), a differential postharvest behavior was observed compared to that in wild-type fruits at room temperature. Interestingly, fruits of the MDH-deficient genotype showed higher malate content and poorer postharvest behavior than non-transformed fruits by losing more water and being more susceptible to opportunistic fungal infections and *Botrytis cinerea* spores. In contrast, the fumarase-deficient genotype, with a relatively low malate content, presented a decrease in water loss and in susceptibility to opportunistic fungi [57]. The mechanism underlying malate's role in postharvest responses could not be clearly explained; however, the authors suggested a role for SSC, which changes in the opposite manner in *MDH*- and *fumarase*-silenced lines, in osmotic potential and subsequent water loss during storage. Another study using recombinant inbred lines originating from the cultivated tomato *Solanum lycopersicum* and the wild-type species *Solanum pimpinellifolium* also pointed out the association between malate content, fruit firmness and shelf-life [58]. A comprehensive polar metabolite profiling was performed by GC–MS and NMR and a combination of neuronal clustering and network construction displayed a strong correlation between glycerate and malate content and postharvest, which also showed a negative correlation with fructose levels [58]. This association between metabolites and agronomic traits such as firmness and storage behavior suggested that malate could be a good biomarker to select genotypes with enhanced quality traits, such as improved postharvest life [58].

By performing a GC–MS metabolic characterization of *S. lycopersicum* cv. 'Plaisance' fruits during ripening and postharvest stages, Oms-Oliu et al. [59] showed that one sugar (mannose) and three organic acids (citramalate, gluconate and keto-gulonate) were strongly increased once the fruit was removed from the vine and that these compounds could be indicators of metabolic shifts during postharvest storage [59] (Table 1). As an example, the enhanced gluconate levels could be a consequence of tartarate biosynthesis from ascorbate degradation or energy balance changes during tomato storage [62,63]. Free mannose levels are generally low, as this monomer usually composes carbohydrate polymers. However, it can be found in a free form as a result of cell wall disassembly and hemicellulose breakdown during fruit senescence, as described in tomato, apple and pear [59,62,64].

Organic acids, particularly citric acid, accumulate at high levels in *Citrus* fruits, such as lemons, oranges, grapefruits or pummelos. A study on 'Hirado Buntan' pummelo focused on the relationship between organic acids, measured by high-performance capillary electrophoresis, and fruit senescence during postharvest storage at both ambient and cold temperatures. The authors observed a general decrease in malate, citrate, aconitate and fumarate during storage, accompanied by important fluctuations in their levels; this decrease was associated with a loss of fruit quality [31]. The combination of transcriptomic analysis paralleled the metabolomic data, suggesting that the *peroxisomal MDH*—the expression of which correlated with malate levels—is responsible for organic acid metabolism regulation during postharvest. This result indicated that the glyoxylate cycle, which occurs in peroxisomes and glyoxysomes, is central to organic acid regulation by supplying succinate for the TCA cycle [31]. Tang et al. [33] also observed a decrease in several organic acids analyzed by GC–MS, such as malate, citrate and α-ketoglutarate, during postharvest storage of 'Powell' oranges at room temperature. In this case, they suggested that malate could be used as a substrate for gluconeogenesis, being converted into phosphoenolpyruvate (PEP) by the action of two enzymes upregulated in oranges kept at room temperature: PEP carboxykinase and pyruvate orthophosphate dikinase (PPDK). Similarly, an increased abundance of PPDK proteins associated with decreased malate content was observed in peaches subjected to heat treatment followed by storage at 20 °C [65]. Another study using GC–MS analysis in different varieties from the *Citrus* genus suggested that a conversion of organic acids to sugars during fruit postharvest senescence at ambient temperature occurs, as negative correlations were frequently observed between metabolites belonging to the two groups and that the SSC/titratable acidity ratio increased during storage [66]. The succinate content increased during pummelo postharvest storage, showing a positive correlation with GABA and glutamine [31]. In addition, GABA increased during the postharvest senescence of Powell oranges, matched by an upregulation of the genes involved in the GABA shunt [33]. In this sense, the GABA shunt was outlined as an important pathway for organic acid catabolism and for balancing organic acid and amino acid levels. Indeed, superfluous citrate can be converted into amino acids via the GABA shunt [33,67]. Moreover, Sun et al. [31] observed an increase in ROS during pummelo storage, which correlated with enhanced mitochondrial damage. Cross-talk between ROS and organic acids could occur during postharvest senescence, as TCA enzymes have been described to be very sensitive to inhibition by ROS [68,69], while organic acids could be involved in the direct ROS scavenging [70,71].

3.3. Amino Acids

Amino acid content is also affected to a large extent by postharvest storage, as these compounds are involved in several pathways induced during fruit ripening [47]. In particular, during senescence, amino acid catabolism can counteract the reduction in electron supply from the TCA cycle [72]. Ubiquitination of proteins controls their degradation to free amino acids, and upregulation of the ubiquitin pathway has been reported in stored peaches that were previously were heat treated [73]. Dopamine, a derivative of the aromatic amino acid tyrosine, has been proposed as a postharvest marker in banana fruit stored at 25 °C [36] (Table 1). Indeed, NMR-based metabolite profiling of the senescence of bananas stored at room temperature showed that dopamine levels were undetectable at the last postharvest stage. Concomitant with dopamine disappearance was the sudden appearance of salsolinol, which has been described to originate from dopamine and acetaldehyde, the latter formed from ethanol, which is also generated in the late postharvest stage [36,74]. The authors concluded that the conversion of dopamine to salsolinol led to a decrease in fruit quality, making bananas less fit for consumption [36].

Additionally, several amino acids play a key role in tolerance to abiotic stresses in fruits during postharvest senescence. Indeed, a GC–MS comparative study between pineapple varieties tolerant and susceptible to CI stored at 10 °C outlined that amino acid increases during chilling stress may be associated with a delay in symptom appearance, such as internal browning, by presumably contributing to the synthesis of enzymes involved in tissue repair and, in the case of cysteine, aspartate and valine, by acting as osmoprotectants [75]. Proline is a well-documented stress-related amino acid and among the main osmolytes that are accumulated during plant stresses, playing important membrane protection and ROS-scavenging functions [76,77] (Table 1). Grape storage in a CO_2-enriched atmosphere resulted in a threefold endogenous proline increase when compared to that in air-stored grapes [60], and proline accumulation is a common trend in postharvest fruits subjected to treatments to attenuate CI, such as zucchinis [61], mangoes [78], bananas [79], pears [80] and loquats [81]. However, the possible role of amino acids in counteracting CI seems to be species dependent, as GABA, aspartate, phenylalanine and proline increase in peach stored at 0 °C for 21 days was not associated with CI protection, since their levels, quantified by GC–MS, were enhanced in both resistant and susceptible genotypes [56].

A recent study in strawberry also outlined the possible role of amino acids in plant defense, as pathogen resistance mechanisms implicated this group of metabolites [82]. The increase in asparagine, aspartic acid, threonine, glutamic acid, glutamine, alanine and glycine in CO_2-treated strawberries compared to control fruits could, at least partially, explain the lower fungal decay observed in the first group [83].

4. Postharvest Impact on Secondary Metabolites

The two main families of secondary metabolites present in fruits are polyphenol and terpenoid compounds, responsible for their appealing color and also important for their organoleptic and nutritional characteristics [84]. Apart from their importance in the human diet, these molecules are involved in plant defense and responses against biotic and abiotic stresses. In particular, metabolomic approaches have helped in deciphering their role during fruit storage and how different postharvest strategies impact on them. Here, impact on polyphenols, including anthocyanins, and carotenoids during fruit shelf-life is discussed in the next paragraphs.

4.1. Polyphenol Compounds

Dynamic metabolite changes, profiled by high-performance LC–MS (HPLC–MS), were observed during grape postharvest ripening and dehydration, the metabolic responses being genotype dependent [85]. A particular feature was the cultivar-specific accumulation of stilbenes, a class of phenylpropanoid compounds, with antifungal activity. On the other hand, anthocyanins and other flavonoids, belonging to another phenylpropanoid class, were depleted along postharvest dehydration [85]. An untargeted HPLC–MS profiling during grape ripening and withering (postharvest drying), combined with transcriptomic and proteomic data integration, also correlated the presence of stress-related secondary compounds (stilbenes and acylated anthocyanins) with the postharvest phase. The synthesis of defense molecules could be a response to abiotic stress (dehydration) or biotic stress (eventual pathogen attack). In addition, three metabolites (two taxifolins and tetrahydroxyflavanone-O-deoxyhexoside), belonging to the flavonoid class, have been proposed as putative markers in order to assess berry fruit quality traits (Table 2) [86]. In grapes, the accumulation of different stilbenes during cold postharvest storage was monitored by UHPLC–MS/MS [87]. This increase was also observed when grape fruits were kept at high CO_2 [87]. In contrast, CA storage has been described to have negative effects on anthocyanin accumulation in strawberry fruits, compounds responsible for the color of the ripe fruit [88]. In this sense, postharvest cold storage is a mandatory strategy to enhance anthocyanin content in some fruits such as blood oranges, some varieties of plums and anthocyanin-rich tomatoes [89,90]. Interestingly, it has been described that tomato anthocyanin-rich lines are able to maintain fruit quality for longer during storage, mainly by reducing their susceptibility to *Botrytis cinerea* [91,92].

In mandarins, heat treatment previous to storage at 12–16 °C positively impacts polyphenol metabolism by increasing flavonoids and lignin content (flavonoids measured by HPLC–MS). The effect of this postharvest strategy can be seen as a modulation of fruit defense against biotic and abiotic stress during postharvest storage, by supplying chemical (flavonoids) and physical barriers against pathogen attack [93]. The relationship between polyphenol content and resistance to postharvest decay caused by *Penicillium expansum* has also been described in apple; indeed, resistant and susceptible apple genotypes could be discriminated based on polyphenol content, measured by UPLC–MS (Table 2) [94]. However, a general polyphenol increase during fruit shelf-life does not always occur, as described by untargeted UHPLC–MS in several mango varieties stored at room temperature during six days, in which it was found that only gallic acid and epicatechin content was enhanced after storage [95].

Table 2. Secondary metabolites (polyphenols and carotenoids) identified as putative biomarkers by metabolomic profiling studies to assess fruit quality changes during postharvest storage. d: day; w: weeks; UHPLC–HRAM MS^n: ultra-high-performance liquid chromatography coupled to high resolution multiple-stage mass spectrometry.

Metabolite	Effect on Fruit	Postharvest Treatment	Behavior during Postharvest	Fruit Species	Metabolomic Platform	Reference
taxifolin deoxyhexoside, taxifolin hexoside tetrahydroxyflavanone-*O*-deoxyhexoside	Antifungal activity, withering stress responses	withering, 91 d	Increase	grape	Untargeted HPLC–MS	[86]
procyanidin B1, epi-catechin	Resistance to *Penicillium expansum*	2 °C storage	Increase	apple	UHPLC–HRAM MS^n	[94]
β-cryptoxanthin	Part of β, β-xanthophyll pool in mature oranges	12 °C up to 7 w	Increase	sweet orange	HPLC	[96]

4.2. Carotenoids

Carotenoids are an important class of terpenoids, responsible for the attractive color of many fruits and vegetables. While their low stability during postharvest storage, mainly due to a rapid turnover of β-carotene, has been described in many staple crops, postharvest accumulation in *Citrus* and tomato seems to be temperature dependent [96,97]. Carotenoid levels in grapefruit, determined by HPLC, stored at 2 and 12 °C established a link between carotenoid content and CI symptom suppression, suggesting that they play a role in preventing cold damage by protecting plastid structures [98,99]. Furthermore, the ratio between 9-Z-violaxanthin (yellow hues) and β-citraurin (orange-red pigments), responsible for the external orange fruit color, was lower in sweet oranges stored at 12 °C than at 2 °C, outlining that this important quality indicator is better maintained at moderate temperatures. Additionally, increased levels of β-cryptoxanthin in orange pulp stored at 12 °C should be pointed out, due to health-beneficial provitamin A activity (Table 2) [96]. Carotenoid content, measured by HPLC, was also drastically increased during postharvest storage of winter squash at 21 °C, even if no induction of the biosynthetic genes could be observed. Starch degradation during winter squash storage, with the concomitant release of soluble sugars which may act as substrates for terpenoid synthesis, and downregulation of genes involved in carotenoid turnover, could be the explanation of their enhanced content [100]. In other fruits, such as green pepper, carotenoid accumulation during postharvest storage has a negative impact on consumer acceptance. Pepper reddening depends on the metabolic dynamic of chlorophyll degradation and active synthesis of carotenoids, such as β-carotene and capsanthin, as depicted by HPLC-based profiling of these pigments [101]. Quantification of chlorophyll by spectrophotometry has also pointed out its breakdown as a deterioration factor occurring during pear or lime shelf-life [102,103]. In this sense, postharvest strategies, such as chlorine dioxide fumigation or hot water treatment, may be effective in downregulating genes involved in chlorophyll-degrading enzymes [101,102].

5. Volatile Profiles during Postharvest and Their Impact on Fruit Aroma

In fruits, there are three major classes of metabolites responsible for flavor: sugars, acids, and volatile. While fruit taste is mostly dependent on the ratio of sugars and acids, it is the volatiles that determinate the unique flavor of fruits. Most volatiles present in mature fruits originate from terpenoid and phenylpropanoid pathways or are fatty and amino acid derivatives [104]. Volatile profiling is typically achieved by extracting them from the headspace (HS), i.e., the airspace around the fruit, and detecting them by GC–MS. Sampling from headspace is most often performed by the adsorption of the volatiles on a stationary phase coated on a fused silica fiber and is known as solid-phase microextraction (SPME) [104]. Another GC–MS-based strategy for volatile profiling is their collection from chopped fruits on a Super Q column, followed by elution with methylene chloride [105]. To overcome metabolite co-elution by one-dimensional GC, GC x GC–MS has been implemented to increase separation efficiency and volatile detection [106,107]. As not all volatiles impact fruit aroma, a complementary approach, known as GC—olfactometry, can be used to determine odor-active compounds [108]. During postharvest, it could be established that important shifts in fruit volatile profiles are normally observed and are often responsible for the decreased sensory acceptability after prolonged storage. For instance, general trends profiled by GC–olfactometry, describe a loss of 'green' or 'fresh' notes and a concomitant increase in 'fruity', 'overripe' or 'musty' aromas [109]. Changes in aroma are a consequence of metabolic pathways that are active during postharvest and, in turn, appear to be largely depend on the storage strategies used by industry. For example, among the symptoms related to CI is the negative impact perceived on aroma production, a phenomenon described in many species, such as strawberries [110], kiwifruit [111], tomatoes [112] and peaches [113]. Tomatoes stored at 5 °C for 7 days were significantly less palatable than fruits recently harvested, and this decrease in consumer acceptance, established by taste panels, was a consequence of changes in volatile emissions [105]. Furthermore, a higher increase in 'musty' and 'damp' aroma notes was observed in tomatoes stored at 10 °C than in those stored at 12.5 °C, suggesting that the latter temperature storage was able to maintain better sensory attributes [114].

Fermentation metabolism and amino acid and fatty acid catabolism are of great importance regarding the production and accumulation of volatiles in harvested fruits. Indeed, the activation of amino acid and fatty acid degradation to generate TCA cycle acetyl-CoA precursors and thus maintain energy production leads to the accumulation of specific substrates for volatile formation. In mandarin, a combination of metabolomic and transcriptomic data outlined the upregulation of genes involved in branched-chain amino acid catabolism, fatty acid cleavage and ethanol fermentation, which suggested that central metabolism modifications are accountable for the increase in branched-chain esters ('fruity', 'overripe' aroma), fatty acid-derived volatiles ('musty' notes) and ethanol [115]. The activation of anaerobic fermentative metabolism due to postharvest abiotic stress is especially important in 'off-aroma' compound generation and has been described in fruits of several species, including strawberries [116–119], apples [120], mandarins [109] and peaches [121], among others. Indeed, the glycolysis end-product pyruvate can alternatively serve as a substrate for anaerobic respiration and ATP production under O_2-limiting conditions, which produces a shift from aerobic respiration to the fermentation pathway [16,120,122,123]. As a consequence, off-aroma volatiles, namely ethanol, acetaldehyde and ethyl acetate, accumulate, playing a key role in fruit quality decline [117,124] (Table 3).

Table 3. Volatile compounds identified as putative biomarkers to evaluate the effects of postharvest storage on fruit aroma. d: day; w: week; GC–O: gas chromatography–olfactometry; GLC–MS: capillary gas–liquid chromatography–mass spectrometry.

Volatile	Effect on Fruit	Postharvest Treatment	Behavior during Postharvest	Fruit Species	Metabolomic Platform	Reference
ethanol, ethyl acetate, acetaldehyde	'Off-aroma' generation, 'alcohol' aroma	3 °C, 3 ws supplemented with different CO_2 concentration	Increase	strawberry	HS-SPME–GC–MS	[117]
ethanol, ethyl acetate, acetaldehyde	'Off-aroma' generation, 'alcohol' aroma	5 °C, 6 ws + 20 °C, 1 w	Increase	mandarin	HS-SPME–GC–MS and GC–O	[109]
ethanol, ethyl acetate, acetaldehyde	'Off-aroma' generation, 'alcohol' aroma	2.5 °C, 7 d + 1 °C followed by two different low oxygen protocols up to 240 d	Increase	apple	HS-SPME–GC–MS	[11]
ethanol, ethyl acetate, acetaldehyde	'Off-aroma' generation, 'alcohol' aroma	0 °C, 6 w + 20 °C, 2 days supplemented with different CO_2 concentration	Increase	grape	HS-SPME–GC–MS	[87]
β-myrcene	Decrease in aroma quality	2 °C or 12 °C, 7 w	Increase at 2 °C, decrease at 12 °C	grapefruit	HS-SPME–GC–MS	[98]
ketone nootkatone	Confers characteristic ripe aroma fragrance	12 °C, 7 w	Increase in 12 °C	grapefruit	HS-SPME–GC–MS	[98]
limonene	Cold-induced responses	2 °C, 7 w	Increase	grapefruit	HS-SPME–GC–MS	[98]
limonene	Cold-induced responses	1 °C, 7 w	Increase	lemon	HS-SPME–GC–MS	[125]
limonene	Cold-induced responses	5 °C, up to 6 w	Increase	mandarin	HS-SPME–GC–MS	[126]
limonene	Senescence predictor	Combination of treatments, including 15 °C, 7 d + 2 °C, 18 d, 13 °C, 17 d	Increase	grapefruit	GLC–MS	[127]
α-farnesene	Correlation with CI symptom development in 0 °C storage	0 °C up to 12 w, with or without ethylene	Increase	lime, mandarin, grapefruit, orange	GC–MS	[128]
linalool	Key component of fruit aroma	10 °C, 10 d + 22 °C until fully ripe	Decrease in low-temperature storage	papaya	HS-SPME–GC–MS	[129]
linalool	Key component of fruit aroma	0, 2, 5, and 10 °C up to 3 months	Decrease in low-temperature storage	Muscat table grapes	HS-SPME–GC–MS	[130]

Fermentative metabolism activation and off-flavor compound formation are mainly associated with low oxygen concentration under CA storage [118,123,124,131]. However, the production of ethanol via fermentation may also be a consequence of a decline in cellular energy status [132]. As long as energy demand is maintained, fermentation can be endured; nevertheless, failure of cellular homeostasis, such as an imbalance in the pH or ROS production, will lead to storage-induced disorders, strongly affecting fruit quality [133]. Understanding how or when fermentation occurs can help to limit ethanol production. Metabolomic approaches using ^{1}H NMR and GC–MS profiling were used to assess metabolite gradients within the fruit, which may be related to in situ hypoxia in the central part of the ripening fruit [134,135]. CA is of special importance for the long-term storage of fruits such as apples, and could maintain a better aroma quality [136,137]. It appears that the low-oxygen pressure employed during CA affects volatile emissions in a genotype-dependent manner [11]. Indeed, a multiplatform metabolomic approach (proton-NMR, GC–MS and HS-SPME–GC–MS) comparing 'Red Delicious' and 'Granny Smith' apple varieties showed strong activation of fermentative metabolism in the former, with ethanol and acetaldehyde accumulation, while the latter dealt with hypoxia by a reconfiguration of nitrogen metabolism through the intensification of alanine levels to prevent excessive accumulation of pyruvate [11]. Low oxygen may induce changes in metabolite concentrations that reflect a decrease in biosynthetic process, inhibition of the TCA cycle, and activation of anaerobic metabolism, which means accumulation of sucrose and organic acids and diversion of pyruvate to ethanol and alanine [134]. Table grapes stored under elevated CO_2 concentrations (5 kPa O_2 and 15 kPa CO_2) showed an upregulation of genes involved in pyruvate synthesis (*pyruvate kinase*, *PEP carboxykinase* and *NADP-dependent malic acid enzyme*) and a concomitant increase in volatiles, detected by HS-SPME–GC–MS, derived from pyruvate degradation—some of which were suspected to generate 'off flavor'. Additionally, the increased expression of a specific *alcohol dehydrogenase* gene (*ADH*) under anaerobic atmospheric conditions enhanced the accumulation of off-aroma volatiles, including ethanol, acetaldehyde and ethyl acetate [87].

Metabolic reconfiguration during postharvest affects volatile patterns beyond the generation of off-aroma compounds, and changes occurring in most important volatile classes are described in the next sections.

5.1. Fatty and Amino Acid-Derived Volatiles

Fatty acid-derived volatiles, responsible for aldehyde, alcohol and ester accumulation, the last being the predominant class of aromatic compounds in fruits of several species, seem to be strongly impacted by low-temperature storage [110,138,139]. Free fatty acids such as linoleic acid and linolenic acid are reduced to aldehydes by the lipoxygenase pathway (LOX). Next, aldehydes are reduced to alcohols followed by alcohols to esters by ADH and alcohol acyltransferase (AAT), respectively (for a review, see [84]). Interestingly, correlations among LOX, ADH and AAT activities, gene expression and decreased volatile production under refrigerated postharvest conditions have been established in several fruit-bearing species [140,141]. In particular, a relationship between a reduction in ADH activity and decreased ester content, monitored by SPME–GC–MS technology, during pear cold storage has been established [138]. In tomatoes, ADH activity was diminished as a consequence of refrigerated conditions at both 10 and 12.5 °C, and storage was associated with an increase in the aldehyde/alcohol ratio at 10 °C [112,114]. Furthermore, a decrease in *ADH2*, *LoxC* and *AAT1* transcripts after 8 days of cold storage (5 °C) was associated with lower levels of C6 and C5 (fatty acid-derived) volatiles in chilled tomatoes [105]. Additionally, low temperature also seems to affect upstream lipid catabolism by downregulating the expression of several genes involved in the formation of the unsaturated free fatty acids linoleic acid and linolenic acid, limiting substrate availability for ester biosynthesis [142]. Furthermore, membrane damage during cold storage has also been suggested to impair ester synthesis, as a relatively high leakage rate, a commonly used marker for membrane permeability, was measured in pears stored for a long time under refrigerated conditions [138]. The impact of cold storage on the aromatic compound profile reaches further than that on the pattern of lipid-derived volatiles.

Branched-chain volatiles derived from the direct precursors of branched-chain amino acids, measured by GC, after methylene chloride extraction, were also shown to decrease during tomato cold storage and are correlated with a lower expression of two branched-chain aminotransferases (*BCAT1* and *BCAT7*) involved in the first step of the catabolism of these amino acids [105]. Additional treatments, such as hot air or UV-C, combined with cold storage could counteract the negative effect on ester biosynthesis by promoting the *LOX* pathway, as has been demonstrated in peaches [143]. Similarly, a pre-chilling heat treatment (52 °C, 5 min) has been shown to alleviate the depletion of important volatiles for tomato aroma quality during its postharvest storage; in this case, the volatiles include amino acid- and carotenoid-derived compounds profiled by HS-SPME–GC–MS [144]. Fatty acid-derived alcohols were also higher under elevated CO_2 concentrations compared to those of recently harvested grapes and cold-stored berries under atmospheric conditions due to the upregulation of the *LOX* pathway, together with *ADH* [87].

Low-oxygen storage has a broader impact on volatile content than ethanol and off-aroma compound generation, as demonstrated by the different content of ethyl esters between 'Granny Smith' and Red Delicious' apples [11]. Indeed, ethanol can serve as a substrate for ethyl esters, enhancing their synthesis [136,145,146] and competitively inhibiting the formation of esters originated from other alcohols [147]. As a consequence, an imbalance between the ratio of ethyl and the remaining esters occurs during postharvest storage of fruits of ethanol-accumulating apple varieties and those of many other fruit-bearing species, most likely affecting aroma perception. The fruits of 'Granny Smith' and 'Royal Gala' apple varieties did not seem to accumulate ethanol under low-oxygen-pressure storage; however, a negative effect on ester synthesis, in particular straight-chain esters, was observed, with the impact proportional to the decrease in O_2 pressure [148–154]. This decrease can be explained by the fact that the LOX pathway requires the presence of oxygen. This effect has also been described in other apple varieties. [137,153]. In contrast, the concentration of branched-chain esters, monitored by HS-SPME–GC–MS, did not seem to be negatively affected by low oxygen, possibly because branched-chain amino acid levels were unaltered [154]. Furthermore, low oxygen suppresses the production of the hormone ethylene, which is involved in ester synthesis, as demonstrated during apple or banana storage in the presence of its antagonist 1-MCP [155–157].

5.2. Terpenoid Volatiles

Several studies in *Citrus* have highlighted important changes in terpenoid volatiles monitored by HS-SPME–GC–MS. These changes were related to CI and tolerance to cold storage. In mandarins, the accumulation of terpenoid volatiles is associated with chilling-sensitive fruits [158]. This increase is temperature dependent, and the authors suggest that it was responsible for decreased fruit palatability, as terpenes can contribute to an unpleasant aroma, providing 'musty', 'resinous' and 'oily' notes.

Another study on volatile emission by intact grapefruits stored at 12 and 2 °C for 7 weeks outlined important differences in the profiles of the terpenoid volatiles as a consequence of temperature [98]. Interestingly, grapefruits stored at 2 °C experienced a strong increase in monoterpene content, particularly in limonene and β-myrcene levels; this group of volatiles was strongly decreased in fruits stored at 12 °C at the beginning of the postharvest period, after which their content remained unchanged. In contrast, sesquiterpene emissions were predominant in the fruits stored at 12 °C [98]. While the accumulation of the monoterpene β-myrcene under 2 °C can negatively impact negatively consumer acceptance, by providing 'musty' and 'wet soil' aroma notes to the fruits [108], sesquiterpene ketone nootkatone levels, an important volatile in grapefruit aroma, seems to be promoted (in term of it content) under moderate–intermediate-temperature storage, but not refrigerated conditions [98,159] (Table 3). Taken together, these data suggest that the aroma quality of grapefruits could be better maintained during intermediate-temperature storage, as has been previously demonstrated in mandarins [126,149].

The trend in the accumulation of the monoterpene limonene and the sesquiterpene α-farnesene, which showed a transient increase after one week of storage under the two different temperatures, could be related to cold-induced responses. Limonene release, measured both by HS-SPME–GC–MS and capillary gas–liquid chromatography–MS, has also been described in other *Citrus* species [125–127,160] and could be a consequence of cell wall and plasmatic membrane disruptions in the oil glands [99,125]. The degree of accumulation of α-farnesene is correlated with the susceptibility of different cold-sensitive *Citrus* species to CI development at 0 °C storage [128]. Interestingly, α-farnesene stopped being emitted by grapefruits after 3 weeks of storage at 12 °C, which coincided with the decrease in the observed CI symptoms, i.e., peel injury. At 2 °C, α-farnesene emissions were maintained during the whole postharvest period, concomitant with the CI symptom progression, confirming the relation between the detection of this volatile and CI manifestation. In this sense, the detection of α-farnesene by metabolome-driven approaches could be of high value as a potential biomarker to assess *Citrus* quality during postharvest (Table 3). Similarly, the monoterpene linalool, a key component of the aroma of papaya, was negatively affected by cold postharvest storage at 10 °C, and a concomitant downregulation of linalool synthase expression was also observed, suggesting that this volatile could be used as a marker to define papaya quality during postharvest storage [129] (Table 3). Linalool is also the predominant compound responsible for flavor in Muscat table grapes, a highly appreciated quality trait [161]. The postharvest storage of 'Shine Muscat' grapes at different temperatures between 0 and 10 °C showed that the decrease in linalool, profiled by HS-SPME–GC–MS, was enhanced at relatively low (0, 2 and 5 °C) temperatures in both fruit skin and flesh. Concomitant with linalool levels, grapes stored for four weeks at 10 °C presented a higher Muscat flavor than grapes stored at 0 °C for the same duration [130]. A possible effect of temperature on linalool synthesis or on the interconversion of free (aroma-producing) linalool and its glycosidically (odorless) bound form could be responsible for the observed differences in its concentration. In this sense, optimal storage at 10 °C for a short period or at relatively low temperatures followed by poststorage conditioning at 10 °C is fundamental for maintaining aroma quality for consumers [130] (Table 3).

The combination of 1-MCP with high O_2 or high CO_2 seemed to favor terpene content, as did the CO_2-enriched atmosphere in lemon [162]. CA storage under an elevated CO_2 atmosphere (8% CO_2 and 2%–3% O_2) could also promote terpene accumulation in mango. Indeed, fruit injury as a result of CA can enhance the activity of glycosidases, releasing monoterpenes, such as linalool or terpineol, from their glycoside-bound forms [163]. However, all tested CA treatments resulted in a reduction in total sesquiterpenes and also enhanced levels of ethanol, acetaldehyde and esters compared to those under atmospheric conditions. In addition, it was established that mangoes should not be stored under 3%–5% O_2 to avoid excess fermentative compound accumulation and maintain fruit aroma quality [163,164].

6. Conclusions

Although fruit responses to postharvest storage conditions are species and even cultivar dependent, making them especially complicated to study, metabolomic approaches alone or combined with transcriptomic/proteomic analyses are highly useful for understanding how metabolic changes affect quality traits. In particular, reconfiguration of fruit metabolism as a consequence of the abiotic/biotic stress encountered during postharvest storage conditions (cold, hypoxia, pathogens, etc.) has a direct impact on the accumulation of taste- and aroma-producing metabolites, which are decisive attributes for consumers and thus for the fruit industry. Even though many molecular mechanisms active during fruit postharvest storage and senescence remain elusive, future omic studies will shed light on them to optimize fruit storage conditions.

Furthermore, the recent advances in metabolomic-driven technology allows the identification of valuable biomarkers that can be employed by the fruit industry to tightly monitor changes in quality attributes during postharvest storage. In this sense, the use of multiplatform approaches offers the possibility to select a set of metabolite markers, which could better depict the impact of postharvest storage on aroma, taste, appearance and nutritional value [165].

Author Contributions: All the authors did the literature searching, writing and original draft preparation. All authors have read and agreed to the submitted version of the manuscript.

Funding: The authors want to acknowledge funding through grants RTI2018-099797-B-100 (Ministerio de Ciencia, Innovación y Universidades, Spain) and UMA18-FEDERJA-179 (FEDER-Junta Andalucía). In addition, we acknowledge partial funding by the University of Malaga (Plan Propio) and the European Union's H2020 Programme (GoodBerry; grant number 679303). D.P. acknowledges the support by the Spanish Ministry of Science and Innovation (BES-2013–062856).

Conflicts of Interest: The authors declare no conflict of interest.

References

1. Qin, G.; Meng, X.; Wang, Q.; Tian, S. Oxidative damage of mitochondrial proteins contributes to fruit senescence: A redox proteomics analysis. *J. Proteome Res.* **2009**, *8*, 2449–2462. [CrossRef] [PubMed]
2. Ornelas-Paz, J.D.J.; Yahia, E.M.; Ramírez-Bustamante, N.; Pérez-Martínez, J.D.; Escalante-Minakata, M.D.P.; Ibarra-Junquera, V.; Acosta-Muñiz, C.; Guerrero-Prieto, V.; Ochoa-Reyes, E. Physical attributes and chemical composition of organic strawberry fruit (*Fragaria x ananassa* Duch, *Cv.* Albion) at six stages of ripening. *Food Chem.* **2013**, *138*, 372–381. [CrossRef] [PubMed]
3. Giovannoni, J. Genetic regulation of fruit development and ripening. *Plant Cell* **2004**, *16*, 170–181. [CrossRef] [PubMed]
4. Zhang, J.; Cheng, D.; Wang, B.; Khan, I.; Ni, Y. Ethylene Control Technologies in Extending Postharvest Shelf Life of Climacteric Fruit. *J. Agric. Food Chem.* **2017**, *65*, 7308–7319. [CrossRef]
5. Yun, Z.; Jin, S.; Ding, Y.; Wang, Z.; Gao, H.; Pan, Z.; Xu, J.; Cheng, Y.; Deng, X. Comparative transcriptomics and proteomics analysis of citrus fruit, to improve understanding of the effect of low temperature on maintaining fruit quality during lengthy post-harvest storage. *J. Exp. Bot.* **2012**, *63*, 2873–2893. [CrossRef]
6. Yang, Z.; Cao, S.; Su, X.; Jiang, Y. Respiratory activity and mitochondrial membrane associated with fruit senescence in postharvest peaches in response to UV-C treatment. *Food Chem.* **2014**, *161*, 16–21. [CrossRef]
7. Lauxmann, M.A.; Brun, B.; Borsani, J.; Bustamante, C.A.; Budde, C.O.; Lara, M.V.; Drincovich, M.F. Transcriptomic Profiling during the Post-Harvest of Heat-Treated Dixiland Prunus persica Fruits: Common and Distinct Response to Heat and Cold. *PLoS ONE* **2012**, *7*, e51052. [CrossRef]
8. Cruz-Mendívil, A.; López-Valenzuela, J.A.; Calderón-Vázquez, C.L.; Vega-García, M.O.; Reyes-Moreno, C.; Valdez-Ortiz, A. Transcriptional changes associated with chilling tolerance and susceptibility in "Micro-Tom" tomato fruit using RNA-Seq. *Postharvest Biol. Technol.* **2015**, *99*, 141–151. [CrossRef]
9. Li, D.; Limwachiranon, J.; Li, L.; Du, R.; Luo, Z. Involvement of energy metabolism to chilling tolerance induced by hydrogen sulfide in cold-stored banana fruit. *Food Chem.* **2016**, *208*, 272–278. [CrossRef]
10. Aghdam, M.S.; Bodbodak, S. Postharvest Heat Treatment for Mitigation of Chilling Injury in Fruits and Vegetables. *Food Bioprocess Technol.* **2014**, *7*, 37–53. [CrossRef]
11. Brizzolara, S.; Santucci, C.; Tenori, L.; Hertog, M.; Nicolai, B.; Stürz, S.; Zanella, A.; Tonutti, P. A metabolomics approach to elucidate apple fruit responses to static and dynamic controlled atmosphere storage. *Postharvest Biol. Technol.* **2017**, *127*, 76–87. [CrossRef]
12. Luo, H.; Song, J.; Toivonen, P.; Gong, Y.; Forney, C.; Campbell Palmer, L.; Fillmore, S.; Pang, X.Q.; Zhang, Z.Q. Proteomic changes in 'Ambrosia' apple fruit during cold storage and in response to delayed cooling treatment. *Postharvest Biol. Technol.* **2018**, *137*, 66–76. [CrossRef]
13. Moreno, A.S.; Perotti, V.E.; Margarit, E.; Bello, F.; Vázquez, D.E.; Podestá, F.E.; Tripodi, K.E.J. Metabolic profiling and quality assessment during the postharvest of two tangor varieties subjected to heat treatments. *Postharvest Biol. Technol.* **2018**, *142*, 10–18. [CrossRef]
14. Perotti, V.E.; Del Vecchio, H.A.; Sansevich, A.; Meier, G.; Bello, F.; Cocco, M.; Garrán, S.M.; Anderson, C.; Vázquez, D.; Podestá, F.E. Proteomic, metabalomic, and biochemical analysis of heat treated Valencia oranges during storage. *Postharvest Biol. Technol.* **2011**, *62*, 97–114. [CrossRef]
15. Serek, M.; Tamari, G.; Sisler, E.C.; Borochov, A. Inhibition of ethylene-induced cellular senescence symptoms by 1-methylcyclopropene, a new inhibitor of ethylene action. *Physiol. Plant.* **1995**, *94*, 229–232. [CrossRef]
16. Pedreschi, R.; Lurie, S. Advances and current challenges in understanding postharvest abiotic stresses in perishables. *Postharvest Biol. Technol.* **2015**, *107*, 77–89. [CrossRef]
17. Obata, T.; Fernie, A.R. The use of metabolomics to dissect plant responses to abiotic stresses. *Cell. Mol. Life Sci.* **2012**, *69*, 3225–3243. [CrossRef]
18. Kosová, K.; Vítámvás, P.; Prášil, I.T.; Renaut, J. Plant proteome changes under abiotic stress—Contribution of proteomics studies to understanding plant stress response. *J. Proteomics* **2011**, *74*, 1301–1322. [CrossRef]

19. Bjerrum, J.T. *Metabonomics*; Springer: New York. NY, USA, 2015; ISBN 9781493923762.
20. Antonio, C. *Plant Metabolomics: Methods and Protocols*; Antonio, C., Ed.; Springer: New York, NY, USA, 2018; ISBN 9781493978182.
21. Begou, O.; Gika, H.G.; Wilson, I.D.; Theodoridis, G. Hyphenated MS-based targeted approaches in metabolomics. *Analyst* **2017**, *142*, 3079–3100. [CrossRef]
22. Azzi-Achkouty, S.; Estephan, N.; Ouaini, N.; Rutledge, D.N. Headspace solid-phase microextraction for wine volatile analysis. *Crit. Rev. Food Sci. Nutr.* **2017**, *57*, 2009–2020. [CrossRef]
23. Kroymann, J. Natural diversity and adaptation in plant secondary metabolism. *Curr. Opin. Plant Biol.* **2011**, *14*, 246–251. [CrossRef] [PubMed]
24. Moco, S.; Bino, R.J.; Vorst, O.; Verhoeven, H.A.; De Groot, J.; van Beek, T.A.; Vervoort, J.; de Vos, R.C.H. A Liquid Chromatography-Mass Spectrometry-Based Metabolome Database for Tomato. *Plant Physiol.* **2006**, *141*, 1205–1218. [CrossRef] [PubMed]
25. Ramautar, R.; Somsen, G.W.; de Jong, G.J. CE-MS for metabolomics: Developments and applications in the period 2010–2012. *Electrophoresis* **2013**, *34*, 86–98. [CrossRef] [PubMed]
26. Monton, M.R.N.; Soga, T. Metabolome analysis by capillary electrophoresis—Mass spectrometry. *J. Chromatogr. A* **2007**, *1168*, 237–246. [CrossRef]
27. Krishnan, P.; Kruger, N.J.; Ratcliffe, R.G. Metabolite fingerprinting and profiling in plants using NMR. *J. Exp. Bot.* **2005**, *56*, 255–265. [CrossRef]
28. Goulas, V.; Minas, I.S.; Kourdoulas, P.M.; Lazaridou, A.; Molassiotis, A.N.; Gerothanassis, I.P.; Manganaris, G.A. ^{1}H NMR Metabolic Fingerprinting to Probe Temporal Postharvest Changes on Qualitative Attributes and Phytochemical Profile of Sweet Cherry Fruit. *Front. Plant Sci.* **2015**, *6*, 1–11. [CrossRef]
29. Zhang, J.; Wang, X.; Yu, O.; Tang, J.; Gu, X.; Wan, X.; Fang, C. Metabolic profiling of strawberry (*Fragaria* × *ananassa* Duch.) during fruit development and maturation. *J. Exp. Bot.* **2011**, *62*, 1103–1118. [CrossRef]
30. Tieman, D.; Bliss, P.; McIntyre, L.M.; Blandon-Ubeda, A.; Bies, D.; Odabasi, A.Z.; Rodríguez, G.R.; Van Der Knaap, E.; Taylor, M.G.; Goulet, C.; et al. The chemical interactions underlying tomato flavor preferences. *Curr. Biol.* **2012**, *22*, 1035–1039. [CrossRef]
31. Sun, X.; Zhu, A.; Liu, S.; Sheng, L.; Ma, Q.; Zhang, L.; Nishawy, E.M.E.; Zeng, Y.; Xu, J.; Ma, Z.; et al. Integration of Metabolomics and Subcellular Organelle Expression Microarray to Increase Understanding the Organic Acid Changes in Post-harvest Citrus Fruit. *J. Integr. Plant Biol.* **2013**, *55*, 1038–1053. [CrossRef]
32. Leisso, R.S.; Hanrahan, I.; Mattheis, J.P.; Rudell, D.R. Controlled atmosphere storage, temperature conditioning, and antioxidant treatment alter postharvest 'honeycrisp' metabolism. *HortScience* **2017**, *52*, 423–431. [CrossRef]
33. Tang, N.; Deng, W.; Hu, N.; Chen, N.; Li, Z. Metabolite and transcriptomic analysis reveals metabolic and regulatory features associated with Powell orange pulp deterioration during room temperature and cold storage. *Postharvest Biol. Technol.* **2016**, *112*, 75–86. [CrossRef]
34. Osorio, S.; Carneiro, R.T.; Lytovchenko, A.; Mcquinn, R.; Sørensen, I.; Vallarino, J.G.; Giovannonni, J.J.; Fernie, A.R.; Rose, J.K.C. Genetic and metabolic effects of ripening mutations and vine detachment on tomato fruit quality. *Plant Biotechnol. J.* **2019**, *18*, 106–118. [CrossRef] [PubMed]
35. Kim, M.J.; Lee, M.Y.; Shon, J.C.; Kwon, Y.S.; Liu, K.-H.; Lee, C.H.; Ku, K.-M. Untargeted and targeted metabolomics analyses of blackberries—Understanding postharvest red drupelet disorder. *Food Chem.* **2019**, *300*, 125169. [CrossRef] [PubMed]
36. Yuan, Y.; Zhao, Y.; Yang, J.; Jiang, Y.; Lu, F.; Jia, Y.; Yang, B. Metabolomic analyses of banana during postharvest senescence by ^{1}H-high resolution-NMR. *Food Chem.* **2017**, *218*, 406–412. [CrossRef] [PubMed]
37. Salzano, A.M.; Renzone, G.; Sobolev, A.P.; Carbone, V.; Petriccione, M.; Capitani, D.; Vitale, M.; Novi, G.; Zambrano, N.; Pasquariello, M.S.; et al. Unveiling Kiwifruit Metabolite and Protein Changes in the Course of Postharvest Cold Storage. *Front. Plant Sci.* **2019**, *10*, 71. [CrossRef]
38. Sheen, J. Master Regulators in Plant Glucose Signaling Networks. *J. Plant Biol.* **2014**, *57*, 67–79. [CrossRef]
39. Kim, J. Sugar metabolism as input signals and fuel for leaf senescence. *Genes and Genomics* **2019**, *41*, 737–746. [CrossRef]
40. Jia, H.; Wang, Y.; Sun, M.; Li, B.; Han, Y.; Zhao, Y.; Li, X.; Ding, N.; Li, C.; Ji, W.; et al. Sucrose functions as a signal involved in the regulation of strawberry fruit development and ripening. *New Phytol.* **2013**, *198*, 453–465. [CrossRef]

41. Liu, Y.H.; Offler, C.E.; Ruan, Y.L. Regulation of fruit and seed response to heat and drought by sugars as nutrients and signals. *Front. Plant Sci.* **2013**, *4*, 1–13. [CrossRef]
42. Li, D.; Mou, W.; Wang, Y.; Li, L.; Mao, L.; Ying, T.; Luo, Z. Exogenous sucrose treatment accelerates postharvest tomato fruit ripening through the influence on its metabolism and enhancing ethylene biosynthesis and signaling. *Acta Physiol. Plant.* **2016**, *38*, 225. [CrossRef]
43. Jia, H.; Jiu, S.; Zhang, C.; Wang, C.; Tariq, P.; Liu, Z.; Wang, B.; Cui, L.; Fang, J. Abscisic acid and sucrose regulate tomato and strawberry fruit ripening through the abscisic acid-stress-ripening transcription factor. *Plant Biotechnol. J.* **2016**, *14*, 2045–2065. [CrossRef] [PubMed]
44. Wingler, A.; Roitsch, T. Metabolic regulation of leaf senescence: Interactions of sugar signalling with biotic and abiotic stress responses. *Plant Biol.* **2008**, *10*, 50–62. [CrossRef] [PubMed]
45. Vallarino, J.G.; Yeats, T.H.; Maximova, E.; Rose, J.K.; Fernie, A.R.; Osorio, S. Postharvest changes in LIN5-down-regulated plants suggest a role for sugar deficiency in cuticle metabolism during ripening. *Phytochemistry* **2017**, *142*, 11–20. [CrossRef] [PubMed]
46. Farcuh, M.; Rivero, R.M.; Sadka, A.; Blumwald, E. Ethylene regulation of sugar metabolism in climacteric and non-climacteric plums. *Postharvest Biol. Technol.* **2018**, *139*, 20–30. [CrossRef]
47. Uarrota, V.G.; Fuentealba, C.; Hernández, I.; Defilippi-Bruzzone, B.; Meneses, C.; Campos-Vargas, R.; Lurie, S.; Hertog, M.; Carpentier, S.; Poblete-Echeverría, C.; et al. Integration of proteomics and metabolomics data of early and middle season Hass avocados under heat treatment. *Food Chem.* **2019**, *289*, 512–521. [CrossRef] [PubMed]
48. Keunen, E.; Peshev, D.; Vangronsveld, J.; Van den Ende, W.; Cuypers, A. Plant sugars are crucial players in the oxidative challenge during abiotic stress. Extending the traditional concept. *Plant Cell Environ.* **2013**, *36*, 1242–1255. [CrossRef]
49. Davik, J.; Koehler, G.; From, B.; Torp, T.; Rohloff, J.; Eidem, P.; Wilson, R.C.; Sønsteby, A.; Randall, S.K.; Alsheikh, M. Dehydrin, alcohol dehydrogenase, and central metabolite levels are associated with cold tolerance in diploid strawberry (*Fragaria* spp.). *Planta* **2013**, *237*, 265–277. [CrossRef]
50. Nishizawa, A.; Yabuta, Y.; Shigeoka, S. Galactinol and Raffinose Constitute a Novel Function to Protect Plants from Oxidative Damage. *Plant Physiol.* **2008**, *147*, 1251–1263. [CrossRef]
51. Korn, M.; Gärtner, T.; Erban, A.; Kopka, J.; Selbig, J.; Hincha, D.K. Predicting Arabidopsis freezing tolerance and heterosis in freezing tolerance from metabolite composition. *Mol. Plant* **2010**, *3*, 224–235. [CrossRef]
52. Lauxmann, M.A.; Borsani, J.; Osorio, S.; Lombardo, V.A.; Budde, C.O.; Bustamante, C.A.; Monti, L.L.; Andreo, C.S.; Fernie, A.R.; Drincovich, M.F.; et al. Deciphering the metabolic pathways influencing heat and cold responses during post-harvest physiology of peach fruit. *Plant Cell Environ.* **2014**, *37*, 601–616. [CrossRef]
53. Wu, Q.; Zhou, Y.; Zhang, Z.; Li, T.; Jiang, Y.; Gao, H.; Yun, Z. Effect of blue light on primary metabolite and volatile compound profiling in the peel of red pitaya. *Postharvest Biol. Technol.* **2020**, *160*. [CrossRef]
54. Xu, J.; Zhang, Y.; Qi, D.; Huo, H.; Dong, X.; Tian, L.; Zhang, X.; Liu, C.; Cao, Y. Postharvest metabolomic changes in Pyrus ussuriensis Maxim. wild accession „Zaoshu Shanli. *J. Sep. Sci.* **2018**, *41*, 4001–4013. [CrossRef] [PubMed]
55. Mack, C.; Wefers, D.; Schuster, P.; Weinert, C.H.; Egert, B.; Bliedung, S.; Trierweiler, B.; Muhle-goll, C.; Bunzel, M.; Luy, B.; et al. Untargeted multi-platform analysis of the metabolome and the non-starch polysaccharides of kiwifruit during postharvest ripening. *Postharvest Biol. Technol.* **2017**, *125*, 65–76. [CrossRef]
56. Bustamante, C.A.; Monti, L.L.; Gabilondo, J.; Scossa, F.; Valentini, G.; Budde, C.O.; Lara, M.V.; Fernie, A.R.; Drincovich, M.F. Differential Metabolic Rearrangements after Cold Storage Are Correlated with Chilling Injury Resistance of Peach Fruits. *Front. Plant Sci.* **2016**, *7*, 1–15. [CrossRef] [PubMed]
57. Centeno, D.C.; Osorio, S.; Nunes-Nesi, A.; Bertolo, A.L.F.; Carneiro, R.T.; Araújo, W.L.; Steinhauser, M.-C.; Michalska, J.; Rohrmann, J.; Geigenberger, P.; et al. Malate Plays a Crucial Role in Starch Metabolism, Ripening, and Soluble Solid Content of Tomato Fruit and Affects Postharvest Softening. *Plant Cell* **2011**, *23*, 162–184. [CrossRef] [PubMed]
58. López, M.G.; Zanor, M.I.; Pratta, G.R.; Stegmayer, G.; Boggio, S.B.; Conte, M.; Bermúdez, L.; Coluccio Leskow, C.; Rodríguez, G.R.; Picardi, L.A.; et al. Metabolic analyses of interspecific tomato recombinant inbred lines for fruit quality improvement. *Metabolomics* **2015**, *11*, 1416–1431. [CrossRef]

59. Oms-Oliu, G.; Hertog, M.L.A.T.M.; Van de Poel, B.; Ampofo-Asiama, J.; Geeraerd, A.H.; Nicolai, B.M. Metabolic characterization of tomato fruit during preharvest development, ripening, and postharvest shelf-life. *Postharvest Biol. Technol.* **2011**, *62*, 7–16. [CrossRef]
60. Vazquez-Hernandez, M.; Navarro, S.; Sanchez-Ballesta, M.T.; Merodio, C.; Escribano, M.I. Short-term high CO2 treatment reduces water loss and decay by modulating defense proteins and organic osmolytes in Cardinal table grape after cold storage and shelf-life. *Sci. Hortic. (Amst.)* **2018**, *234*, 27–35. [CrossRef]
61. Palma, F.; Carvajal, F.; Jiménez-Muñoz, R.; Pulido, A.; Jamilena, M.; Garrido, D. Exogenous γ-aminobutyric acid treatment improves the cold tolerance of zucchini fruit during postharvest storage. *Plant Physiol. Biochem.* **2019**, *136*, 188–195. [CrossRef]
62. Pedreschi, R.; Franck, C.; Lammertyn, J.; Erban, A.; Kopka, J.; Hertog, M.; Verlinden, B.; Nicolaï, B. Metabolic profiling of ' Conference ' pears under low oxygen stress. *Postharvest* **2009**, *51*, 123–130. [CrossRef]
63. DeBolt, S.; Cook, D.R.; Ford, C.M. L-tartaric acid synthesis from vitamin C in higher plants. *Proc. Natl. Acad. Sci. USA* **2006**, *103*, 5608–5613. [CrossRef]
64. Hatoum, D.; Annaratone, C.; Hertog, M.L.A.T.M.; Geeraerd, A.H.; Nicolai, B.M. Targeted metabolomics study of "Braeburn" apples during long-term storage. *Postharvest Biol. Technol.* **2014**, *96*, 33–41. [CrossRef]
65. Lara, M.V.; Borsani, J.; Budde, C.O.; Lauxmann, M.A.; Lombardo, V.A.; Murray, R.; Andreo, C.S.; Drincovich, M.F. Biochemical and proteomic analysis of "Dixiland" peach fruit (*Prunus persica*) upon heat treatment. *J. Exp. Bot.* **2009**, *60*, 4315–4333. [CrossRef] [PubMed]
66. Ding, Y.; Chang, J.; Ma, Q.; Chen, L.; Liu, S.; Jin, S.; Han, J. Network Analysis of Postharvest Senescence Process in Citrus Fruits Revealed by Transcriptomic and Metabolomic Profiling. *Plant Physiol.* **2015**, *168*, 357–376. [CrossRef] [PubMed]
67. Degu, A.; Hatew, B.; Nunes-Nesi, A.; Shlizerman, L.; Zur, N.; Katz, E.; Fernie, A.R.; Blumwald, E.; Sadka, A. Inhibition of aconitase in citrus fruit callus results in a metabolic shift towards amino acid biosynthesis. *Planta* **2011**, *234*, 501–513. [CrossRef] [PubMed]
68. Tretter, L.; Adam-Vizi, V. Inhibition of Krebs Cycle Enzymes by Hydrogen Peroxide: A Key Role of alpha-Ketoglutarate Dehydrogenase in Limiting NADH Production under Oxidative Stress. *J. Neurosci.* **2000**, *20*, 8972–8979. [CrossRef]
69. Nulton-Persson, A.C.; Szweda, L.I. Modulation of Mitochondrial Function by Hydrogen Peroxide. *J. Biol. Chem.* **2001**, *276*, 23357–23361. [CrossRef]
70. Nath, K.A.; Ngo, E.; Hebbel, R.P.; Croatt, J.; Zhou, B.; Nutter, L.M.; Hebbel, P.; Croatt, J.; Zhou, B.; Nutter, M. Alpha-Ketoacids scavenge H202 in vitro and in vivo and reduce menadione-induced DNA injury and cytotoxicity. *Am. J. Physiol.* **1995**, *268*, 227–236. [CrossRef]
71. Varma, S.D.; Devamanoharan, P.S.; Morris, S.M. Prevention of Cataracts by Nutritional and Metabolic Antioxidants. *Crit. Rev. Food Sci. Nutr.* **1995**, *35*, 111–129. [CrossRef]
72. Araujo, W.L.; Nunes-Nesi, a.; Osorio, S.; Usadel, B.; Fuentes, D.; Nagy, R.; Balbo, I.; Lehmann, M.; Studart-Witkowski, C.; Tohge, T.; et al. Antisense Inhibition of the Iron-Sulphur Subunit of Succinate Dehydrogenase Enhances Photosynthesis and Growth in Tomato via an Organic Acid-Mediated Effect on Stomatal Aperture. *Plant Cell* **2011**, *23*, 600–627. [CrossRef]
73. Zhang, L.; Yu, Z.; Jiang, L.; Jiang, J.; Luo, H.; Fu, L. Effect of post-harvest heat treatment on proteome change of peach fruit during ripening. *J. Proteom.* **2011**, *74*, 1135–1149. [CrossRef] [PubMed]
74. De los Ángeles Juricic, M.; Berríos-Cárcamo, P.A.; Acevedo, M.L.; Israel, Y.; Almodóvar, I.; Cassels, B.K. Salsolinol and isosalsolinol: Condensation products of acetaldehyde and dopamine. Separation of their enantiomers in the presence of a large excess of dopamine. *J. Pharm. Biomed. Anal.* **2012**, *63*, 170–174. [CrossRef] [PubMed]
75. Luengwilai, K.; Beckles, D.M.; Roessner, U.; Dias, D.A.; Lui, V.; Siriphanich, J. Identification of physiological changes and key metabolites coincident with postharvest internal browning of pineapple (*Ananas comosus* L.) fruit. *Postharvest Biol. Technol.* **2018**, *137*, 56–65. [CrossRef]
76. Szabados, L.; Savouré, A. Proline: A multifunctional amino acid. *Trends Plant Sci.* **2010**, *15*, 89–97. [CrossRef]
77. Guy, C.; Kaplan, F.; Kopka, J.; Selbig, J.; Hincha, D.K. Metabolomics of temperature stress. *Physiol. Plant.* **2008**, *132*, 220–235. [CrossRef]
78. Zhang, Z.; Zhu, Q.; Hu, M.; Gao, Z.; An, F.; Li, M.; Jiang, Y. Low-temperature conditioning induces chilling tolerance in stored mango fruit. *Food Chem.* **2017**, *219*, 76–84. [CrossRef]

79. Luo, Z.; Li, D.; Du, R.; Mou, W. Hydrogen sulfide alleviates chilling injury of banana fruit by enhanced antioxidant system and proline content. *Sci. Hortic. (Amst.)* **2015**, *183*, 144–151. [CrossRef]
80. Li, D.; Cheng, Y.; Dong, Y.; Shang, Z.; Guan, J. Effects of low temperature conditioning on fruit quality and peel browning spot in 'Huangguan' pears during cold storage. *Postharvest Biol. Technol.* **2017**, *131*, 68–73. [CrossRef]
81. Zhang, Y.; Jin, P.; Huang, Y.; Shan, T.; Wang, L.; Li, Y.; Zheng, Y. Effect of hot water combined with glycine betaine alleviates chilling injury in cold-stored loquat fruit. *Postharvest Biol. Technol.* **2016**, *118*, 141–147. [CrossRef]
82. Rojas, C.M.; Senthil-Kumar, M.; Tzin, V.; Mysore, K.S. Regulation of primary plant metabolism during plant-pathogen interactions and its contribution to plant defense. *Front. Plant Sci.* **2014**, *5*, 1–12. [CrossRef]
83. Bang, J.; Lim, S.; Yi, G.; Lee, J.G.; Lee, E.J. Integrated transcriptomic-metabolomic analysis reveals cellular responses of harvested strawberry fruit subjected to short-term exposure to high levels of carbon dioxide. *Postharvest Biol. Technol.* **2019**, *148*, 120–131. [CrossRef]
84. Pott, D.M.; Osorio, S.; Vallarino, J.G. From Central to Specialized Metabolism: An Overview of Some Secondary Compounds Derived From the Primary Metabolism for Their Role in Conferring Nutritional and Organoleptic Characteristics to Fruit. *Front. Plant Sci.* **2019**, *10*, 835. [CrossRef] [PubMed]
85. Zenoni, S.; Fasoli, M.; Guzzo, F.; Dal Santo, S.; Amato, A.; Anesi, A.; Commisso, M.; Herderich, M.; Ceoldo, S.; Avesani, L.; et al. Disclosing the Molecular Basis of the Postharvest Life of Berry in Different Grapevine Genotypes. *Plant Physiol.* **2016**, *172*, 1821–1843. [CrossRef] [PubMed]
86. Zamboni, A.; Di Carli, M.; Guzzo, F.; Stocchero, M.; Zenoni, S.; Ferrarini, A.; Tononi, P.; Toffali, K.; Desiderio, A.; Lilley, K.S.; et al. Identification of Putative Stage-Specific Grapevine Berry Biomarkers and Omics Data Integration into Networks. *Plant Physiol.* **2010**, *154*, 1439–1459. [CrossRef]
87. Maoz, I.; De Rosso, M.; Kaplunov, T.; Vedova, A.D.; Sela, N.; Flamini, R.; Lewinsohn, E.; Lichter, A. Metabolomic and transcriptomic changes underlying cold and anaerobic stresses after storage of table grapes. *Sci. Rep.* **2019**, *9*, 1–14. [CrossRef]
88. Li, D.; Zhang, X.; Li, L.; Aghdam, M.S.; Wei, X.; Liu, J.; Xu, Y.; Luo, Z. Elevated CO_2 delayed the chlorophyll degradation and anthocyanin accumulation in postharvest strawberry fruit. *Food Chem.* **2019**, *285*, 163–170. [CrossRef]
89. Carmona, L.; Alquézar, B.; Marques, V.V.; Peña, L. Anthocyanin biosynthesis and accumulation in blood oranges during postharvest storage at different low temperatures. *Food Chem.* **2017**, *237*, 7–14. [CrossRef]
90. Wang, L.; Sang, W.; Xu, R.; Cao, J. Alteration of flesh color and enhancement of bioactive substances via the stimulation of anthocyanin biosynthesis in 'Friar' plum fruit by low temperature and the removal. *Food Chem.* **2020**, *310*, 125862. [CrossRef]
91. Petric, T.; Kiferle, C.; Perata, P.; Gonzali, S. Optimizing shelf life conditions for anthocyanin-rich tomatoes. *PLoS ONE* **2018**, *13*, e0205650. [CrossRef]
92. Zhang, Y.; Butelli, E.; De Stefano, R.; Schoonbeek, H.J.; Magusin, A.; Pagliarani, C.; Wellner, N.; Hill, L.; Orzaez, D.; Granell, A.; et al. Anthocyanins double the shelf life of tomatoes by delaying overripening and reducing susceptibility to gray mold. *Curr. Biol.* **2013**, *23*, 1094–1100. [CrossRef]
93. Yun, Z.; Gao, H.; Liu, P.; Liu, S.; Luo, T.; Jin, S.; Xu, Q.; Xu, J.; Cheng, Y. Comparative proteomic and metabolomic profiling of citrus fruit with enhancement of disease resistance by postharvest heat treatment. *BMC Plant Biol.* **2013**, *13*, 44. [CrossRef] [PubMed]
94. Sun, J.; Janisiewicz, W.J.; Nichols, B.; Jurick, W.M.; Chen, P. Composition of phenolic compounds in wild apple with multiple resistance mechanisms against postharvest blue mold decay. *Postharvest Biol. Technol.* **2017**, *127*, 68–75. [CrossRef]
95. Monribot-Villanueva, J.L.; Elizalde-Contreras, J.M.; Aluja, M.; Segura-Cabrera, A.; Birke, A.; Guerrero-Analco, J.A.; Ruiz-May, E. Endorsing and extending the repertory of nutraceutical and antioxidant sources in mangoes during postharvest shelf life. *Food Chem.* **2019**, *285*, 119–129. [CrossRef] [PubMed]
96. Carmona, L.; Zacarías, L.; Rodrigo, M.J. Stimulation of coloration and carotenoid biosynthesis during postharvest storage of "Navelina" orange fruit at 12 °C. *Postharvest Biol. Technol.* **2012**, *74*, 108–117. [CrossRef]
97. Javanmardi, J.; Kubota, C. Variation of lycopene, antioxidant activity, total soluble solids and weight loss of tomato during postharvest storage. *Postharvest Biol. Technol.* **2006**, *41*, 151–155. [CrossRef]

98. Lado, J.; Gurrea, A.; Zacarías, L.; Rodrigo, M.J. Influence of the storage temperature on volatile emission, carotenoid content and chilling injury development in Star Ruby red grapefruit. *Food Chem.* **2019**, *295*, 72–81. [CrossRef]
99. Lado, J.; Rodrigo, M.J.; Cronje, P.; Zacarías, L. Involvement of lycopene in the induction of tolerance to chilling injury in grapefruit. *Postharvest Biol. Technol.* **2015**, *100*, 176–186. [CrossRef]
100. Zhang, M.K.; Zhang, M.P.; Mazourek, M.; Tadmor, Y.; Li, L. Regulatory control of carotenoid accumulation in winter squash during storage. *Planta* **2014**, *240*, 1063–1074. [CrossRef]
101. Wei, F.; Fu, M.; Li, J.; Yang, X.; Chen, Q.; Tian, S. Chlorine dioxide delays the reddening of postharvest green peppers by affecting the chlorophyll degradation and carotenoid synthesis pathways. *Postharvest Biol. Technol.* **2019**, *156*, 110939. [CrossRef]
102. Kaewsuksaeng, S.; Tatmala, N.; Srilaong, V.; Pongprasert, N. Postharvest heat treatment delays chlorophyll degradation and maintains quality in Thai lime (*Citrus aurantifolia* Swingle cv. Paan) fruit. *Postharvest Biol. Technol.* **2015**, *100*, 1–7. [CrossRef]
103. Zhang, H.; Wang, R.; Wang, T.; Fang, C.; Wang, J. Methyl salicylate delays peel yellowing of 'Zaosu' pear (*Pyrus bretschneideri*) during storage by regulating chlorophyll metabolism and maintaining chloroplast ultrastructure. *J. Sci. Food Agric.* **2019**, *99*, 4816–4824. [CrossRef] [PubMed]
104. Vallarino, J.G.; Erban, A.; Fehrle, I.; Fernie, A.R.; Kopka, J.; Osorio, S. Acquisition of Volatiles Compounds by Gas Chromatography–Mass Spectrometry (GC-MS). In *Plant Metabolomics: Methods and Protocols*; Antonio, C., Ed.; Humana Press: New York, NY, USA, 2018; Volume 1778, pp. 225–239. ISBN 9781493978199.
105. Zhang, B.; Tieman, D.M.; Jiao, C.; Xu, Y.; Chen, K.; Fe, Z.; Giovannoni, J.J.; Klee, H.J. Chilling-induced tomato flavor loss is associated with altered volatile synthesis and transient changes in DNA methylation. *Proc. Natl. Acad. Sci. USA* **2016**, *113*, 12580–12585. [CrossRef] [PubMed]
106. Aith Barbará, J.; Primieri Nicolli, K.; Souza-Silva, É.A.; Camarão Telles Biasoto, A.; Welke, J.E.; Alcaraz Zini, C. Volatile profile and aroma potential of tropical Syrah wines elaborated in different maturation and maceration times using comprehensive two-dimensional gas chromatography and olfactometry. *Food Chem.* **2020**, *308*, 125552. [CrossRef] [PubMed]
107. Uekane, T.M.; Nicolotti, L.; Griglione, A.; Bizzo, H.R.; Rubiolo, P.; Bicchi, C.; Rocha-Leão, M.H.M.; Rezende, C.M. Studies on the volatile fraction composition of three native Amazonian-Brazilian fruits: Murici (*Byrsonima crassifolia* L., Malpighiaceae), bacuri (*Platonia insignis* M., Clusiaceae), and sapodilla (*Manilkara sapota* L., Sapotaceae). *Food Chem.* **2017**, *219*, 13–22. [CrossRef]
108. Tietel, Z.; Plotto, A.; Fallik, E.; Lewinsohn, E.; Porat, R. Taste and aroma of fresh and stored mandarins. *J. Sci. Food Agric.* **2011**, *91*, 14–23. [CrossRef] [PubMed]
109. Tietel, Z.; Porat, R.; Weiss, K.; Ulrich, D. Identification of aroma-active compounds in fresh and stored "Mor" mandarins. *Int. J. Food Sci. Technol.* **2011**, *46*, 2225–2231. [CrossRef]
110. Ayala-Zavala, J.F.; Wang, S.Y.; Wang, C.Y.; González-Aguilar, G.A. Effect of storage temperatures on antioxidant capacity and aroma compounds in strawberry fruit. *LWT Food Sci. Technol.* **2004**, *37*, 687–695. [CrossRef]
111. Günther, C.S.; Marsh, K.B.; Winz, R.A.; Harker, R.F.; Wohlers, M.W.; White, A.; Goddard, M.R. The impact of cold storage and ethylene on volatile ester production and aroma perception in "Hort16A" kiwifruit. *Food Chem.* **2015**, *169*, 5–12. [CrossRef]
112. Díaz de León-Sánchez, F.; Pelayo-Zaldívar, C.; Rivera-Cabrera, F.; Ponce-Valadez, M.; Ávila-Alejandre, X.; Fernández, F.J.; Escalona-Buendía, H.B.; Pérez-Flores, L.J. Effect of refrigerated storage on aroma and alcohol dehydrogenase activity in tomato fruit. *Postharvest Biol. Technol.* **2009**, *54*, 93–100. [CrossRef]
113. Zhang, B.; Xi, W.P.; Wei, W.W.; Shen, J.; Ferguson, I.; Chen, K. Changes in aroma-related volatiles and gene expression during low temperature storage and subsequent shelf-life of peach fruit. *Postharvest Biol. Technol.* **2011**, *60*, 7–16. [CrossRef]
114. Ponce-Valadez, M.; Escalona-Buendía, H.B.; Villa-Hernández, J.M.; de León-Sánchez, F.D.; Rivera-Cabrera, F.; Alia-Tejacal, I.; Pérez-Flores, L.J. Effect of refrigerated storage (12.5 °C) on tomato (*Solanum lycopersicum*) fruit flavor: A biochemical and sensory analysis. *Postharvest Biol. Technol.* **2016**, *111*, 6–14. [CrossRef]
115. Tietel, Z.; Feldmesser, E.; Lewinsohn, E.; Fallik, E.; Porat, R. Changes in the transcriptome of "Mor" mandarin flesh during storage: Emphasis on molecular regulation of fruit flavor deterioration. *J. Agric. Food Chem.* **2011**, *59*, 3819–3827. [CrossRef] [PubMed]

116. Ponce-Valadez, M.; Fellman, S.M.; Giovannoni, J.; Gan, S.S.; Watkins, C.B. Differential fruit gene expression in two strawberry cultivars in response to elevated CO_2 during storage revealed by a heterologous fruit microarray approach. *Postharvest Biol. Technol.* **2009**, *51*, 131–140. [CrossRef]
117. Almenar, E.; Hernández-Muñoz, P.; Lagarón, J.M.; Catalá, R.; Gavara, R. Controlled atmosphere storage of wild strawberry fruit (*Fragaria vesca* L.). *J. Agric. Food Chem.* **2006**, *54*, 86–91. [CrossRef]
118. Ke, D.; Goldstein, L.; O'Mahony, M.; Kader, A.A. Effects of Short-term Exposure to Low O_2 and High CO_2 Atmospheres on Quality Attributes of Strawberries. *J. Food Sci.* **1991**, *56*, 50–54. [CrossRef]
119. Larsen, M.; Watkins, C.B. Firmness and concentrations of acetaldehyde, ethyl acetate and ethanol in strawberries stored in controlled and modified atmospheres. *Postharvest Biol. Technol.* **1995**, *5*, 39–50. [CrossRef]
120. Mellidou, I.; Buts, K.; Hatoum, D.; Ho, Q.T.; Johnston, J.W.; Watkins, C.B.; Schaffer, R.J.; Gapper, N.E.; Giovannoni, J.J.; Rudell, D.R.; et al. Transcriptomic events associated withinternal browning of apple during postharvest storage. *BMC Plant Biol.* **2014**, *14*, 1–17. [CrossRef]
121. Wu, X.; Jiang, L.; Yu, M.; An, X.; Ma, R.; Yu, Z. Proteomic analysis of changes in mitochondrial protein expression during peach fruit ripening and senescence. *J. Proteomics* **2016**, *147*, 197–211. [CrossRef]
122. Tadege, M.; Dupuis, I.; Kuhlemeier, C. Ethanolic fermentation: New functions for an old pathway. *Trends Plant Sci.* **1999**, *4*, 320–325. [CrossRef]
123. Blanch, M.; Rosales, R.; Palma, F.; Sanchez-Ballesta, M.T.; Escribano, M.I.; Merodio, C. CO_2-driven changes in energy and fermentative metabolism in harvested strawberries. *Postharvest Biol. Technol.* **2015**, *110*, 33–39. [CrossRef]
124. Pérez, A.G.; Sanz, C. Effect of High-Oxygen and High-Carbon-Dioxide Atmospheres on Strawberry Flavor and Other Quality Traits. *J. Agric. Food Chem.* **2001**, *49*, 2370–2375. [CrossRef] [PubMed]
125. Obenland, D.M.; Margosan, D.A.; Houck, L.G.; Aung, L.H. Essential Oils and Chilling Injury in Lemon. *HortScience* **1997**, *32*, 108–111. [CrossRef]
126. Obenland, D.; Collin, S.; Sievert, J.; Arpaia, M.L. Mandarin flavor and aroma volatile composition are strongly influenced by holding temperature. *Postharvest Biol. Technol.* **2013**, *82*, 6–14. [CrossRef]
127. Biolatto, A.; Vazquez, D.E.; Sancho, A.M.; Carduza, F.J.; Pensel, N.A. Effect of commercial conditioning and cold quarantine storage treatments on fruit quality of "Rouge La Toma" grapefruit (*Citrus paradisi* Macf.). *Postharvest Biol. Technol.* **2005**, *35*, 167–176. [CrossRef]
128. Yuen, C.M.C.; Tridjaja, N.O.; Wills, R.B.H.; Wild, B.L. Chilling injury development of 'Tahitian' lime, 'Emperor' mandarin, 'Marsh' grapefruit and 'Valencia' orange. *J. Sci. Food Agric.* **1995**, *67*, 335–339. [CrossRef]
129. Gomes, B.L.; Fabi, J.P.; Purgatto, E. Cold storage affects the volatile profile and expression of a putative linalool synthase of papaya fruit. *Food Res. Int.* **2016**, *89*, 654–660. [CrossRef]
130. Matsumoto, H.; Ikoma, Y. Effect of postharvest temperature on the muscat flavor and aroma volatile content in the berries of "Shine Muscat" (*Vitis labruscana* Baily × V. *vinifera* L.). *Postharvest Biol. Technol.* **2016**, *112*, 256–265. [CrossRef]
131. Blanch, M.; Rosales, R.; Mateos, R.; Perez-Gago, M.B.; Sanchez-Ballesta, M.T.; Escribano, M.I.; Merodio, C. Effects of high CO_2 levels on fermentation, peroxidation, and cellular water stress in Fragaria vesca stored at low temperature in conditions of unlimited O_2. *J. Agric. Food Chem.* **2015**, *63*, 761–768. [CrossRef]
132. Wright, A.H.; Delong, J.M.; Arul, J.; Prange, R.K. The trend toward lower oxygen levels during apple (*Malus × domestica* Borkh) storage—A review. *J. Hortic. Sci. Biotechnol.* **2015**, *90*, 1–13. [CrossRef]
133. Veltman, R.H.; Lenthéric, I.; Van Der Plas, L.H.W.; Peppelenbos, H.W. Internal browning in pear fruit (*Pyrus communis* L. cv Conference) may be a result of a limited availability of energy and antioxidants. *Postharvest Biol. Technol.* **2003**, *28*, 295–302. [CrossRef]
134. Biais, B.; Allwood, J.W.; Deborde, C.; Xu, Y.; Maucourt, M.; Beauvoit, B.; Dunn, W.B.; Jacob, D.; Goodacre, R.; Rolin, D.; et al. 1H NMR, GC-EI-TOFMS, and data set correlation for fruit metabolomics: Application to spatial metabolite analysis in melon. *Anal. Chem.* **2009**, *81*, 2884–2894. [CrossRef] [PubMed]
135. Aprea, E.; Gika, H.; Carlin, S.; Theodoridis, G.; Vrhovsek, U.; Mattivi, F. Metabolite profiling on apple volatile content based on solid phase microextraction and gas-chromatography time of flight mass spectrometry. *J. Chromatogr. A* **2011**, *1218*, 4517–4524. [CrossRef] [PubMed]
136. Thewes, F.R.; Brackmann, A.; de Oliveira Anese, R.; Bronzatto, E.S.; Schultz, E.E.; Wagner, R. Dynamic controlled atmosphere storage suppresses metabolism and enhances volatile concentrations of 'Galaxy' apple harvested at three maturity stages. *Postharvest Biol. Technol.* **2017**, *127*, 1–13. [CrossRef]

137. López, M.L.; Villatoro, C.; Fuentes, T.; Graell, J.; Lara, I.; Echeverría, G. Volatile compounds, quality parameters and consumer acceptance of "Pink Lady®" apples stored in different conditions. *Postharvest Biol. Technol.* **2007**, *43*, 55–66. [CrossRef]
138. Shi, F.; Zhou, X.; Zhou, Q.; Tan, Z.; Yao, M.M.; Wei, B.D.; Ji, S.J. Membrane lipid metabolism changes and aroma ester loss in low-temperature stored Nanguo pears. *Food Chem.* **2018**, *245*, 446–453. [CrossRef]
139. Lama, K.; Alkalai-Tuvia, S.; Perzelan, Y.; Fallik, E. Nutritional qualities and aroma volatiles of harvested red pepper fruits stored at suboptimal temperatures. *Sci. Hortic. (Amst.)* **2016**, *213*, 42–48. [CrossRef]
140. Fellman, J.K.; Miller, T.W.; Mattinson, D.S.; Mattheis, J.P. Factors that influence biosynthesis of volatile flavor compounds in apple fruits. *HortScience* **2000**, *35*, 1026–1033. [CrossRef]
141. Bai, J.; Baldwin, E.A.; Imahori, Y.; Kostenyuk, I.; Burns, J.; Brecht, J.K. Chilling and heating may regulate C6 volatile aroma production by different mechanisms in tomato (*Solanum lycopersicum*) fruit. *Postharvest Biol. Technol.* **2011**, *60*, 111–120. [CrossRef]
142. Shi, F.; Zhou, X.; Yao, M.M.; Zhou, Q.; Ji, S.J.; Wang, Y. Low-temperature stress-induced aroma loss by regulating fatty acid metabolism pathway in 'Nanguo' pear. *Food Chem.* **2019**, *297*, 124927. [CrossRef]
143. Zhou, D.; Sun, Y.; Li, M.; Zhu, T.; Tu, K. Postharvest hot air and UV-C treatments enhance aroma-related volatiles by simulating the lipoxygenase pathway in peaches during cold storage. *Food Chem.* **2019**, *292*, 294–303. [CrossRef]
144. Wang, L.; Baldwin, E.; Luo, W.; Zhao, W.; Brecht, J.; Bai, J. Key tomato volatile compounds during postharvest ripening in response to chilling and pre-chilling heat treatments. *Postharvest Biol. Technol.* **2019**, *154*, 11–20. [CrossRef]
145. Liu, W.W.; Qi, H.Y.; Xu, B.H.; Li, Y.; Tian, X.B.; Jiang, Y.Y.; Xu, X.F.; Lv, D.Q. Ethanol treatment inhibits internal ethylene concentrations and enhances ethyl ester production during storage of oriental sweet melons (*Cucumis melo* var. *makuwa* Makino). *Postharvest Biol. Technol.* **2012**, *67*, 75–83. [CrossRef]
146. Wang, L.; Luo, Z.; Li, J.; Yang, M.; Yan, J.; Lu, H.; Li, D.; Chen, C.; Aghdam, M.S.; Wu, B.; et al. Morphological and quality characterization of grape berry and rachis in response to postharvest 1-methylcyclopropene and elevated oxygen and carbon dioxide atmospheres. *Postharvest Biol. Technol.* **2019**, *153*, 107–117. [CrossRef]
147. Dixon, J.; Hewett, E.W. Exposure to hypoxia conditions alters volatile concentrations of apple cultivars. *J. Sci. Food Agric.* **2001**, *81*, 22–29. [CrossRef]
148. Miszczak, A.; Forney, C.F.; Prange, R.K. Development of aroma volatiles and color during postharvest ripening of "kent" strawberries. *J. Am. Soc. Hortic. Sci.* **1995**, *120*, 650–655. [CrossRef]
149. Obenland, D.; Collin, S.; Mackey, B.; Sievert, J.; Arpaia, M.L. Storage temperature and time influences sensory quality of mandarins by altering soluble solids, acidity and aroma volatile composition. *Postharvest Biol. Technol.* **2011**, *59*, 187–193. [CrossRef]
150. Ji, Y.; Hu, W.; Jiang, A.; Xiu, Z.; Liao, J.; Yang, X.; Guan, Y.; Saren, G.; Feng, K. Effect of ethanol treatment on the quality and volatiles production of blueberries after harvest. *J. Sci. Food Agric.* **2019**, *99*, 6296–6306. [CrossRef]
151. Forney, C.F.; Kalt, W.; Jordan, M.A. The composition of strawberry aroma is influenced by cultivar, maturity, and storage. *HortScience* **2000**, *35*, 1022–1026. [CrossRef]
152. Lumpkin, C.; Fellman, J.K.; Rudell, D.R.; Mattheis, J.P. "Fuji" apple (*Malus domestica* Borkh.) volatile production during high pCO_2 controlled atmosphere storage. *Postharvest Biol. Technol.* **2015**, *100*, 234–243. [CrossRef]
153. Echeverría, G.; Fuentes, T.; Graell, J.; Lara, I.; López, M.L. Aroma volatile compounds of "Fuji" apples in relation to harvest date and cold storage technology: A comparison of two seasons. *Postharvest Biol. Technol.* **2004**, *32*, 29–44. [CrossRef]
154. Both, V.; Brackmann, A.; Thewes, F.R.; Ferreira, D.D.F.; Wagner, R. Effect of storage under extremely low oxygen on the volatile composition of "Royal Gala" apples. *Food Chem.* **2014**, *156*, 50–57. [CrossRef] [PubMed]
155. Defilippi, B.G.; Kader, A.A.; Dandekar, A.M. Apple aroma: Alcohol acyltransferase, a rate limiting step for ester biosynthesis, is regulated by ethylene. *Plant Sci.* **2005**, *168*, 1199–1210. [CrossRef]
156. Ciesa, F.; Dalla Via, J.; Wisthaler, A.; Zanella, A.; Guerra, W.; Mikoviny, T.; Märk, T.D.; Oberhuber, M. Discrimination of four different postharvest treatments of "Red Delicious" apples based on their volatile organic compound (VOC) emissions during shelf-life measured by proton transfer reaction mass spectrometry (PTR-MS). *Postharvest Biol. Technol.* **2013**, *86*, 329–336. [CrossRef]

157. Zhu, X.; Shen, L.; Fu, D.; Si, Z.; Wu, B.; Chen, W.; Li, X. Effects of the combination treatment of 1-MCP and ethylene on the ripening of harvested banana fruit. *Postharvest Biol. Technol.* **2015**, *107*, 23–32. [CrossRef]
158. Tietel, Z.; Lewinsohn, E.; Fallik, E.; Porat, R. Importance of storage temperatures in maintaining flavor and quality of mandarins. *Postharvest Biol. Technol.* **2012**, *64*, 175–182. [CrossRef]
159. Ortuño, A.; García-Puig, D.; Fuster, M.D.; Pérez, M.L.; Sabater, F.; Porras, I.; García-Lidón, A.; Del Río, J.A. Flavanone and Nootkatone Levels in Different Varieties of Grapefruit and Pummelo. *J. Agric. Food Chem.* **1995**, *43*, 1–5. [CrossRef]
160. Obenland, D.M.; Aung, L.H.; Bridges, D.L.; Mackey, B.E. Volatile emissions of navel oranges as predictors of freeze damage. *J. Agric. Food Chem.* **2003**, *51*, 3367–3371. [CrossRef]
161. Marais, J. Terpenes in the Aroma of Grapes and Wines: A Review. *South African J. Enol. Vitic.* **1983**, *4*, 49–58. [CrossRef]
162. Ma, Y.; Li, S.; Yin, X.; Xing, Y.; Lin, H.; Xu, Q.; Bi, X.; Chen, C. Effects of controlled atmosphere on the storage quality and aroma compounds of lemon fruits using the designed automatic control apparatus. *Biomed. Res. Int.* **2019**, *2019*. [CrossRef]
163. Lalel, H.J.D.; Singh, Z. Controlled atmosphere storage of "Delta R2E2" mango fruit affects production of aroma volatile compounds. *J. Hortic. Sci. Biotechnol.* **2006**, *81*, 449–457. [CrossRef]
164. Ntsoane, M.L.; Luca, A.; Zude-Sasse, M.; Sivakumar, D.; Mahajan, P.V. Impact of low oxygen storage on quality attributes including pigments and volatile compounds in 'shelly' mango. *Sci. Hortic. (Amst.)* **2019**, *250*, 174–183. [CrossRef]
165. Pott, D.M.; de Abreu e Lima, F.; Soria, C.; Willmitzer, L.; Fernie, A.R.; Nikoloski, Z.; Osorio, S.; Vallarino, J.G. Metabolic reconfiguration of strawberry physiology in response to postharvest practices. *Food Chem.* **2020**, *321*, 126747. [CrossRef] [PubMed]

metabolites

MDPI

Article

Network Analysis Provides Insight into Tomato Lipid Metabolism

Anastasiya Kuhalskaya [1,2], Micha Wijesingha Ahchige [1], Leonardo Perez de Souza [1], José Vallarino [1], Yariv Brotman [1,2] and Saleh Alseekh [1,3,*]

[1] Max Planck Institute of Molecular Plant Physiology, 14476 Potsdam-Golm, Germany; Kuhalskaya@mpimp-golm.mpg.de (A.K.); Wijesingha@mpimp-golm.mpg.de (M.W.A.); LPerez@mpimp-golm.mpg.de (L.P.d.S.); Vallarino@mpimp-golm.mpg.de (J.V.); brotman@mpimp-golm.mpg.de (Y.B.)

[2] Department of Life Sciences, Ben Gurion University of the Negev, 84105 Beersheva, Israel

[3] Centre of Plant Systems Biology and Biotechnology, 4000 Plovdiv, Bulgaria

* Correspondence: Alseekh@mpimp-golm.mpg.de; Tel.: +49-(0)331-567-8211

Received: 14 February 2020; Accepted: 11 April 2020; Published: 14 April 2020

Abstract: Metabolic correlation networks have been used in several instances to obtain a deeper insight into the complexity of plant metabolism as a whole. In tomato (*Solanum lycopersicum*), metabolites have a major influence on taste and overall fruit quality traits. Previously a broad spectrum of metabolic and phenotypic traits has been described using a *Solanum pennellii* introgression-lines (ILs) population. To obtain insights into tomato fruit metabolism, we performed metabolic network analysis from existing data, covering a wide range of metabolic traits, including lipophilic and volatile compounds, for the first time. We provide a comprehensive fruit correlation network and show how primary, secondary, lipophilic, and volatile compounds connect to each other and how the individual metabolic classes are linked to yield-related phenotypic traits. Results revealed a high connectivity within and between different classes of lipophilic compounds, as well as between lipophilic and secondary metabolites. We focused on lipid metabolism and generated a gene-expression network with lipophilic metabolites to identify new putative lipid-related genes. Metabolite–transcript correlation analysis revealed key putative genes involved in lipid biosynthesis pathways. The overall results will help to deepen our understanding of tomato metabolism and provide candidate genes for transgenic approaches toward improving nutritional qualities in tomato.

Keywords: lipophilic compounds; lipid-related genes; lipid metabolism

---

1. Introduction

Plants produce a wide variety of biochemical compounds, starting at the central or primary metabolism which generates compounds absolutely vital for plant survival and continuing through the pathways of specialized or secondary metabolism [1–5]. Specialized metabolites display a tremendous diversity, are often specific to certain plant lineages, and play many different roles in adaptation to the environment [6]. Volatile organic compounds (VOCs) are often discussed as a subgroup of secondary metabolites, with their low molecular weight enabling movement across cell membranes and release into the surrounding environment [7]. Secondary metabolites can act as direct or indirect defense agents by deterring herbivores, fending off pathogens, and/or attracting predators or pollinators [8]. When it comes to human consumption of plants, secondary metabolites also fulfill an important role, since they can have beneficial health effects and some of them contribute to flavor [9,10].

Since lipids may either be primary (e.g., glycerolipids, phospholipids) or secondary (derived from the isoprenoid pathway) metabolites, they are often discussed in the literature separately from other metabolites [11–13]. Lipids fulfill many different functions, ranging from carbon storage via cell-membrane components to signaling molecules [14].

The cultivated tomato (*Solanum lycopersicum*) has been widely used for metabolomic studies [15]. With production of over 150 million metric tonnes in 2017, tomato is the second most consumed vegetable in Europe after potato and the first by market value (www.fao.org/faostat/en/#home). The drought-tolerant green-fruited relative, *Solanum pennellii*, has been successfully hybridized with the cultivated tomato, and the offspring from that cross has been used to identify a wide range of phenotypic and metabolic quantitative trait loci (QTL) [16–18]. The importance of tomatoes for human diet, combined with the availability of genetic diversity from wild relative species, makes tomato an optimal model crop for studying different aspects of plant physiology [19]. In recent years, tomato metabolism has been intensively studied. Tomato's specialized metabolites in particular have received much attention, since many compounds are known to have positive health effects [20], serving among others as antioxidants. Many efforts have been undertaken to understand the genetic basis of their biosynthesis and to increase the production of these compounds in fruits [21–26].

However, due to its immense complexity and interconnection, it is a great challenge to understand plant metabolism as a whole. QTL mapping is a commonly used approach to dissect plant metabolism and identify genes in the corresponding pathways [27]. Although QTL mapping is useful for the elucidation of individual pathways, this approach is still not sufficient for dissecting metabolism in its entirety due to the latter's complexity [28]. Thus, metabolite correlation network analysis has been suggested as an additional method for elucidating novel connections in plant metabolism [29]. Several studies have used a network approach to display the correlation between metabolic compounds. Schauer et al. (2006) combined mQTL and network analysis to elucidate the relationship between metabolic and yield-related traits over two seasons of a tomato introgression-line (IL) population [30]. The study revealed a modular network with intra-modular connections of amino acids, sugars, and phenotypic traits, and highlighted the connections between metabolic and phenotypic traits. A similar approach was used to obtain several novel insights concerning the interrelation of seed primary [31] and fruit secondary metabolites with yield-associated traits [22]. Production of seed oils with desirable fatty-acid content raised the interest of oilseed breeders and biotechnologists as a result of the growing demand for vegetable oil [13,32–35]. Studies also focused on tomato seeds because their oil is an excellent source of important fatty acids involved in plant growth [36,37]. Following research has also been extended to other tissues, e.g., tomato leaves [31,38].

Despite the large-scale tomato cultivation and the number of studies related to the fruit's chemical composition and its quality, the majority of tomato metabolome research has focused on polar (non-lipid) primary and secondary compounds: sugars and their derivatives, amino acids, and organic acids. Traits such as aroma, color, and nutritional values, all deriving from secondary metabolites, have been extensively investigated, too [1,39]. However, lipid metabolism is still poorly studied in tomato, with several investigations focusing on cuticular waxes [40–43] and the characterization of genes involved in fatty-acid production [44].

In this study we performed a global correlation-based network analysis of metabolic traits, including lipids, volatiles, primary, and secondary metabolites from tomato fruit pericarp, and in addition plant phenotypic traits across a *S. pennellii* population in three field studies. This allowed us to expose the connectivity network between lipids and other branches of metabolism and key agronomical traits. We further generated a correlation network between the levels of lipophilic compounds and the expression of 1431 lipid-related genes in fruit and leaf. We were able to identify 123 potential lipid-related genes and obtained several novel insights concerning lipid metabolism.

2. Results

2.1. The Overall Metabolic Network in S. pennellii ILs

S. pennellii ILs have contributed over the years to the investigation of more than 2500 QTLs associated with plant morphology and to our understanding of the genetic regulation of primary, secondary metabolism, and lipids [31]. Here, in order to decipher the underlying regulatory organization of lipid metabolism together with other metabolic traits, we created a correlation-based network using already existing data of primary metabolites [45], secondary metabolites [22], lipophilic compounds [44], volatile compounds [10], and yield-related traits [30] across *S. pennellii* ILs in three independent seasons: 2001, 2003, and 2004 (see Section 4, Supplementary material Table S1). We correlated all metabolic and phenotypic traits against each other using Spearman's rank correlation coefficient. Figure 1 shows an overall view of a correlation-based network of all collected data in *S. pennellii* ILs. As expected, high correlations occur more often within the same group of metabolites or phenotypic traits than between different groups. The vast majority of overall correlations in the network were positive (84.4%) (Supplementary material Figure S1 and Table S2).

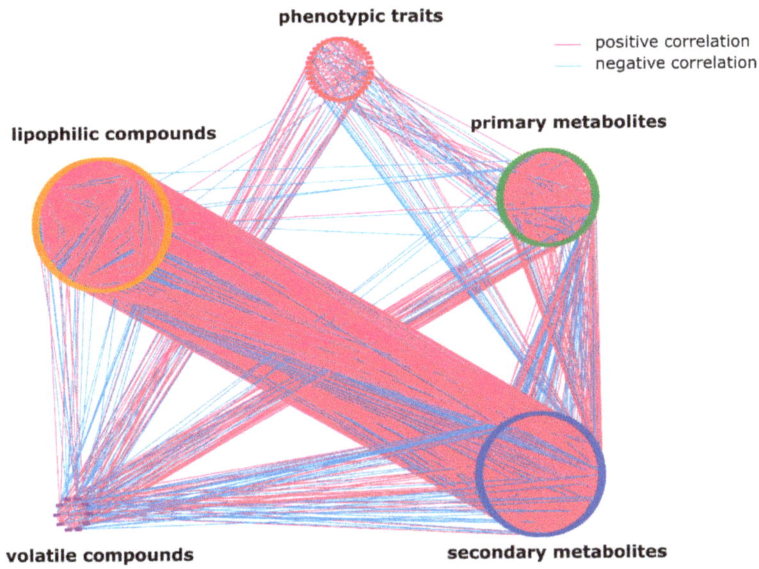

Figure 1. Overall metabolic network. Each node represents a metabolite or a whole plant phenotypic trait; edges connecting two nodes show an association between two traits. In total, the network is composed of 455 nodes and about 15,000 edges assembled into five large groups: lipophilic metabolites comprise of 171 nodes, primary and secondary metabolites have 89 and 147 nodes, respectively, phenotypic traits have 38 nodes, and the smallest group consists of 10 nodes and represents volatile compounds (Supplementary material Table S1).

2.2. Lipophilic Compounds Expose Weak Correlation with Phenotypic Traits and Primary Metabolites

We initially focused on correlations between lipids and phenotypic traits, volatiles, primary, and secondary metabolites (see Section 4).

Few correlations between different lipid classes and yield-related traits such as brix and plant weight were identified (Figure 2a). Specifically, galactolipids were correlated negatively to plant weight. Our results showed correlations between phenotypic traits and other metabolic classes, with 46% of traits being connected to primary metabolites. Brix revealed more connections to primary metabolites (primarily sugars) than to phenotypic traits. Brix reflects the total amount of soluble solids and therefore cannot be truly regarded as a phenotypic trait.

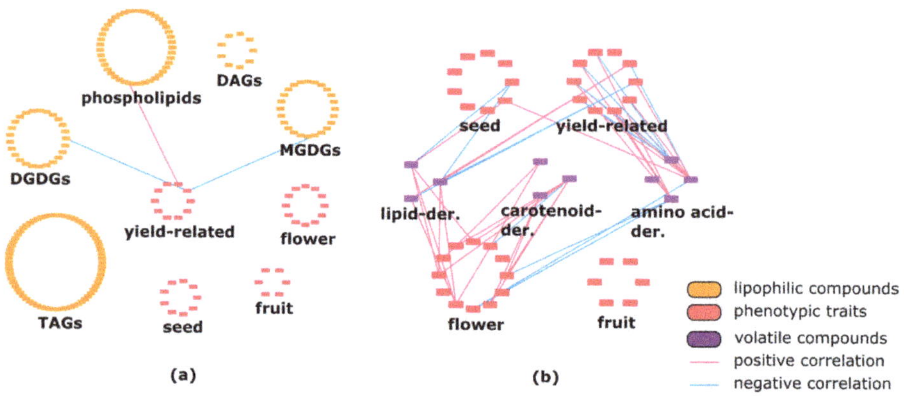

Figure 2. (**a**) Correlation between phenotypic traits and lipophilic compounds in *Solanum pennellii* introgression-lines (ILs). Yield-related traits showed three connection with several lipid classes such as phospho- and galactolipids. DAG—diacylglycerol; TAG—triacylglycerol; DGDG—digalactosyldiacylglycerol; MGDG—monogalactosyldiacylglycerol. (**b**) Correlation between phenotypic traits and volatiles in *S. pennellii* ILs. Various phenotypic traits are linked to each volatile class.

The network showed negative and positive correlations among phenotypic traits and various volatile compounds (Figure 2b). Lipid-, amino-acid-, and carotenoid-derived volatiles showed 69.2%, 55%, and 90% positive correlations with several phenotypic traits, respectively. Lipid-derived volatile compounds (hexanal, trans-2-hexenal, cis-3-hexen-1-ol) showed some positive correlations with flower-, seed-, and yield-related traits. In addition, yield-related traits showed a high number of negative correlations with amino-acid-derived volatile isovaleronitrile. Two secondary metabolites (calystegine A_3 and calystegine B_2) showed several correlations with phenotypic traits (Supplementary material Table S2).

Our data show strong, mainly positive (92.1%) correlations within the class of primary metabolites. In agreement with previous studies, we observed a highly interconnected amino-acid module (Gly, Ile, Val, Thr, and Ser) [31]. However, primary metabolites such as galactinol and guanidine highly connected with specialized metabolites. In addition, two alkaloids (calystegine A3 and calystegine B2), belonging to the class of secondary metabolites, presented a high number of correlations with primary metabolites.

We observed seven links between primary metabolites and lipophilic compounds such as phospholipids, glycerolipids, and galactolipids.

Analysis of primary metabolites against volatile compounds indicated no significant correlations between amino-acid-derived volatiles like isovaleronitrile, benzaldehyde, 3-methylbutanal, and 2-phenylethanol and their upstream biosynthesis substrates such as leucine and phenylalanine. However, guanidine showed correlations with all amino-acid-derived volatiles. In total, amino-acid-derived volatiles revealed 80.8% positive correlations with primary metabolites. The majority of correlations between lipid-derived volatiles (trans-2-hexenal, hexanal, cis-3-hexen-1-ol) and carotenoid-derived volatiles (geranylacetone and b-ionone) with primary metabolites were positive.

2.3. Lipids Are Highly Correlated with Specialized Metabolites and Show Some Negative Correlations with Volatiles

Our results showed a high number of correlations between various lipophilic compounds and secondary metabolites, with 4588 correlations representing 30.8% of all edges shown in the network

(Supplementary Material Table S2). Our data introduced for the first time, a high correlation between specialized metabolites and lipophilic compounds. For example, phospholipids were correlated to almost all secondary-metabolite subclasses. Glycerolipids showed 1725 positive and 335 negative correlations with secondary metabolites. A high number of positive correlations (96%) was also found between galactolipids and secondary metabolites (Table 1, Supplementary material Figure S2).

Table 1. Percentage of positive correlations between lipids and secondary metabolites.

	MGDGs	DGDGs	Phospholipids	TAGs	DAGs
Acyl sugars	100	96	50	78	100
Flavonoids	99	93	51	88	97
Hydroxycinnamate derivatives	93	94	48	82	95
N-containing compounds	93	92	49	78	90
Unspecified	100	99	50	85	100
Amino acid	100	100	45	71	100
Glycoalkaloids	100	100	50	78	100
Others (phenolics)	98	98	53	79	100

We identified 3249 correlations among various groups of secondary metabolites. Only 2% of all identified correlations were negative. Results indicated a high connection between and within different specialized-compound subclasses.

Correlation analysis revealed 10 positive and five negative interactions between lipid-derived volatiles and secondary metabolites. Correlations between amino-acid-derived volatiles and secondary metabolites showed an almost equal amount of positive and negative connections. Our observations revealed 58.7% (27 out of 46) of negative connections between carotenoid-derived volatiles and specialized metabolites (Figure 3a, Supplementary material Table S2). Additionally, a high number of negative correlations between triacylglycerols and volatiles was observed. Other lipid classes were connected to volatiles mostly positively (Figure 3b, Supplementary material Table S2).

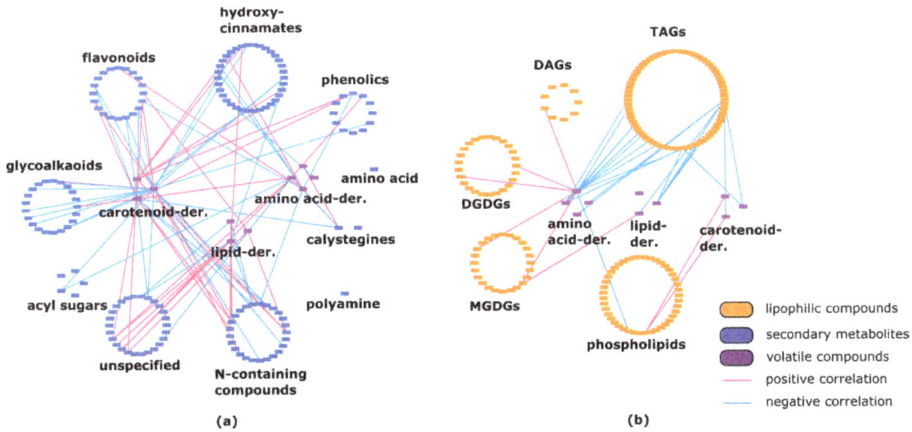

Figure 3. (**a**) Connection between secondary metabolites and volatiles in *Solanum pennellii* ILs. Volatile compounds are linked to almost all classes of specialized metabolites. (**b**) Correlation between lipophilic compounds and volatiles in *Solanum pennellii* ILs. Lipid class of TAGs showed numerous negative connections with all types of volatile compounds. DAG—diacylglycerol; TAG—triacylglycerol; DGDG—digalactosyldiacylglycerol; MGDG—monogalactosyldiacylglycerol.

Within all lipid classes we identified 5841 correlations, representing 39.2% of all edges shown in the network, with 1012 (17.3%) being negative. Lipids belonging to the same subclass showed strong, mainly positive connections with other members of the same class (Table 2). For example, among different triacylglycerols (TAGs), out of 1229 identified correlations 1102 (89.7%) were positive. There was also a minor number of negative correlations within the monogalactosyldiacylglycerol (MGDG) and digalactosyldiacylglycerol (DGDG) subclasses. Out of all identified correlations within the phospholipids subclass, 78.1% were positive. Moreover, there are no negative correlations between diacylglycerols (DAGs) and DGDGs. The highest percentage of negative correlations was observed between phospholipids and TAGs (Table 2, Supplementary material Figure S3).

Table 2. Percentage of positive correlations between different lipid subclasses.

	MGDGs	DGDGs	Phospholipids	TAGs	DAGs
MGDGs	99	99	75	96	90
DGDGs	99	99	70	87	100
Phospholipids	75	70	78	60	65
TAGs	96	87	60	90	85
DAGs	90	100	65	85	100

2.4. *Fruit-Specific Lipid-Related Genes Show a Mainly Positive Pattern of Change with Lipophilic Compounds*

For further exploration of lipid biosynthesis in tomato fruit pericarp, we correlated lipid profiles across 76 introgression lines [44] with multiple next-generation-sequencing gene-expression datasets from the same introgression lines. We used data consisting of 188 and 117 different lipophilic compounds from leaves and fruits of 74 ILs, respectively, and gene-expression data of lipid-related genes (1431) to perform the correlation analysis.

Using Spearman rank correlation coefficient, we identified 59 positive connections out of 173 in fruits and 33 positive correlations out of 73 in leaves. Figure 4 shows the significant correlation values between gene-expression levels and different lipid concentrations in fruits (Figure 4a) and leaves (Figure 4b).

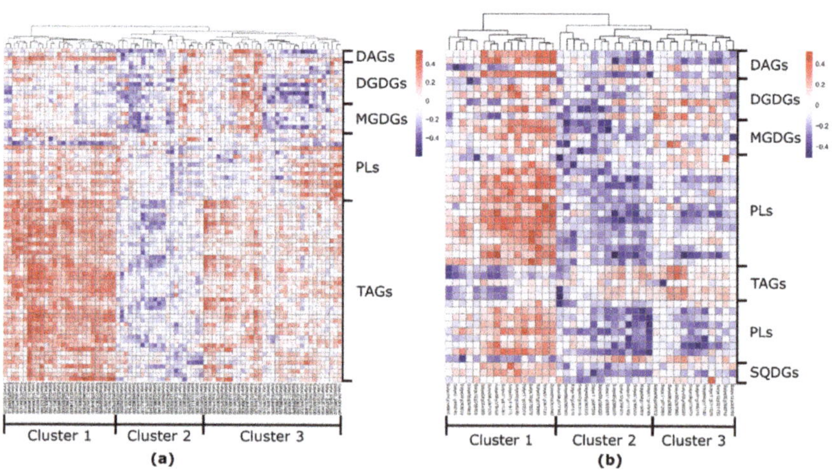

Figure 4. Heatmap of correlation values between lipid-related genes expression and lipid levels in tomato (**a**) fruits and (**b**) leaves. The number of lipid-related genes is higher in the fruit dataset compared to the leaf dataset (81 and 42, respectively). In tomato fruit, lipid-related genes linked mostly to TAGs, while in tomato leaf predominantly to phospho- and galactolipids. DAG—diacylglycerol; TAG—triacylglycerol; DGDG—digalactosyldiacylglycerol; MGDG—monogalactosyldiacylglycerol; SQDG—sulfoquinovosyldiacylglycerol.

Our analysis identified 123 potential lipid-related genes in both fruit and leaf datasets (Supplementary material Table S3). In agreement with previous results [44], we observed that lipase (*Solyc12g055730)* showed a significant high correlation with different levels of TAGs (Figure 4a).

Our fruit data showed that the expression of the gene for 1-acyl-sn-glycerol-3-phosphate acyltransferase (GPAT) (*Solyc11g065890*) is positively correlated with 18 unsaturated TAGs, four of which showing significant mQTLs in IL 11-2 and IL 11-3 [44]. Further, expression analysis revealed that in IL 11-2 this gene is expressed 1.5 times lower compared to M82, and that in IL 11-3 expression of the gene is 0.69 times higher than in M82. Moreover, various TAGs display correlations with lipid-related genes putatively annotated as phospholipase D (*Solyc01g103910*), non-specific lipid transfer protein (*Solyc10g075150*), acyl-ACP thioesterase (*Solyc12g006930*), and lipoxygenase (*Solyc03g122340*), whereas another lipoxygenase (*Solyc01g099210*) showed five positive correlations with various phospholipids. We also observed connections of galactolipids to several lipid-related genes (Supplementary material Table S3).

Finally, we identified strong correlations between lipophilic compounds and the expression of a class III lipase (*Solyc09g091050*) located on chromosome 9. The gene was negatively correlated with DGDG 36:4 (−0.46). Our previous QTL data showed a significant mQTL for galactolipids and phospholipids in this region. The QTL was mapped to a narrow overlapping region of IL 9-3, IL 9-3-1, and IL 9-3-2. The levels of DAGs, DGDGs and MGDGs, and TAGs were significantly decreased in IL 9-3, IL 9-3-1, and IL 9-3-2 compared to M82. Furthermore, gene expression was 4.6-fold higher in cultivated tomato compared to introgression lines harboring the gene (Figure 5).

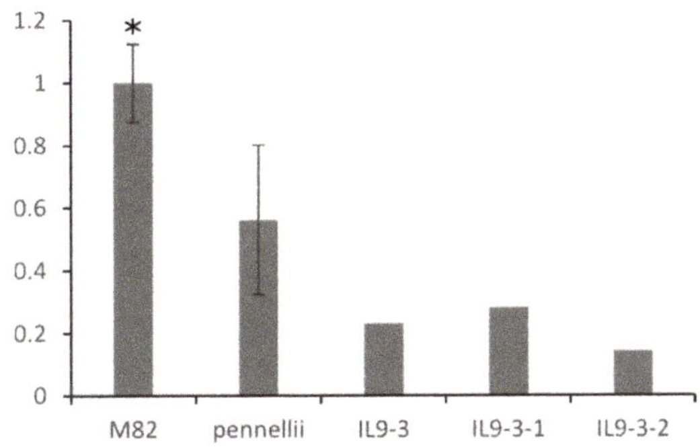

Figure 5. Transcript level of class III lipase (*Solyc09g091050*) in red ripe fruits of M82, *Solanum pennellii*, IL 9-3, IL 9-3-1, IL 9-3-2. Expression of the gene in wild tomato species *Solanum pennellii* as well as in ILs carrying the same allele version is lower in comparison to cultivated tomato variety M82. Asterisks indicate significant differences (* $p < 0.05$).

To provide additional support for the observed results, we performed genomic sequence analysis of the promoter region of the lipase (*Solyc09g091050*) and compared the promoter sequences of *S. lycopersicum cv. M82* and *S. pennellii* by genome alignment [17]. Results showed several small deletions and nucleotide substitutions in the promotor region, while the coding region showed 99% similarity between *M82* and *S. pennellii*.

Additionally, we investigated several significant correlations between different lipid classes and other lipid-related genes in tomato fruits (Supplementary material Table S3).

2.5. Leaf-Specific Lipid-Related Genes Show Many Negative Correlations with Lipophilic Compounds

Similar to the above, we extended our analysis and combined leaf lipid profiling of ILs [44] with transcriptomic data from the same lines [46,47]. Using Spearman rank correlation coefficient, we identified 45.2% (33 out of 73) and 54.8% (40 out of 73) positive and negative correlations, respectively (Figure 4a). Our results show correlations between lipid transfer protein (*Solyc03g079880*),

3-ketoacyl CoA thiolase 1 (*Solyc09g061840*), and glycerophosphoryl diester phosphodiesterase (*Solyc11g045040*) with different DAGs and phospholipids. Phospholipids additionally were linked to phospholipid-translocating flippase (*Solyc01g011100*), diacylglycerol kinase (*Solyc01g096500*), and lipid transfer proteins (*Solyc03g119210* and *Solyc10g075070*). Moreover, one transfer protein, *Solyc03g119210*, was correlated with galactolipids (DGDG 32:3 and SQDG 32:1). Many other significant correlations between different lipid classes and lipid-related gene candidates were identified (Supplementary material Table S3).

The cholesterol acyltransferase gene (*Solyc05g050710*), located on chromosome 5 (IL 5-3), predicted to take part in lipid catabolism, exhibited significant positive correlation with PC 32:0 (0.4). In the same region, several significant mQTLs for different lipid classes were identified [44]. In addition, the *Solyc05g050710* expression is almost 20-fold higher in *S. pennellii* compared to the cultivated variety (M82) (Figure 6a). Comparison of the coding sequence between *S. pennellii* and M82 revealed 99% identity. However, there are many differences in the promotor region, including a 17-bp deletion in *S. pennellii* compared to the cultivated variety, and additionally a 47-bp deletion in M82 compared to the allele derived from the wild species. This may account for the difference in the expression levels of *Solyc05g050710* between M82 and its wild relative *S. pennellii* (Figure 6b).

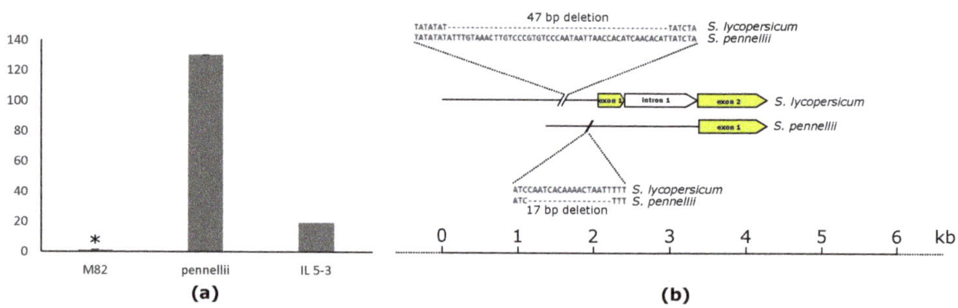

Figure 6. (**a**) Transcript levels of the cholesterol acyltransferase gene (*Solyc05g050710)* in red ripe fruits of M82, *Solanum pennellii*, IL5-3. Expression of the gene in wild tomato species *Solanum pennellii*, similar to IL 5-3 expressing the same allele version, is higher compared to cultivated tomato variety M82. (**b**) Comparison between the promoter regions of allelic versions of *Solyc05g050710* derived from cultivated tomato variety M82 and wild tomato *Solanum pennellii.* Deletions of 47 and 17 bp can be found in the promotor sequence of *Solanum lycorepsicum* and *Solanum pennellii* compared to the respective other. Asterisks indicate significant differences (* $p < 0.001$).

3. Discussion

3.1. Network Analysis: Correlation between Metabolic and Phenotypic Traits

Numerous efforts have been made to identify and characterize metabolic quantitative trait loci (mQTLs) in tomato, with a focus on primary and secondary metabolites [10,22,30,44,48,49]. In addition, QTLs for volatile organic compounds from tomato fruit [24,50–52] and acyl sugars in tomato leaf trichomes have been defined [53,54], with further studies focusing on natural variation [54–58] and cuticle composition [59]. A series of genome-wide association studies (GWAS) contributed to assessing the effects of domestication and crop improvement on the fruit metabolome [15,46,47,59–62]. Despite all these studies, relatively few investigations have hitherto aimed at understanding the genetic basis of lipid composition in tomato fruit, with few studies mainly focused on cuticular lipids [17,59,63,64].

Here, we investigated correlations between different classes of metabolites and phenotypic traits in fruits, combined with expression analysis of lipid-related genes in tomato fruits and leaves. The generated network partially validates previously discovered correlations and presents new ones. For example, a QTL for brix (*Brix9-2-5*) deriving from green-fruited tomato species increases glucose and fructose contents in cultivated tomato fruits [65]. We identified in our network significant positive correlations of brix with different sugars (Supplementary material Table S2). Additionally, we observed strong correlations between amino acids (Gly, Ile, Val, Thr, and Ser), with an average *r*-value of 0.78 and small percentage of negative interactions of correlations within the class of secondary metabolites, in agreement with previous studies [22,31,65].

Large classes of lipophilic compounds, similarly as specialized metabolites, display a considerable diversity across different plant species, e.g., isoprenoid-derived compounds are considered "secondary" metabolites produced in a cell-specific manner and are not directly involved in cell growth and development [13]. Our results showed a high number of significant correlations (30.8% of all) between lipophilic and specialized compounds. The number of significant correlations between lipids and other metabolic traits is smaller.

Several yield-related traits showed correlations to galactolipids and to one phospholipid. Tomato fruit as an organ, unlike maize cobs, for example, does not store much lipids. Unlike seeds, tomato fruit and leaf cells do not accumulate high amounts of storage lipids. Lipophilic compounds in tomato fruits and leaves participating mainly in signaling, membrane structure, and development [66–69].

Furthermore, lipids, similarly to secondary metabolites, were less correlated with phenotypic traits (Figure 1). It has been shown experimentally that the variability in secondary metabolites does not impact morphological and yield-related traits [21,70,71]. Therefore, the ability to change lipid composition or levels in tomato fruits would be a valuable tool for improving fruit quality and flavor. For example, the composition of fatty acids can be significantly changed without altering the overall plant morphology [72]. Nevertheless, versatility of lipophilic compounds might indirectly affect traits like shelf-life, if cuticle lipids are changed [73]. Until now, the influence of lipophilic compounds on overall plant phenotype remains unclear.

The connection between secondary metabolites and lipids seems to be more direct since changes in one compound class can have an effect on the other. Experimental evidence has already highlighted the close connection between secondary metabolites and cuticle lipids [42], however, no experimental validation exists of the relationship between those two metabolic classes in fruit pericarp.

In our data, we identified several negative correlations between lipid-derived volatiles and various TAGs. The observations here confirmed previous finding in tomato fruit describing linkage between decreasing levels of TAGs and simultaneously increasing levels of volatiles originating from lipids [44]. In tomato fruit some volatiles are linked to overall liking and flavor intensity [25].

In our network, we observed strong correlations within the lipid class. Lipophilic compounds correlate mostly positively between each other. For example, most of the correlations between galactolipids and glycerolipids are positive. However, phospholipids showed numerous negative correlations with all other classes. These results indicate that phospholipids are functionally the most distinct lipid class.

3.2. Network Analysis: Combining Metabolite Profiling and Expression for Gene Discovery

Plant lipid metabolism is constantly under investigation. In the well-studied model plant *A. thaliana* only around 40% of the 700 genes putatively annotated as lipid related are functionally characterized. In other plant species like tomato, that number is much smaller [74].

Recently we applied a quantitative genetic strategy to a *S. pennellii* IL population and mapped more than 160 various lipid species belonging to 10 different classes, with a total of 1528 and 428 mQTLs in fruit and leaf, respectively [44]. Here, we combined lipid profiling in leaf and fruit tissues across 76 ILs with gene expression analysis in order to identify genes involved in lipid biosynthesis.

To validate our approach, we checked whether our data confirmed previously proven correlations. We identified connections between a class III triacylglycerol lipase (*Solyc12g055730*) and various TAGs (TAG 48:2, TAG 58:0, TAG 48:3). It has been suggested that the enzyme catalyzes TAGs for further volatiles biosynthesis [44].

Our results highlighted several other lipid-related candidate genes in fruits and leaves (Figure 4). In tomato fruit, for example, the eQTL of lipase (*Solyc09g091050*) and mQTL of DGDGs and phospholipids confirmed high correlation between the gene and DGDG 36:4 (–0.46). Further validation of the gene's function is required.

Furthermore, we identified a high number of lipid-related genes that correlated positively with TAGs. For example, the gene putatively annotated as acyl-ACP thioesterase (*Solyc12g006930*) correlates positively with nine TAGs with an average *r*-value of 0.43. The enzyme is essential in the process of chain termination during de novo fatty-acid synthesis [75]. Another example is the 1-acyl-sn-glycerol-3-phosphate acyltransferase gene (GPAT) (*Solyc11g065890*), which correlated with 18 unsaturated TAGs. The *A. thaliana* ortholog (*At3g57650*) was shown to be involved in phospholipid and TAG biosynthesis [76,77]. In tomato, GPAT catalyzes acylation at the *sn*-1 position of glycerol-3-phosphate to produce lysophosphatidic acid (LPA) with subsequent TAG synthesis [78]. For the same gene we found a correlation between level of expression from leaf dataset and MGDG 32:6. Differences in connections depending on the tissue type could suggest altered function of the same gene in fruits and leaves.

Interestingly, our data revealed mainly positive correlations between expression of lipid-related genes and levels of lipophilic compounds in tomato fruit, whereas in tomato leaves these were mostly negative (Supplementary material Table S3). This may suggest that the genes involved in biosynthesis and regulation of lipid metabolism are generally different between fruits and leaves. These results are supported by a higher number of identified mQTLs in fruit compared to leaves [44]. This may indicate that lipid-related genes were less affected in leaves than in fruits in the context of the domestication process [79]. This could further mean that tomato fruits as sink tissues, which are dependent on carbon supply from source tissues, might need a tighter regulation of lipid production, compared to tissues like leaves where carbon is assimilated, making a flux to lipid metabolism "shorter" and more flexible [80].

In our leaf dataset, a high number of lipid-related genes were found to be correlated mainly with phospholipids and galactolipids compared to other subclasses. Galactolipids, for instance, represent the most abundant lipid class in thylakoid membranes, organelles specifically in leaves [81]. For instance, in our study we identified correlations between phospholipids and phospholipid-translocating flippase (*Solyc01g011100*), diacylglycerol kinase (*Solyc01g096500*), and lipid transfer proteins (*Solyc03g119210*, *Solyc10g075070*). Moreover, *Solyc03g119210* correlates with galactolipids. Another lipid-related gene candidate expressed in tomato leaves—3-ketoacyl CoA thiolase 1 (*Solyc09g061840*)—exposes correlations with DAGs and phospholipids [82]. The gene ortholog in *A. thaliana* was suggested to be involved in fatty-acid beta-oxidation [83]. Several other lipid-related genes such as particle serine esterase (*Solyc04g077180*) [84], cyclopropane-fatty-acyl-phospholipid synthase (*Solyc04g056450*) [85], acyl-CoA-binding protein (*Solyc08g075690*) [86], and long-chain fatty alcohol dehydrogenase (*Solyc09g090350*) [87] showed correlations with various phospholipids, galactolipids, and glycerolipids (Supplementary material Table S3).

It has been observed that lipid metabolism could be genetically regulated on intra-class and inter-class levels [44]. We have identified several examples of genes following the pattern of intra-class level regulation such as GPAT (*Solyc11g065890*), which correlates with 18 TAGs. In contrast, the pattern of inter-class regulation was followed by lipid transfer protein gene (*Solyc03g079880*) or 3-ketoacyl CoA thiolase 1 (*Solyc09g061840*), which contributes to regulation of two different lipid subclasses simultaneously.

In this study, we evaluated metabolic trait correlations and performed global analysis of trait associations across a *S. pennellii* IL population. This is by no means the first time that network analysis has been used for evaluation of the relationships between traits in wide genetic populations, with many previous examples in Arabidopsis, potato, and maize [88–90]. Besides, the analysis has been applied for a range of different traits and tissue types in tomato populations [30,31,45]. However, here we included for the first time three major different lipid classes and revealed several insights concerning the interrelation of traits from yield-associated traits, primary, and secondary metabolism, volatiles with lipids.

Our network using correlation between gene expression and metabolite levels combined with DNA sequence analysis highlighted several candidate genes putatively involved in lipid biosynthesis or regulation. The presented results complement previous studies regarding metabolic traits in a *S. pennellii* IL population [10,22,30,44,45] and can be used for expanding the knowledge of lipid metabolism in tomato.

4. Materials and Methods

4.1. Plant Material

Data used in this study were based on *S. pennellii* ILs. The *S. pennellii* IL population was created by replacement of marker-defined genetic regions of the wild species *S. pennellii* with homologous fragments of the cultivated tomato *S. lycopersicum* (M82), representing whole wild-genome coverage of *S. pennellii* [16].

We used already available data for primary and secondary metabolites, lipids, and phenotypic traits. The data were obtained using a population grown in the Western Galilee Experimental Station in Akko, Israel, in a completely randomized design with one plant per m^2. The field was irrigated with 320 m^3 of water per 1000 m^2 of field area throughout the season. The harvest of fruit was done when 80%–100% of tomatoes were red [16]. All the data were obtained from peeled-off fruit pericarp. Primary and secondary metabolites were available for three independent seasons: 2001, 2003, and 2004 [30,45]. Lipids data were available for seasons 2001 and 2003 [44]. Yield-related traits were available for seasons 2001 and 2004 [30], and flower-, seed- and fruit-related phenotypic traits—for season 2004 [30,91].

The data for volatile compounds were obtained from an *S. pennellii* IL population [10]. All lines were grown in randomized, replicated plots in three different sites (Gainesville, Citra, and Live Oak, Florida) over the seasons of 2002 to 2004. Volatile data we focused on in our study were consistently available only for season 2003. We also used available data for volatiles of interest for season 2004. Plants were grown using standard commercial practices in raised plastic mulched beds. Fruits from all plants for each line were combined and analyzed as they reached the red ripe stage [10].

Transcriptomic data across *S. pennellii* ILs RNA-seq of 1431 lipid-related genes from young leaf [47] and fruit peeled-off pericarp were obtained under http://ted.bti.cornell.edu [46]. We selected all expressed lipid-related genes across the tomato genome (based on GO). We extracted 647 and 786 lipid-related genes from leaf and fruit datasets, respectively.

4.2. Correlation Analysis

All metabolite and transcript values used for correlation analysis correspond to the standard scores of the log transformed data. Spearman correlation matrices were calculated in R (R Development Core Team, 2010) using the *cor* function of the *stats* package (https://www.rdocumentation.org/packages/stats). For the trait/trait correlations we used a critical p-value of 0.05, since it is a commonly used threshold for statistical analysis. Correlation p-values were obtained by performing a permutation test based on Spearman correlations using the using the perm.cor.test function of *jmuOutlier* package (https://www.rdocumentation.org/packages/jmuOutlier/versions/2.2/topics/perm.cor.test), set for 20,000 permutations. To select the most meaningful correlations for the network analysis, arbitrary cut-offs were set to an absolute correlation coefficient higher than 0.3 and 0.4 for trait/trait and trait/transcript networks, respectively. We set up the cut-off for trait/trait correlations so that known correlations would be incorporated in our network. Approved significant correlation between brix and sugars were shown before [65]. In our network the average correlation between brix and sugars were 0.378. Additionally, we used a relatively relaxed correlation coefficient threshold of 0.3, because we were integrating data from different platforms. For trait/transcript cut-off 0.3 we reported in total 1537 correlations with lipid classes. To get more insight on some of these correlations we decided to use a stronger cut-off of correlation coefficient (≥0.40).

Depending on the measurement or dataset, we had different amounts of replicates for the different traits in the introgression lines. For GC-MS replicate number was between 1 and 11, for LC-MS between 1 and 12, for lipids between 3 and 4, for phenotypic traits between 1 and 12 and for volatiles always 1. The number of replicates for M82 wild type was usually much higher.

The metabolite/metabolite network plot was produced using Cytoscape version 3.6.1 with nodes representing different metabolites and phenotypic traits and edges representing pairwise correlation above the set threshold. All metabolites and transcripts exhibiting at least one pairwise correlation above the metabolite/transcript correlation network cut-off were selected to be represented in the heatmap produced using the *pheatmap* (https://CRAN.R-project.org/package=pheatmap) package in R.

4.3. Promoter Analysis

Promoter analysis of *Solyc05g050710* and *Solyc09g091050* was performed on the accessions *S. lycopersicum* (M82) and *S. pennellii* (LA0716). Alignment of the promoter region of *Solyc05g050710* and *Solyc09g091050* was done with CLUSTALW (http://www.genome.jp/tools/clustalw/).

4.4. Trait Classes Used for Correlations

Phenotypic traits were divided to flower traits (anther length, anther width, anther length/width ratio, ovary length, ovary width, ovary length/width ratio, style length, style width, style length/width ratio), seed traits (seed length, seed width, seed length/width ratio, seed weight, seed number per fruit, seed weight per fruit, seed number per plant, seed weight per plant, seed number per fruit unit, inflorescence, flowers per inflorescence, flowers per plant), fruit-related (fruit length, fruit width, fruit length/width ratio, fruit pericarp thickness, fruit length/pericarp thickness ratio, fruit width/pericarp thickness ratio, fruit locule number) and yield-related ((brix (BX), brix yield (BY), plant weight (PW), total yield (TY), harvest index (HI), biomass (BM), fruit number (FN), red fruit weight (RED), earliness (EA), mean fruit weight (FW)) subgroups. Specialized metabolites were divided to flavonoids, glycoalkaloids, phenolics, N-containing compounds, hydroxycinnamate derivatives, acyl sugars, polyamines, and unspecified compounds subgroups, lipophilic compounds to trialycglycerols, diacylglycerols, phospholipids, digalactoyldiacylglycerols, monogalactoyldiacylglycerols subgroups, and volatile compounds to carotenoid-, lipid-, and amino acid-derived subgroups. Volatile compounds were divided to carotenoid-, lipid-, and amino acid-derived subgroups.

Supplementary Materials: The following are available online http://www.mdpi.com/2218-1989/10/4/152/s1; Figure S1: Percentage of positive and negative correlations in overall metabolic network, Figure S2: Percentage of positive and negative correlations between lipophilic compounds and secondary metabolites, Figure S3: Percentage of positive and negative correlations within the class of lipophilic compounds. Table S1: classes and subclasses of metabolic and phenotypic traits used for the network analysis, Table S2: All identified correlations between metabolic and phenotypic traits used for the network analysis. Table S3: correlation between lipid-related genes in fruits and leaves.

Author Contributions: Conceptualization, S.A.;Y.B.; methodology, L.P.d.S., A.K., M.W.A.; formal analysis, A.K.M., W.A.; writing—original draft preparation, S.A., A.K., M.W.A.; writing—review and editing, S.A., Y.B., L.P.d.S., J.V., A.K., M.W.A.; visualization, A.K., M.W.A.; supervision, S.A., Y.B. All authors have read and agreed to the published version of the manuscript.

Funding: This work was in part supported by the PlantaSYST project by the European Union's Horizon 2020 research and innovation programme (SGA-CSA No 664621 and No 739582 under FPA No. 664620), and ISRAEL SCIENCE FOUNDATION (grant No. 859/19).

Acknowledgments: We thank Dr. Hezi Tenenboim for his kind editorial assistance.

Conflicts of Interest: The authors declare no conflict of interest.

References

1. Pott, D.M.; Osorio, S.; Vallarino, J.G. From Central to Specialized Metabolism: An Overview of Some Secondary Compounds Derived from the Primary Metabolism for Their Role in Conferring Nutritional and Organoleptic Characteristics to Fruit. *Front. Plant Sci.* **2019**, *10*, 835. [CrossRef] [PubMed]
2. Dixon, R.A.; Strack, D. Phytochemistry meets genome analysis, and beyond. *Phytochemistry* **2003**, *62*, 815–816. [CrossRef]
3. Rai, A.; Saito, K.; Yamazaki, M. Integrated omics analysis of specialized metabolism in medicinal plants. *Plant J.* **2017**, *90*, 764–787. [CrossRef] [PubMed]
4. Fang, C.; Fernie, A.R.; Luo, J. Exploring the Diversity of Plant Metabolism. *Trends Plant Sci.* **2019**, *24*, 83–98. [CrossRef]
5. Fernie, A.R.; Trethewey, R.N.; Krotzky, A.J.; Willmitzer, L. Metabolite profiling: From diagnostics to systems biology. *Nat. Rev. Mol. Cell Biol.* **2004**, *5*, 763–769. [CrossRef]
6. Maeda, H.A. Evolutionary Diversification of Primary Metabolism and Its Contribution to Plant Chemical Diversity. *Front. Plant Sci.* **2019**, *10*, 881. [CrossRef]
7. Dudareva, N.; Klempien, A.; Muhlemann, J.K.; Kaplan, I. Biosynthesis, function and metabolic engineering of plant volatile organic compounds. *New Phytol.* **2013**, *198*, 16–32. [CrossRef]
8. Kessler, A.; Kalske, A. Plant Secondary Metabolite Diversity and Species Interactions. *Annu. Rev. Ecol. Evol. Syst.* **2018**, *49*, 115–138. [CrossRef]
9. Martin, C. The interface between plant metabolic engineering and human health. *Curr. Opin. Biotechnol.* **2013**, *24*, 344–353. [CrossRef]
10. Tieman, D.M.; Zeigler, M.; Schmelz, E.A.; Taylor, M.G.; Bliss, P.; Kirst, M.; Klee, H.J. Identification of loci affecting flavour volatile emissions in tomato fruits. *J. Exp. Bot.* **2006**, *57*, 887–896. [CrossRef]
11. Nakabayashi, R.; Saito, K. Metabolomics for unknown plant metabolites. *Anal. Bioanal. Chem.* **2013**, *405*, 5005–5011. [CrossRef] [PubMed]
12. Zhao, J.; Li, L.; Zhao, Y.; Zhao, C.; Chen, X.; Liu, P.; Zhou, H.; Zhang, J.; Hu, C.; Chen, A.; et al. Metabolic changes in primary, secondary, and lipid metabolism in tobacco leaf in response to topping. *Anal. Bioanal. Chem.* **2018**, *410*, 839–851. [CrossRef] [PubMed]
13. Ohlrogge, J.; Browse, J. Lipid biosynthesis. *Plant Cell* **1995**, *7*, 957–970. [PubMed]
14. Aid, F. Plant Lipid Metabolism. *Lipid Metabolism* **2019**. [CrossRef]
15. Zhu, G.; Wang, S.; Huang, Z.; Zhang, S.; Liao, Q.; Zhang, C.; Lin, T.; Qin, M.; Peng, M.; Yang, C.; et al. Rewiring of the Fruit Metabolome in Tomato Breeding. *Cell* **2018**, *172*, 249–261. [CrossRef]
16. Eshed, Y.; Zamir, D. An Introgression Line Population of Lycopersicon Pennellii in the Cultivated Tomato Enables the Identification and Fine Mapping of Yield-Associated Qtl. *Genetics* **1995**, *141*, 1147–1162.
17. Bolger, A.; Scossa, F.; Bolger, M.E.; Lanz, C.; Maumus, F.; Tohge, T.; Quesneville, H.; Alseekh, S.; Sorensen, I.; Lichtenstein, G.; et al. The genome of the stress-tolerant wild tomato species Solanum pennellii. *Nat. Genet.* **2014**, *46*, 1034–1038. [CrossRef]
18. Lippman, Z.B.; Semel, Y.; Zamir, D. An integrated view of quantitative trait variation using tomato interspecific introgression lines. *Curr. Opin. Genet. Dev.* **2007**, *17*, 545–552. [CrossRef]
19. Kimura, S.; Sinha, N. Tomato (Solanum lycopersicum): A Model Fruit-Bearing Crop. *CSH Protoc.* **2008**. [CrossRef]
20. Raiola, A.; Rigano, M.M.; Calafiore, R.; Frusciante, L.; Barone, A. Enhancing the Health-Promoting Effects of Tomato Fruit for Biofortified Food. *Mediators Inflamm.* **2014**, *2014*. [CrossRef]
21. Butelli, E.; Titta, L.; Giorgio, M.; Mock, H.-P.; Matros, A.; Peterek, S.; Schijlen, E.G.W.M.; Hall, R.D.; Bovy, A.G.; Luo, J.; et al. Enrichment of tomato fruit with health-promoting anthocyanins by expression of select transcription factors. *Nat. Biotechnol.* **2008**, *26*, 1301–1308. [CrossRef] [PubMed]
22. Alseekh, S.; Tohge, T.; Wendenberg, R.; Scossa, F.; Omranian, N.; Li, J.; Kleessen, S.; Giavalisco, P.; Pleban, T.; Mueller-Roeber, B.; et al. Identification and mode of inheritance of quantitative trait loci for secondary metabolite abundance in tomato. *Plant Cell* **2015**, *27*, 485–512. [CrossRef] [PubMed]
23. Mutschler, M.A.; Doerge, R.W.; Liu, S.C.; Kuai, J.P.; Liedl, B.E.; Shapiro, J.A. QTL analysis of pest resistance in the wild tomato Lycopersicon pennellii: QTLs controlling acylsugar level and composition. *Theor. Appl. Genet.* **1996**, *92*, 709–718. [CrossRef] [PubMed]

24. Mathieu, S.; Cin, V.D.; Fei, Z.; Li, H.; Bliss, P.; Taylor, M.G.; Klee, H.J.; Tieman, D.M. Flavour compounds in tomato fruits: Identification of loci and potential pathways affecting volatile composition. *J. Exp. Bot.* **2009**, *60*, 325–337. [CrossRef]
25. Tieman, D.; Zhu, G.; Resende, M.F.R.; Lin, T.; Nguyen, C.; Bies, D.; Rambla, J.L.; Beltran, K.S.O.; Taylor, M.; Zhang, B.; et al. A chemical genetic roadmap to improved tomato flavor. *Science* **2017**, *355*, 391–394. [CrossRef]
26. Adato, A.; Mandel, T.; Mintz-Oron, S.; Venger, I.; Levy, D.; Yativ, M.; Domínguez, E.; Wang, Z.; Vos, R.C.H.D.; Jetter, R.; et al. Fruit-Surface Flavonoid Accumulation in Tomato Is Controlled by a SlMYB12-Regulated Transcriptional Network. *PLoS Genet.* **2009**, *5*, e1000777. [CrossRef]
27. Wu, S.; Alseekh, S.; Cuadros-Inostroza, A.; Fusari, C.M.; Mutwil, M.; Kooke, R.; Keurentjes, J.B.; Fernie, A.R.; Willmitzer, L.; Brotman, Y. Combined Use of Genome-Wide Association Data and Correlation Networks Unravels Key Regulators of Primary Metabolism in Arabidopsis thaliana. *PLoS Genet.* **2016**, *12*, e1006363. [CrossRef]
28. Kruger, N.J.; Ratcliffe, R.G. Pathways and fluxes: Exploring the plant metabolic network. *J. Exp. Bot.* **2012**, *63*, 2243–2246. [CrossRef]
29. Toubiana, D.; Fernie, A.R.; Nikoloski, Z.; Fait, A. Network analysis: Tackling complex data to study plant metabolism. *Trends Biotechnol.* **2013**, *31*, 29–36. [CrossRef]
30. Schauer, N.; Semel, Y.; Roessner, U.; Gur, A.; Balbo, I.; Carrari, F.; Pleban, T.; Perez-Melis, A.; Bruedigam, C.; Kopka, J.; et al. Comprehensive metabolic profiling and phenotyping of interspecific introgression lines for tomato improvement. *Nat. Biotechnol.* **2006**, *24*, 447–454. [CrossRef]
31. Toubiana, D.; Semel, Y.; Tohge, T.; Beleggia, R.; Cattivelli, L.; Rosental, L.; Nikoloski, Z.; Zamir, D.; Fernie, A.R.; Fait, A. Metabolic Profiling of a Mapping Population Exposes New Insights in the Regulation of Seed Metabolism and Seed, Fruit, and Plant Relations. *PLoS Genet.* **2012**, *8*, e1002612. [CrossRef] [PubMed]
32. Sales-Campos, H.; de Souza, P.R.; Peghini, B.C.; da Silva, J.S.; Cardoso, C.R. An overview of the modulatory effects of oleic acid in health and disease. *Mini Rev. Med. Chem.* **2013**, *13*, 201–210. [PubMed]
33. Farvid, M.S.; Ding, M.; Pan, A.; Sun, Q.; Chiuve, S.E.; Steffen, L.M.; Willett, W.C.; Hu, F.B. Dietary linoleic acid and risk of coronary heart disease: A systematic review and meta-analysis of prospective cohort studies. *Circulation* **2014**, *130*, 1568–1578. [CrossRef]
34. Weselake, R.J.; Taylor, D.C.; Rahman, M.H.; Shah, S.; Laroche, A.; McVetty, P.B.E.; Harwood, J.L. Increasing the flow of carbon into seed oil. *Biotechnol. Adv.* **2009**, *27*, 866–878. [CrossRef] [PubMed]
35. Baud, S.; Lepiniec, L. Physiological and developmental regulation of seed oil production. *Prog. Lipid Res.* **2010**, *49*, 235–249. [CrossRef] [PubMed]
36. Aluyor, E.O.; Ori-Jesu, M. The use of antioxidants in vegetable oils–A review. *Afr. J. Biotechnol.* **2008**, *7*, 4836–4842.
37. Botineștean, C.; Hădărugă, N.G.; Hădărugă, D.I.; Jianu, I. Fatty Acids Composition by Gas Chromatography – Mass Spectrometry (GC-MS) and most important physical- chemicals parameters of Tomato Seed Oil. *J. Agroaliment. Process. Technol.* **2012**, *18*, 89–94.
38. Nunes-Nesi, A.; Alseekh, S.; de Oliveira Silva, F.M.; Omranian, N.; Lichtenstein, G.; Mirnezhad, M.; González, R.R.R.; Sabio y Garcia, J.; Conte, M.; Leiss, K.A.; et al. Identification and characterization of metabolite quantitative trait loci in tomato leaves and comparison with those reported for fruits and seeds. *Metabolomics* **2019**, *15*, 46. [CrossRef]
39. Tohge, T.; Fernie, A.R. Metabolomics-Inspired Insight into Developmental, Environmental and Genetic Aspects of Tomato Fruit Chemical Composition and Quality. *Plant Cell Physiol.* **2015**, *56*, 1681–1696. [CrossRef]
40. Vallarino, J.G.; Yeats, T.H.; Maximova, E.; Rose, J.K.; Fernie, A.R.; Osorio, S. Postharvest changes in LIN5-down-regulated plants suggest a role for sugar deficiency in cuticle metabolism during ripening. *Phytochemistry* **2017**, *142*, 11–20. [CrossRef]
41. Mintz-Oron, S.; Mandel, T.; Rogachev, I.; Feldberg, L.; Lotan, O.; Yativ, M.; Wang, Z.; Jetter, R.; Venger, I.; Adato, A.; et al. Gene Expression and Metabolism in Tomato Fruit Surface Tissues. *Plant Physiol.* **2008**, *147*, 823–851. [CrossRef] [PubMed]
42. España, L.; Heredia-Guerrero, J.A.; Reina-Pinto, J.J.; Fernández-Muñoz, R.; Heredia, A.; Domínguez, E. Transient Silencing of CHALCONE SYNTHASE during Fruit Ripening Modifies Tomato Epidermal Cells and Cuticle Properties1[C][W]. *Plant Physiol.* **2014**, *166*, 1371–1386. [CrossRef] [PubMed]

43. Heredia, A.; Heredia-Guerrero, J.A.; Domínguez, E. CHS silencing suggests a negative cross-talk between wax and flavonoid pathways in tomato fruit cuticle. *Plant Signal. Behav.* **2015**, *10*, e1019979. [CrossRef] [PubMed]
44. Garbowicz, K.; Liu, Z.; Alseekh, S.; Tieman, D.; Taylor, M.; Kuhalskaya, A.; Ofner, I.; Zamir, D.; Klee, H.J.; Fernie, A.R.; et al. Quantitative Trait Loci Analysis Identifies a Prominent Gene Involved in the Production of Fatty Acid-Derived Flavor Volatiles in Tomato. *Mol. Plant* **2018**, *11*, 1147–1165. [CrossRef]
45. Schauer, N.; Semel, Y.; Balbo, I.; Steinfath, M.; Repsilber, D.; Selbig, J.; Pleban, T.; Zamir, D.; Fernie, A.R. Mode of Inheritance of Primary Metabolic Traits in Tomato. *Plant Cell* **2008**, *20*, 509–523. [CrossRef]
46. Tomato Functional Genomics Database (TFGD). Available online: http://ted.bti.cornell.edu/ (accessed on 13 February 2020).
47. Chitwood, D.H.; Kumar, R.; Headland, L.R.; Ranjan, A.; Covington, M.F.; Ichihashi, Y.; Fulop, D.; Jiménez-Gómez, J.M.; Peng, J.; Maloof, J.N.; et al. A Quantitative Genetic Basis for Leaf Morphology in a Set of Precisely Defined Tomato Introgression Lines. *Plant Cell* **2013**, *25*, 2465–2481. [CrossRef]
48. Do, P.T.; Prudent, M.; Sulpice, R.; Causse, M.; Fernie, A.R. The Influence of Fruit Load on the Tomato Pericarp Metabolome in a Solanum chmielewskii Introgression Line Population. *Plant Physiol.* **2010**, *154*, 1128–1142. [CrossRef]
49. Ballester, A.-R.; Tikunov, Y.; Molthoff, J.; Grandillo, S.; Viquez-Zamora, M.; de Vos, R.; de Maagd, R.A.; van Heusden, S.; Bovy, A.G. Identification of Loci Affecting Accumulation of Secondary Metabolites in Tomato Fruit of a Solanum lycopersicum × Solanum chmielewskii Introgression Line Population. *Front. Plant Sci.* **2016**, *7*, 1428. [CrossRef]
50. Tieman, D.; Taylor, M.; Schauer, N.; Fernie, A.R.; Hanson, A.D.; Klee, H.J. Tomato aromatic amino acid decarboxylases participate in synthesis of the flavor volatiles 2-phenylethanol and 2-phenylacetaldehyde. *Proc. Natl. Acad. Sci. USA* **2006**, *103*, 8287–8292. [CrossRef]
51. Mageroy, M.H.; Tieman, D.M.; Floystad, A.; Taylor, M.G.; Klee, H.J. A Solanum lycopersicum catechol-O-methyltransferase involved in synthesis of the flavor molecule guaiacol. *Plant J.* **2012**, *69*, 1043–1051. [CrossRef]
52. Rambla, J.L.; Tikunov, Y.M.; Monforte, A.J.; Bovy, A.G.; Granell, A. The expanded tomato fruit volatile landscape. *J. Exp. Bot.* **2014**, *65*, 4613–4623. [CrossRef] [PubMed]
53. Schilmiller, A.; Shi, F.; Kim, J.; Charbonneau, A.L.; Holmes, D.; Daniel Jones, A.; Last, R.L. Mass spectrometry screening reveals widespread diversity in trichome specialized metabolites of tomato chromosomal substitution lines. *Plant J.* **2010**, *62*, 391–403. [CrossRef] [PubMed]
54. Schilmiller, A.L.; Charbonneau, A.L.; Last, R.L. Identification of a BAHD acetyltransferase that produces protective acyl sugars in tomato trichomes. *Proc. Natl. Acad. Sci. USA* **2012**, *109*, 16377–16382. [CrossRef] [PubMed]
55. Sallaud, C.; Rontein, D.; Onillon, S.; Jabès, F.; Duffé, P.; Giacalone, C.; Thoraval, S.; Escoffier, C.; Herbette, G.; Leonhardt, N.; et al. A novel pathway for sesquiterpene biosynthesis from Z,Z-farnesyl pyrophosphate in the wild tomato Solanum habrochaites. *Plant Cell* **2009**, *21*, 301–317. [CrossRef]
56. Gonzales-Vigil, E.; Hufnagel, D.E.; Kim, J.; Last, R.L.; Barry, C.S. Evolution of TPS20-related terpene synthases influences chemical diversity in the glandular trichomes of the wild tomato relative Solanum habrochaites. *Plant J.* **2012**, *71*, 921–935. [CrossRef]
57. Matsuba, Y.; Nguyen, T.T.H.; Wiegert, K.; Falara, V.; Gonzales-Vigil, E.; Leong, B.; Schäfer, P.; Kudrna, D.; Wing, R.A.; Bolger, A.M.; et al. Evolution of a complex locus for terpene biosynthesis in solanum. *Plant Cell* **2013**, *25*, 2022–2036. [CrossRef]
58. Tikunov, Y.M.; Molthoff, J.; de Vos, R.C.H.; Beekwilder, J.; van Houwelingen, A.; van der Hooft, J.J.J.; Nijenhuis-de Vries, M.; Labrie, C.W.; Verkerke, W.; van de Geest, H.; et al. Non-smoky glycosyltransferase1 prevents the release of smoky aroma from tomato fruit. *Plant Cell* **2013**, *25*, 3067–3078. [CrossRef]
59. Yeats, T.H.; Buda, G.J.; Wang, Z.; Chehanovsky, N.; Moyle, L.C.; Jetter, R.; Schaffer, A.A.; Rose, J.K.C. The fruit cuticles of wild tomato species exhibit architectural and chemical diversity, providing a new model for studying the evolution of cuticle function. *Plant J.* **2012**, *69*, 655–666. [CrossRef]
60. Sauvage, C.; Segura, V.; Bauchet, G.; Stevens, R.; Do, P.T.; Nikoloski, Z.; Fernie, A.R.; Causse, M. Genome-Wide Association in Tomato Reveals 44 Candidate Loci for Fruit Metabolic Traits1. *Plant Physiol.* **2014**, *165*, 1120–1132. [CrossRef]

61. Ye, J.; Li, W.; Ai, G.; Li, C.; Liu, G.; Chen, W.; Wang, B.; Wang, W.; Lu, Y.; Zhang, J.; et al. Genome-wide association analysis identifies a natural variation in basic helix-loop-helix transcription factor regulating ascorbate biosynthesis via D-mannose/L-galactose pathway in tomato. *PLoS Genet.* **2019**, *15*, e1008149. [CrossRef]
62. Ye, J.; Wang, X.; Hu, T.; Zhang, F.; Wang, B.; Li, C.; Yang, T.; Li, H.; Lu, Y.; Giovannoni, J.J.; et al. An InDel in the Promoter of Al-ACTIVATED MALATE TRANSPORTER9 Selected during Tomato Domestication Determines Fruit Malate Contents and Aluminum Tolerance. *Plant Cell* **2017**, *29*, 2249–2268. [CrossRef] [PubMed]
63. Fernandez-Moreno, J.-P.; Levy-Samoha, D.; Malitsky, S.; Monforte, A.J.; Orzaez, D.; Aharoni, A.; Granell, A. Uncovering tomato quantitative trait loci and candidate genes for fruit cuticular lipid composition using the Solanum pennellii introgression line population. *J. Exp. Bot.* **2017**, *68*, 2703–2716. [CrossRef] [PubMed]
64. Ofner, I.; Lashbrooke, J.; Pleban, T.; Aharoni, A.; Zamir, D. Solanum pennellii backcross inbred lines (BILs) link small genomic bins with tomato traits. *Plant J.* **2016**, *87*, 151–160. [CrossRef] [PubMed]
65. Fridman, E.; Pleban, T.; Zamir, D. A recombination hotspot delimits a wild-species quantitative trait locus for tomato sugar content to 484 bp within an invertase gene. *Proc. Natl. Acad. Sci. USA* **2000**, *97*, 4718–4723. [CrossRef]
66. Lin, W.; Oliver, D.J. Role of triacylglycerols in leaves. *Plant Sci.* **2008**, *175*, 233–237. [CrossRef]
67. Chapman, K.D.; Dyer, J.M.; Mullen, R.T. Biogenesis and functions of lipid droplets in plants. *J. Lipid Res.* **2012**, *53*, 215–226. [CrossRef]
68. Chapman, K.D.; Dyer, J.M.; Mullen, R.T. Commentary: Why don't plant leaves get fat? *Plant Sci.* **2013**, *207*, 128–134. [CrossRef]
69. Dowhan, W. Understanding phospholipid function: Why are there so many lipids? *J. Biol. Chem.* **2017**, *292*, 10755–10766. [CrossRef]
70. Luo, J.; Butelli, E.; Hill, L.; Parr, A.; Niggeweg, R.; Bailey, P.; Weisshaar, B.; Martin, C. AtMYB12 regulates caffeoyl quinic acid and flavonol synthesis in tomato: Expression in fruit results in very high levels of both types of polyphenol. *Plant J.* **2008**, *56*, 316–326. [CrossRef]
71. Zhang, Y.; Butelli, E.; Alseekh, S.; Tohge, T.; Rallapalli, G.; Luo, J.; Kawar, P.G.; Hill, L.; Santino, A.; Fernie, A.R.; et al. Multi-level engineering facilitates the production of phenylpropanoid compounds in tomato. *Nat. Commun.* **2015**, *6*, 8635. [CrossRef]
72. Cook, D.; Grierson, D.; Jones, C.; Wallace, A.; West, G.; Tucker, G. Modification of fatty acid composition in tomato (Lycopersicon esculentum) by expression of a borage delta6-desaturase. *Mol. Biotechnol.* **2002**, *21*, 123–128. [CrossRef] [PubMed]
73. Lara, I.; Heredia, A.; Domínguez, E. Shelf Life Potential and the Fruit Cuticle: The Unexpected Player. *Front. Plant Sci.* **2019**, *10*, 1–18. [CrossRef] [PubMed]
74. Li-Beisson, Y.; Shorrosh, B.; Beisson, F.; Andersson, M.X.; Arondel, V.; Bates, P.D.; Baud, S.; Bird, D.; DeBono, A.; Durrett, T.P.; et al. Acyl-Lipid Metabolism. *Arabidopsis Book* **2013**, *11*, e0161. [CrossRef] [PubMed]
75. Acyl-[acyl-carrier-protein] hydrolase - Solanum lycopersicum (Tomato). Available online: https://www.uniprot.org/uniprot/A0A3Q7J279 (accessed on 20 January 2020).
76. Angkawijaya, A.E.; Nguyen, V.C.; Nakamura, Y. Enhanced root growth in phosphate-starved Arabidopsis by stimulating de novo phospholipid biosynthesis through the overexpression of LYSOPHOSPHATIDIC ACID ACYLTRANSFERASE 2 (LPAT2). *Plant Cell Environ.* **2017**, *40*, 1807–1818. [CrossRef] [PubMed]
77. Misra, N.; Panda, P.K.; Parida, B.K. Genome-wide identification and evolutionary analysis of algal LPAT genes involved in TAG biosynthesis using bioinformatic approaches. *Mol. Biol. Rep.* **2014**, *41*, 8319–8332. [CrossRef]
78. Chen, X.; Snyder, C.L.; Truksa, M.; Shah, S.; Weselake, R.J. sn-Glycerol-3-phosphate acyltransferases in plants. *Plant Signal. Behav.* **2011**, *6*, 1695–1699. [CrossRef]
79. Lytovchenko, A.; Eickmeier, I.; Pons, C.; Osorio, S.; Szecowka, M.; Lehmberg, K.; Arrivault, S.; Tohge, T.; Pineda, B.; Anton, M.T.; et al. Tomato Fruit Photosynthesis Is Seemingly Unimportant in Primary Metabolism and Ripening but Plays a Considerable Role in Seed Development. *Plant Physiol.* **2011**, *157*, 1650–1663. [CrossRef]
80. Osorio, S.; Ruan, Y.-L.; Fernie, A.R. An update on source-to-sink carbon partitioning in tomato. *Front. Plant Sci.* **2014**, *5*, 516. [CrossRef]

81. Hölzl, G.; Witt, S.; Kelly, A.A.; Zähringer, U.; Warnecke, D.; Dörmann, P.; Heinz, E. Functional differences between galactolipids and glucolipids revealed in photosynthesis of higher plants. *Proc. Natl. Acad. Sci. USA* **2006**, *103*, 7512–7517. [CrossRef]
82. Uncharacterized protein–Solanum lycopersicum (Tomato). Available online: https://www.uniprot.org/uniprot/A0A3Q7I392 (accessed on 20 January 2020).
83. AT2G33150(PKT3). Available online: https://www.arabidopsis.org/servlets/TairObject?type=locus&name=At2g33150 (accessed on 18 March 2020).
84. Sol Genomics Network. Available online: https://www.solgenomics.net/locus/22428/view (accessed on 16 March 2020).
85. Sol Genomics Network. Available online: https://www.solgenomics.net/locus/21871/view (accessed on 16 March 2020).
86. Sol Genomics Network. Available online: https://www.solgenomics.net/locus/32411/view (accessed on 16 March 2020).
87. Sol Genomics Network. Available online: https://www.solgenomics.net/locus/35352/view (accessed on 16 March 2020).
88. Rowe, H.C.; Hansen, B.G.; Halkier, B.A.; Kliebenstein, D.J. Biochemical networks and epistasis shape the Arabidopsis thaliana metabolome. *Plant Cell* **2008**, *20*, 1199–1216. [CrossRef]
89. Lisec, J.; Römisch-Margl, L.; Nikoloski, Z.; Piepho, H.-P.; Giavalisco, P.; Selbig, J.; Gierl, A.; Willmitzer, L. Corn hybrids display lower metabolite variability and complex metabolite inheritance patterns. *Plant J.* **2011**, *68*, 326–336. [CrossRef] [PubMed]
90. Carreno-Quintero, N.; Acharjee, A.; Maliepaard, C.; Bachem, C.W.B.; Mumm, R.; Bouwmeester, H.; Visser, R.G.F.; Keurentjes, J.J.B. Untargeted metabolic quantitative trait loci analyses reveal a relationship between primary metabolism and potato tuber quality. *Plant Physiol.* **2012**, *158*, 1306–1318. [CrossRef] [PubMed]
91. Semel, Y.; Nissenbaum, J.; Menda, N.; Zinder, M.; Krieger, U.; Issman, N.; Pleban, T.; Lippman, Z.; Gur, A.; Zamir, D. Overdominant quantitative trait loci for yield and fitness in tomato. *Proc. Natl. Acad. Sci. USA* **2006**, *103*, 12981–12986. [CrossRef] [PubMed]

metabolites

MDPI

Article

Metabolomics to Exploit the Primed Immune System of Tomato Fruit

Estrella Luna [1], Amélie Flandin [2,3], Cédric Cassan [2,3], Sylvain Prigent [2,3], Chloé Chevanne [2], Camélia Feyrouse Kadiri [2], Yves Gibon [2,3] and Pierre Pétriacq [2,3,*]

1 School of Biosciences, Uni. Birmingham, Birmingham B15 2TT, UK; e.lunadiez@bham.ac.uk
2 UMR BFP, University Bordeaux, INRAE, 33882 Villenave d'Ornon, France; amelie.flandin@inrae.fr (A.F.); cedric.cassan@inrae.fr (C.C.); sylvain.prigent@inrae.fr (S.P.); chloe.chevanne@etu.u-bordeaux.fr (C.C.); camelia.kadiri@etu.u-bordeaux.fr (C.F.K.); yves.gibon@inrae.fr (Y.G.)
3 Bordeaux Metabolome, MetaboHUB, PHENOME-EMPHASIS, 33140 Villenave d'Ornon, France
* Correspondence: pierre.petriacq@inrae.fr

Received: 14 February 2020; Accepted: 4 March 2020; Published: 6 March 2020

Abstract: Tomato is a major crop suffering substantial yield losses from diseases, as fruit decay at a postharvest level can claim up to 50% of the total production worldwide. Due to the environmental risks of fungicides, there is an increasing interest in exploiting plant immunity through priming, which is an adaptive strategy that improves plant defensive capacity by stimulating induced mechanisms. Broad-spectrum defence priming can be triggered by the compound ß-aminobutyric acid (BABA). In tomato plants, BABA induces resistance against various fungal and bacterial pathogens and different methods of application result in durable protection. Here, we demonstrate that the treatment of tomato plants with BABA resulted in a durable induced resistance in tomato fruit against *Botrytis cinerea*, *Phytophthora infestans* and *Pseudomonas syringae*. Targeted and untargeted metabolomics were used to investigate the metabolic regulations that underpin the priming of tomato fruit against pathogenic microbes that present different infection strategies. Metabolomic analyses revealed major changes after BABA treatment and after inoculation. Remarkably, primed responses seemed specific to the type of infection, rather than showing a common fingerprint of BABA-induced priming. Furthermore, top-down modelling from the detected metabolic markers allowed for the accurate prediction of the measured resistance to fruit pathogens and demonstrated that soluble sugars are essential to predict resistance to fruit pathogens. Altogether, our results demonstrate that metabolomics is particularly insightful for a better understanding of defence priming in fruit. Further experiments are underway in order to identify key metabolites that mediate broad-spectrum BABA-induced priming in tomato fruit.

Keywords: tomato; metabolomics; biochemical phenotyping; priming; BABA; *Botrytis cinerea*; *Phytophthora infestans*; *Pseudomonas syringae*

---

1. Introduction

The increase in world food demand and the indiscriminate use of chemical fertilisation highlight the need to adopt sustainable crop production strategies. Given the major threat of phytopathogenic microbes to food production [1] and ecosystem stability worldwide [2], novel practices are needed to combat these threats. Tomatoes are a highly consumed fruit that represent the eleventh largest commodity, with nearly 183 million tons produced in 4 million hectares in 2017 [3]. Crop yields are strongly affected by filamentous and bacterial pathogens, including the fungus *Botrytis cinerea*, the oomycete *Phytophthora infestans* and the bacterium *Pseudomonas syringae*. These pathogens can claim the complete loss of the crop within days of exposure [4–6]. Currently, strategies of control against these biological threats are based on the use of chemical pesticides applied at a pre-harvest stage.

However, up to 50% of tomato losses occur at a post-harvest stage [7] due to, among many reasons, the inability to use chemicals at this stage due to residue toxicity. Therefore, new methods of disease control are needed in order to control infections by pathogenic microbes. Exploiting the plant immune system can represent an effective strategy to provide sustainable disease protection [8,9].

Plants are able to defend themselves against pathogens thanks to their innate immune system [10]. In addition, plants are able to sensitise their immune system to protect themselves against biotic threats. This is known as the priming of defence, which is commonly referred as the adaptive part of the plant immune system [11]. Priming occurs after the perception of stimuli that lead to an enhanced responsiveness of defence mechanisms upon subsequent attack [12]. Among stimuli, the chemical agent β-aminobutyric acid (BABA) has been widely studied for its capacity to result in broad-spectrum-induced resistance (IR) in a broad range of plant species [13]. BABA is a non-protein amino acid that has been demonstrated to be a plant product [14]. The work done in the model plant *Arabidopsis* (*A.*) *thaliana* revealed that this outstanding performance is due to the priming activity of multiple signalling pathways [13]. BABA primes salicylic acid (SA)-dependent defences and the deposition of callose, which result in effective protection against biotrophic and necrotrophic pathogens, respectively [15,16]. Importantly, it has been reported that, in many plant species, treatments with BABA result in a stress phenotype that manifests as changes in plant development (e.g., growth, yield, seed production) [17–19]. The discovery of the molecular receptor of BABA in *A. thaliana* sheds light into the reasons behind the stress response associated with this chemical: BABA binds to an aspartyl-tRNA synthetase and blocks the enzyme, consequently triggering the accumulation of its canonical substrate, uncharged tRNA, which leads to the activation of the stress response associated with amino acid imbalance in the plant [20]. Moreover, high concentrations of BABA or in specific plant species such as potatoes, can also lead to stress, as the chemical directly activates defence mechanisms, which is a costly trade-off in terms of energy resources to the plant [19,21].

In the tomato system, BABA is known to be able to induce resistance against at least 10 different pests and pathogens, including *B. cinerea*, *P. infestans* and *P. syringae* [13]. Similarly to what has been described in *A. thaliana*, priming of SA-dependent mechanisms has been reported [13]. However, it is likely that further priming mechanisms are responsible for its capacity to induce broad-spectrum resistance. Moreover, studies have demonstrated that priming of BABA is long-lasting [22]. For instance, it has been reported that, after the treatment of tomato seedlings, BABA-IR against *B. cinerea* is maintained for weeks in leaves [18]. Moreover, analysis of the resistance phenotypes in tomato fruit demonstrated that BABA-IR reaches the fruiting stages, as tomatoes from plants that had been treated with BABA at the seedling stage were more resistant to *B. cinerea* than the control [23].

Treatments of plants with BABA did not impact yield or fruit size but resulted in delayed fruit production and ripening [23]. In fruit, durable induced resistance has been linked to the accumulation of specific metabolites. For example, it was reported that, after treatment of seedlings with BABA, there was an accumulation of metabolites associated with alkaloid, terpenoid or jasmonate pathways [23]. It was therefore speculated that these metabolites could be responsible for the enhanced resistance, and therefore could mark the priming fingerprint in tomato fruit [24]. Importantly, however, the metabolites responsible for expression of priming of defence mechanisms after infection still remain unknown. Here, we aimed to determine the metabolic shifts that underpin the BABA priming of immune responses against pathogens of a different nature that infect fruit, including necrotrophic and biotrophic microbes. A combination of targeted biochemical phenotyping for major plant compounds involved in central metabolism and untargeted metabolomics could unveil discriminant metabolic biomarkers that respond to BABA priming and inoculations. To the best of our knowledge, this is the first metabolomic study with multiple fruit pathosystems in relation to the priming of immune responses.

2. Results

2.1. Effect of BABA on Broad-Spectrum Resistance in Fruit

We determined whether the treatment of tomato plants with BABA resulted in a durable induced resistance in tomato fruit against the fungal necrotrophic pathogen *B. cinerea* (*Bot*), the biotrophic oomycete *P. infestans* (*Phy*) and the hemibiotrophic bacteria *P. syringae* pv. *tomato* (*Pst*). Box-plots showing sample datapoints and mean of the scored symptoms (n = 11) revealed that BABA-treated plants produced fruit that were statistically more resistant to *Bot*, *Phy* and *Pst* ($P < 0.01$) than the controls (Figure 1). This indicates that BABA is able to induce resistance in tomato plants for a durable fruit protection against pathogens that have different infection strategies.

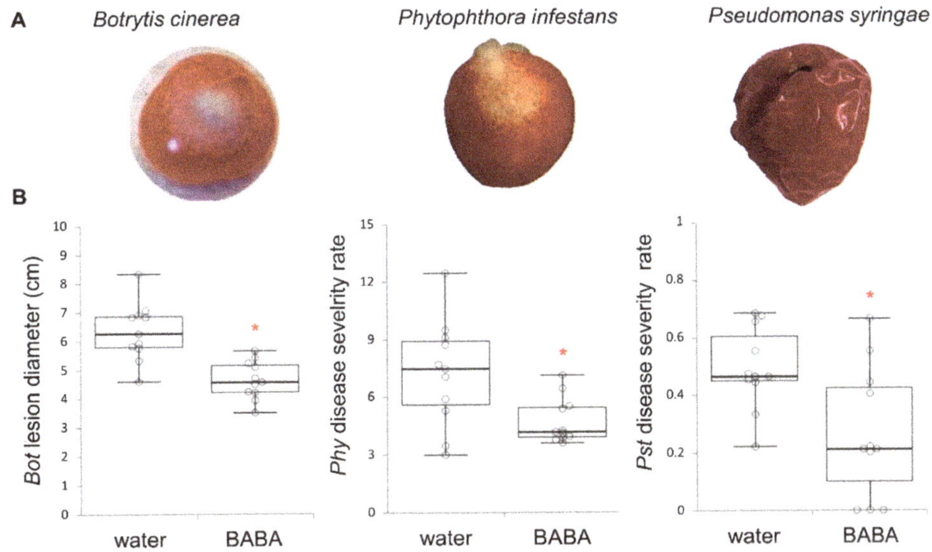

Figure 1. ß-aminobutyric acid (BABA) primes tomato fruit for a durable disease resistance against three different pathogenic microbes. (**A**) photographs showing tomato fruit 2 days after infection with *Botrytis cinerea* (*Bot*, left), *Phytophthora infestans* (*Phy*, middle) or *Pseudomonas syringae* pv. *tomato* (*Pst*, right). (**B**) box-plots of disease symptoms from fruit that originated from plants treated with water or BABA (500 µM) then inoculated with water as a mock control or with *Bot* (left), *Phy* (middle) and *Pst* (right). Symptoms were scored 2 days after inoculation from 11 biologically replicated fruit (n = 11). Asterisks indicate statistically significant differences between water- and BABA-treated plants (*t*-test, $P < 0.01$).

2.2. Effect of BABA on Fruit Yield and Development

We investigated whether the treatment of tomato seedlings with BABA could impact fruit development and yield. As described previously [23], no differences were found in fruit size (Figure 2A), but delayed fruit production (Figure 2B) and fruit ripening (Figure 2C,D) were reported. However, it was observed that, at 13 weeks of growth, the proportion of green fruit was statistically significantly higher in BABA-treated plants (40%) compared to control plants (33%) (Figure 2D), indicating that fruit production did not slowdown in BABA-treated plants after ripening processes had begun (Figure 2D). This might suggest a positive trade-off for BABA treatment on fruit development.

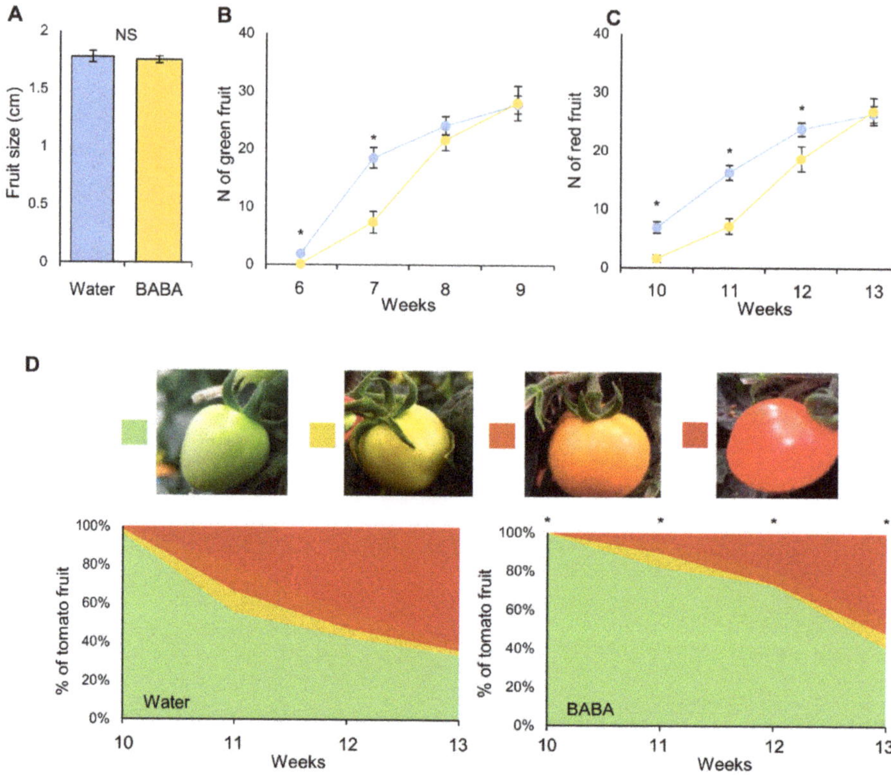

Figure 2. Effect of BABA on fruit yield and development. (**A**) fruit size (cm) from plants treated with water (blue) or 500 μM BABA (yellow). NS: not statistically significant (*t*-test, $P > 0.05$). (**B**) number of green fruit produced from water (blue)- or BABA (yellow)-treated plants after weeks of growth. Asterisks indicate statistically significant differences (*t*-test, $P > 0.05$). (**C**) number of ripped fruit produced from water (blue)- or BABA (yellow)-treated plants after weeks of growth. Asterisks indicate statistically significant differences (*t*-test, $P < 0.05$). (**D**) Proportion of fruit at different stages of fruit ripening in control and BABA-treated plants, expressed as percentage of occurrence per treatment at different weeks of growth. Asterisks indicate statistically significant differences between distributions at specific timepoints (Chi-square test, $P < 0.05$).

2.3. Global Metabolomics after BABA Treatment and After Inoculation

In order to substantiate the disease resistance after BABA treatment (Figure 1), we investigated fruit metabolism from ethanol extracts of freeze-dried tomato pericarps ($n = 4$) using *i*) targeted biochemical profiling of several major compounds involved in the central metabolism [25,26], and *ii*) untargeted metabolomics of semi-polar metabolites, including specialised compounds, via ultra-high-performance liquid chromatography coupled to electrospray ionisation orbitrap high-resolution mass spectrometry (thereafter referred to as LCMS). For this, the first, second and third fruit developed in the plants were used for pathoassays with *B. cinerea (Bot)*, *P. infestans (Phy)* and *P. syringae (Pst)*, respectively. Unbiased processing of LCMS data, followed by filtering of the most reliable variables, generated 6887 metabolomic features (see Materials and Methods). A global overview of metabolic profiles was visualised by an unsupervised multivariate statistical method, Principal Component Analysis (PCA), for all combinations of priming and pathosystems (Figure 3).

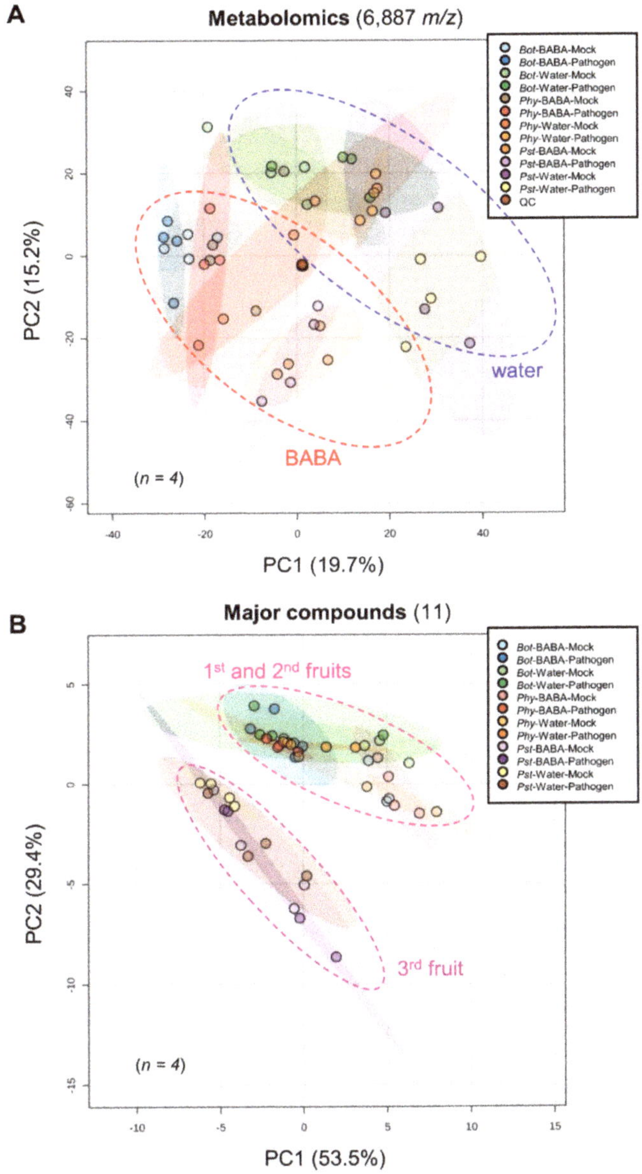

Figure 3. Global metabolomic changes after BABA treatment and after pathogen inoculation. Principle component analysis (PCA) score plots ($n = 4$) of 6887 LCMS-based metabolomics features (**A**) and 11 major compounds analysed by targeted biochemical phenotyping (**B**). Maximal variance explained by each PC is given in brackets.

Firstly, for the 6887 metabolomic signals (Figure 3A), PCA explained 35% of the maximal variance of the dataset and resulted in a clear differentiation of the water and BABA treatments, thus suggesting a greater impact on metabolomic profiles for direct BABA application as compared to other conditions. This was confirmed by a univariate statistical method through a two-factor ANOVA ($P < 0.05$), which quantitatively resulted in more statistically significant markers for the BABA factor (3052; 44%) than for the inoculation factor (2309; 33%) or the interaction (401; 6%) (Table 1).

Table 1. Univariate statistical analysis of the metabolomic features and major compounds.

Two factors ANOVA ($P < 0.05$)		BABA (Water vs. BABA)	Inoculation (Mock vs. *Bot/Phy/Pst*)	BABA × Inoculation
Metabolomic features	Total 6887	3052	2309	401
Major compounds	Total 11	0	4	0

Secondly, we tested 10 major compounds involved in central metabolism (sucrose, fructose, glucose, starch, fructose-6-P, glucose-6-P, glutamate, malate, fumarate and total proteins), as well as total polyphenols. PCA explained 83% of the maximal variance in the dataset and resulted in a separation of fruit by developmental characteristics (i.e., the first and second fruit versus the third fruit) rather than by pathosystems (Figure 3B). Hence, this multivariate differentiation indicates that the profiles of primary metabolites mostly respond to the developmental stage of the fruit, which supports the idea that central metabolism is tuned to fruit growth [27–29]. Complementarily, two-factor ANOVA ($P < 0.05$) only generated significant markers for the inoculation factor (4; 36%), including sucrose, fructose, glutamate and fumarate (Table 1 and Figure S1). Interestingly, such markers dropped upon *Bot* and *Phy* infections, while they were not drastically affected by *Pst* infection (Figure S1). Besides, fructose pools remained low across all treatments within the *Pst* pathoassay (i.e., third fruit). Altogether, this indicates that fruit infection affects the pools of central metabolites and those changes depend on the pathosystem.

Furthermore, a partial segregation of pathogen inoculations was observed on the PCA score plots obtained for each pathosystem from a dataset combining LCMS and targeted analyses (Figure 4). This was further exemplified by a supervised Partial Least Square Discriminant Analysis (PLS-DA) allowing a better differentiation of pathosystems and priming treatments (Figure S2). Two-factor ANOVA ($P < 0.05$) for each pathosystem not only confirmed that the BABA factor quantitatively outweighed the inoculation factor and the interaction, but also showed that all these factors were substantial (Table 2). Hence, this indicates that microbial challenges elicit distinct metabolic profiles. Overall, these results reveal metabolic shifts in fruit upon BABA priming and after pathogen inoculation, notably towards semi-polar biochemicals potentially involved in plant stress mitigation (i.e., specialised metabolites).

Table 2. Univariate statistical analysis for each pathosystem.

Two Factors ANOVA ($P < 0.05$)		BABA	Inoculation	BABA × Inoculation
		Water vs. BABA	Mock vs. Pathogen	
Botrytis cinerea	Total 6998	2297	724	482
Phytophthora infestans	Total 6998	1728	805	674
Pseudomonas syringae	Total 6998	2097	322	313

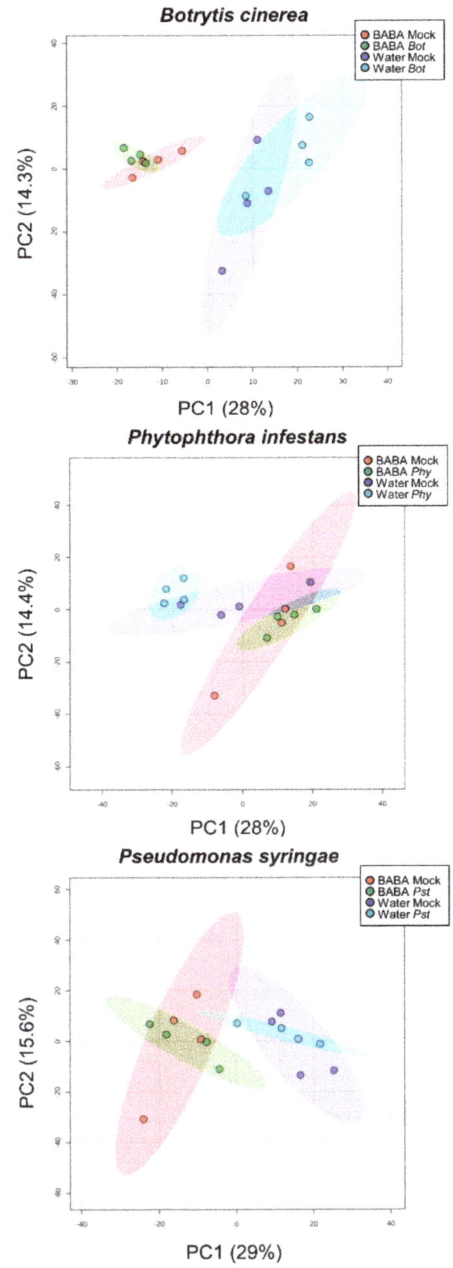

Figure 4. Partial segregation of pathogen inoculations for each fruit pathosystem. PCA score plots (n = 4) of 6898 features (6887 electrospray ionisation orbitrap high-resolution mass spectrometry (LCMS) variables + 11 major compounds) showing metabolomics overview between the three different pathosystems. Maximal variance explained by each PC are given in brackets.

2.4. Primed Responses to Specific Pathogenic Microbes

To gain more insight into BABA priming upon different fruit infections, we next performed quantitative binary comparisons of metabolic markers for each pathosystem by comparing water-treated, mock-inoculated fruit versus *i*) BABA-treated, mock-inoculated fruit, *ii*) water-treated, pathogen-inoculated fruit, and *iii*) BABA-treated, pathogen-inoculated fruit. The resulting statistically significant metabolic markers (t-test, $P < 0.01$) were used to construct Venn diagrams showing common and specific markers (Figure 5A). Very few overlaps were observed between BABA (red) and pathogen (blue) conditions (2, 2 and 0 for *Bot*, *Phy* and *Pst*, respectively) and between pathogen and BABA priming (green) conditions (7, 4 and 4 for *Bot*, *Phy* and *Pst*, respectively). Instead, several markers overlapped between BABA treatment and BABA priming (118, 12 and 158 for *Bot*, *Phy*, and *Pst* respectively), and most markers were found either for BABA treatment (*Phy* and *Pst*) or for BABA priming (*Bot*) (Figure 5A). This suggests that BABA results largely in the accumulation of metabolites that could be used during the expression of priming. In addition, metabolic markers that specifically responded to BABA priming in the different fruit pathosystems were compared through a Venn diagram in order to reveal the common metabolic signatures of BABA priming against the three different pathogens (Figure 5B). Strikingly, no common markers were found upon the three infections, although very few markers were observed between *Pst* and *Bot* (10), *Pst* and *Phy* (3), and *Bot* and *Phy* (1). Hence, the primed responses are likely tailored to the encountered pathogenic microbes.

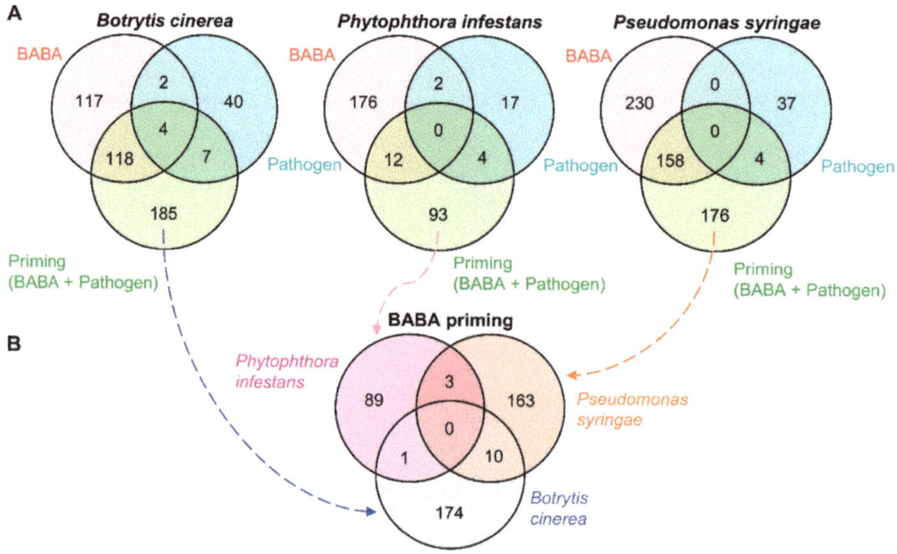

Figure 5. The primed responses are tailored to the encountered pathogenic microbes. (**A**) Venn diagrams showing quantitative binary comparisons were performed for each pathosystem (t-test, $n = 4$, $P < 0.01$) between water-treated, mock-inoculated fruit versus BABA-treated, mock-inoculated fruit (BABA, red); water-treated, pathogen-inoculated fruit (pathogen, blue); BABA-treated, pathogen-inoculated fruit (priming, green). (**B**) Venn diagrams showing the resulting priming clusters for each fruit pathosystem.

2.5. Putative Annotation of Metabolic Markers

We then conducted a tentative annotation of the 14 metabolic markers that were common to *Pst* and *Bot* (10), *Pst* and *Phy* (3), and *Bot* and *Phy* (1) based on their detected *m/z* by high-resolution orbitrap-MS (Table 3). A Kruskal–Wallis test with correction for false rate discovery (Benjamini–Hochberg, $P < 0.05$) confirmed that these 14 metabolic markers showed statistically significant variations that were visualised by bar charts (Figure S3). Putative prediction of compounds and pathways indicated several markers that belonged to the plant defence metabolism, including stress hormones and flavonoids. Interestingly, fungal pathogens (*Bot*, *Phy*) were associated with the induction of the putative marker jasmonoyl–isoleucine. Besides this, (hemi)biotrophic microbes (*Pst* and *Phy*) triggered the accumulation of putative salicylic derivatives and flavonoids (Figure S3). The *Pst*-related primed response further correlated with the depletion of a putative cytokinine. Hence, our results suggest that BABA priming against three different fruit pathogens rely on the induction of pathways involved in the defence hormonal metabolism. Further analytical studies are required to confirm the putative annotation of these priming markers.

2.6. Modelling of Resistance to Multiple Fruit Pathogens

Using a predictive biology approach based on generalised linear models [30], we aimed to determine whether resistance to fruit pathogens could be predicted by the detected metabolic markers (Figure 6). Based on those models (Figure 6A), good correlations were observed between measured and predicted values (mean = 0.87), and were statistically different from correlations based on randomly generated resistance (*t*-test, $P < 2.2 \times 10^{-16}$), which indicated the robustness of the predictions. Furthermore, according to the occurrence of metabolic markers in the models (Figure 6B), fructose appeared to be the best positively correlated predictor (appearing in 99% of the models), as well as sucrose, to a lesser extent (appearing in 41% of the models) (Table S1). Most predictors (32 out of 34) also showed a high statistical significance from a Kruskal–Wallis test with correction for false rate discovery (Benjamini–Hochberg, $P < 0.05$, Table S1). This corroborates the outcome from the two-way ANOVA method (Tables 1 and 2, and Figure S1). Hence, this indicates that soluble sugars involved in the central metabolism are essential to predict resistance to fruit pathogens. The analysis of such compounds is therefore critical for studies involving fruit–pathogen interactions. In addition to sugars, other metabolic predictors appearing in more than 25% of the models showed positive (19 markers) and negative correlations (15 markers) (Figure 6B and Table S1). Further analytical studies are required to annotate and/or identify such markers. Nonetheless, a tentative annotation of the top 15 predictors based on their detected *m/z* by high-resolution orbitrap-MS is presented in Table S1. Unsurprisingly, the resulting putative metabolites belonged to defence pathways (i.e., phenolics, flavonoids, terpenes, amino acid conjugates) and lipids. This suggests that immune perception and signalling seem pivotal in predicting resistance to fruit pathogens.

Table 3. Putative annotation of the primed response markers.

Primed Response	Detected m/z (Da) [1]	RT (min) [1]	P Value [2]	ESI Mode	Putative Adduct	Predicted m/z	Δppm	Putative Compound	Predicted Formula	Putative Pathway
Bot and *Phy*	346.1975	4.1	5.3×10^{-3}	+	[M+Na]+	323.2097	4	Jasmonoyl-isoleucine	C18H29NO4	Jasmonates
Bot and *Pst*	450.1119	7.5	6.7×10^{-3}	-	[M+F]-	431.1206	15	Ribosylzeatin phosphate	C15H22N5O8P	Cytokinines
	289.0896	2.5	5.6×10^{-3}	+	[M+H-2H_2O]+	324.0998	8	5,6-Dimethoxy-[2″,3″:7,8]furanoflavanone	C19H16O5	Flavonoids
	380.1489	3.6	5.3×10^{-3}	+	[M+ACN+H]+	338.1154	0	3,5,7-Trihydroxy-6-prenylflavone	C20H18O5	Flavonoids
	451.1238	7.4	8.4×10^{-3}	+	[M+H]+	450.1162	0	3,4,2′,4′,6′-Pentahydroxychalcone 2′-glucoside	C21H22O11	Flavonoids
	437.2213	7.1	3.8×10^{-2}	+	[M+ACN+Na]+	373.1889	37	Jasmonoyl-tyrosine	C21H27NO5	Jasmonates
	535.3117	9.1	1.6×10^{-2}	-	[M-H]-	536.3114	14	Phosphoglycerolipid (20:2(11Z,14Z)/0:0)	C26H49O9P	Lipids
	268.2271	5.4	2.5×10^{-2}	+	[M+NH4]+	250.1933	0	C16:3n-6,9,12	C16H26O2	Lipids
	652.4048	6.2	3.7×10^{-2}	+	[M+NH4]+	634.3870	24	3-trans-p-Coumaroyl-rotundic acid	C39H54O7	Phenolics
	272.0644	6.0	1.4×10^{-2}	-	[M+F]-	253.0586	27	Salicyloyl-aspartic acid	C11H11NO6	Salicylic derivatives
	154.0216	6.0	1.2×10^{-2}	+	-	-	-	-	-	Unknown
Pst and *Phy*	479.1402	4.3	4.2×10^{-2}	-	[M+Na-2H]-	458.1577	16	7-Hydroxy-5,4′-dimethoxy-8-methylisoflavone 7-O-rhamnoside	C24H26O9	Flavonoids
	525.1456	4.2	2.6×10^{-2}	+	[M+NH4]+	507.1139	3	Delphinidin 3-(acetylglucoside)	C23H23O13	Flavonoids
	452.1954	4.0	8.2×10^{-3}	-	[M+CH3COO]-	393.1940	27	Diphenhydramine salicylate	C24H27NO4	Salicylic derivatives

[1]: metabolomic parameters detected by LCMS. [2]: *P* indicating the statistical significance from a Kruskal–Wallis test followed by correction for false discovery rate using the Benjamini–Hochberg method.

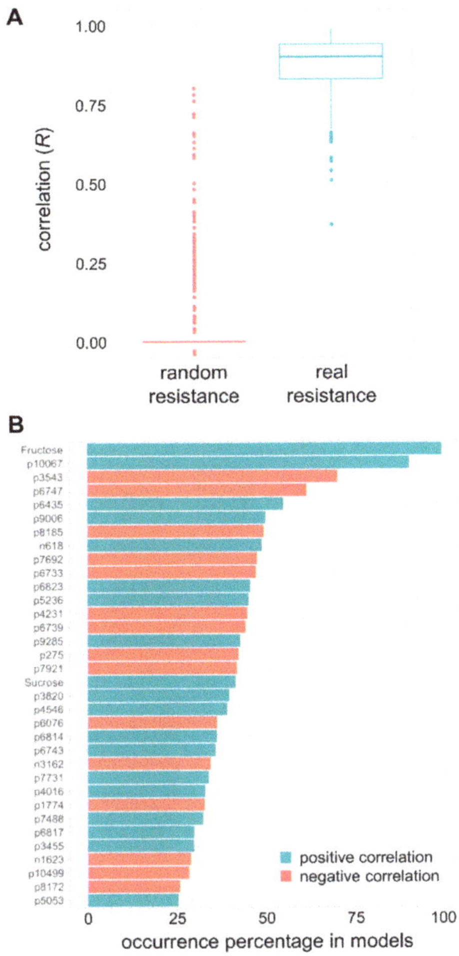

Figure 6. Prediction of biotic resistance from metabolic markers. (**A**) correlation between predicted and measured resistance based on generalised linear models. (**B**) occurrence (%) of the metabolic markers in the models that showed a positive or negative correlation with the resistance to fruit pathogens. Details of untargeted markers are presented in Table S1.

3. Discussion

In the present study, we evaluated the metabolic composition of tomato fruit in relation to the BABA-priming of young tomato plants and the infection of three different pathogens at the fruit stage. To the best of our current knowledge, this is the first metabolomic study on three different fruit pathosytems interacting with BABA priming.

Firstly, untargeted metabolomic profiling indicated a great impact of BABA treatment on metabolic profiles (Figure 3A and Table 1). Hence, the treatment of young tomato plants with BABA metabolically primes fruit tissues, and this stimulation was likely more critical than it was for the pathogen inoculations. This might result from the hormonal nature of BABA, which deeply affects plant metabolism, or from stress-related responses that are activated by the chemical itself, as it has been reported previously for high concentrations of BABA or in another *Solanum* species (i.e., potato) [21].

Secondly, targeted analyses of compounds involved in central metabolism demonstrated that the primary metabolic pools responded to the pathosystem inoculations, which reflected the developmental stage of the fruit, as exemplified by the multivariate distinction between the first/second fruit and the third fruit (Figure 3B). In complement, fructose, sucrose, fumarate and glutamate showed statistically significant variations upon inoculation (Table 1). Since fruit of slightly different ages harbour different profiles of primary compounds, we could assume that central metabolism is tuned to fruit growth, more specifically soluble sugars, amino and organic acids. This agrees with previous phenotyping and modelling studies on tomato that demonstrate metabolic shifts in carbon metabolism in the growing fruit [25,28,29,31]. Furthermore, it has been recently confirmed through transcriptomics and proteomics that the developing fruit not only undergoes metabolic shifts in central pathways, but also redox metabolism, such as for pyridine nucleotides that are detrimental to energy homeostasis [27,32,33]. However, major questions remain regarding the nature and dynamics of shifts in central metabolism upon pathogen inoculation. It is reasonable to expect that further investigations involving a more comprehensive view of fruit primary metabolism and how microbial challenges dynamically affect such pathways might significantly improve our understanding of the relationships between central metabolism and fruit–pathogen interactions. In turn, this should provide novel strategies to obtain fruit of better quality and stress resilience [32].

Upon pathogen challenge, while BABA is effective in leaf tissue, very little is known about its contribution in fruit. According to our fruit pathoassays (Figure 1), the treatment of tomato seedlings with BABA resulted in a broad-spectrum resistance against microbes that have different infection strategies, including necrotrophic or biotrophic, and fungal, oomycete or bacterial pathogens. Further, the primed responses are tailored to the encountered pathogen, as exemplified by the little overlap between the different primed states of the three pathosystems (Figure 5). This implies that the induced resistance state is very specific, which strongly suggests that BABA primes multiple signalling pathways through which such different microbes are resisted in the fruit. Among those metabolic responses, hormonal regulations appear detrimental to BABA-induced immunity [23,24,34]. Accordingly, putative annotation of metabolomic markers indicates that hormone conjugates, including salicylic and jasmonic derivatives, and other defence compounds (i.e., flavonoids), are induced upon infection and BABA treatment (Table 3). Given the diverse set of immune responses that the fruit deploys against different microbial stresses, our study highlights the adaptability of priming as a "stimulus-dependent plasticity of response traits" [35]. For this reason, the exact underlying molecular mechanisms of priming are difficult to describe precisely and their description requires further research [36].

Whilst BABA treatment in many plant species results in a stress phenotype that manifests through developmental alterations (e.g., growth, yield, seed production) [17–19], we found no differences upon BABA application in fruit size, but observed delayed fruit production or fruit ripening (Figure 2), as described previously [23]. However, after the number of ripened fruit had equalised between both treatments, BABA-treated plants continued producing fruit at a much faster rate than the water-treated plants (Figure 2). Seemingly, through its induction of immune responses, BABA thus provides a positive fitness element for tomato plants. This trade-off might emerge, in part, from the stimulation of various signalling pathways, more specifically the ones that link to the central metabolism, such as amino acids or carbohydrates [34]. As a result, BABA-treated plants would perform particularly well. This agrees with what we know about plant perception of BABA in *A. thaliana*. The binding of BABA by an aspartyl-tRNA synthetase blocks the enzyme, consequently triggering the accumulation of its canonical substrate, uncharged tRNA, which leads to changes in amino acid pools in the plant, therefore affecting primary metabolism [20]. Subsequent signalling modulations might result from an alteration in amino acid precursors (e.g., ethylene, auxin) that would alter fruit production [37].

Despite its economic importance, the molecular mechanisms underlying the pathogenicity of *B. cinerea*, *P. infestans* and *P. syringae* are poorly understood in fruit. From a computational systems biology perspective, the study of plant–pathogen interactions involved structural and comparative genomics, transcriptomics, and protein–protein interactions [38]. Further, high-resolution metabolome data and sufficient datapoints over time are essential to calculate metabolite coefficients and thus predict metabolic fluxes [39]. Recently, genome-scale metabolic models of *Solanum* species (i.e., potato, tomato) and *Phy* have been integrated to simulate the metabolic fluxes that occur during infection [40,41]. These studies yield insights into the molecular aspects of photosynthesis suppression by *Phy* via the flux of carboxylation to oxygenation reactions, or the nutrient intakes by *Phy* during different phases of the infection cycle. Interestingly, stage-specific profiles embedded in the joint metabolism of the host and pathogen could potentially be refined by integrating the high-resolution metabolome data of tomato infection [41]. Such elegant works involve leaf tissues. Here, we show that fruit metabolomics and modelling can assist in addressing fruit–pathogen interactions. Using top-down modelling based on the construction of generalised linear models [30], we demonstrate that metabolomics data can be used to accurately predict the measured resistance to various fruit pathogens (Figure 6A). Besides, through the evaluation of the occurrence of best predictors, our data indicate that soluble sugars, more specifically fructose [42], and defence metabolites are pivotal to predict the resistance to fruit pathogens (Table S1). Clearly, a more global systems biology approach based on a higher level of variation in the conditions (e.g., multiple genotypes or priming treatments, various growth stages of the fruit, several infection points) will shed some light on the underlying mechanisms of fruit–pathogen interactions.

Overall, our study validates the value of metabolomics and modelling approaches in the field of phytopathological investigations. This work provides a great perspective for the structural elucidation of the key metabolites involved in broad-spectrum BABA-induced priming in tomato fruit. Although it was only possible to tentatively annotate metabolic biomarkers on the basis of detected HR-accurate *m/z* (Table 3 and Table S1), our analytical and statistical approach can be further optimised for, e.g., metabolite identification through structural elucidation by NMR or targeted MS/MS analyses. A combination of LCMS with purification steps (e.g., SPE cartridge, fractionation) could prove useful for de novo identification.

4. Materials and Methods

4.1. Tomato Cultivation

Tomato (*Solanum lycopersicum*) Micro-Tom was used for all experiments described in this publication. Seeds were incubated for 4 days in wet paper at 28 °C to promote homogeneous germination. Germinates were then planted in individual 80 mL pots containing M3 soil. Plants were grown in a controlled-environment greenhouse chamber with 16h of light, at 26 °C, and 8 h of darkness at 21 °C, and 200 μM.m^{-1}.s^{-1} light intensity. Experiments were performed from November 2017 until May 2018 in the United Kingdom.

4.2. Biochemicals, Reagents and Treatments

All solvents and reagents used in this study were of analytical or MS grades. B-aminobutyric acid (BABA) was obtained from Sigma-Aldrich (A4420-7). Treatments with BABA were performed entirely as described in [23]. Briefly, 2 week-old tomato seedlings were soil-drenched with 8 mL per pot of either water or 5 mM BABA solution, to generate a final concentration of 0.5 mm in the soil. One week post-treatment, roots were carefully washed under running tap water and then plants were transplanted into individual 2.2 L pots containing untreated M3 soil. Plants were allowed to grow for between 9 and 12 weeks until the fruit turned red, at which point they were harvested and infected with the different pathogens.

4.3. Fitness Parameters of Tomato Fruit

Growth and yield were assessed entirely as described in [23]. Assessment of fruit ripening was done as described in [29], by classifying fruit in different levels of maturity by colour.

4.4. Pathogens and Inoculations

Cultivations of *Botrytis cinerea* strain R16 [43], *Phytophthora infestans* 88,069 [44] and *Pseudomonas syringae* pv. *tomato DC3000* (*Pst DC3000*) [45] were done as described in the corresponding publications. For inoculations with *B. cinerea*, the first fruit were used. Inoculations were performed entirely as described in [23]. For infections with *P. infestans*, the second fruit were used. Inoculations were performed by placing 10 µl drops of a spore concentration of 5×10^4 spores/mL onto the needle-wounded tip of the tomato fruit. After infection, fruit were kept at 20 °C in the dark. For *P. syringae* infections, the third fruit were used. Inoculations were done by spraying bacteria onto the fruit in a concentration of 10^8 cells/mL in 10 mM $MgSO_4$ and 0.05% (v/v) Silwet L-77. Infected fruit were kept in the dark at 25 °C. Mock inoculations were performed by following the exact same protocols but without pathogens in the solutions. Fruit were 56 days post-anthesis (dpa), 63 dpa and 70 dpa for the first, second and third fruit, respectively.

Scoring of *B. cinerea* symptoms were performed entirely as described in [23]. Scoring of *P. infestans* disease was done by classifying lesions into different categories of fruit colonization: Class 0; healthy, Class I; necrosis associated with the lesion, Class II; necrosis and mycelium associated with the lesion, Class III; necrosis and mycelium spread in the fruit. Scoring of *Pst DC3000* disease was done by classifying lesions into different categories of fruit damage: Class 0; healthy, Class I; turgent but cracking fruit, Class II; cracked fruit, Class III; fruit tissue collapse. Disease severity rates were calculated from the nominal lesion categories of four fruit per plant (n = 11), as described in [46]. Statistical analysis of disease phenotypes was performed as described in [23].

4.5. Metabolite Extraction

For metabolome analysis, the first, second and third fruit developed by plants were used for pathoassays with *B. cinerea (Bot)*, *P. infestans (Phy)* and *P. syringae (Pst)*, respectively. Experiments on each type of fruit were separated by one week. Infections were performed as described above when the corresponding fruit were fully ripened. Two days after inoculation with the different pathogens, fresh pericarps were rapidly collected into 2 mL-microtubes, then flash-frozen in liquid nitrogen and freeze-dried for 72 h (Pilote Compact, SARL CRYOTEC, Saint-Gély-du-Fesc, France). Fine grinding of dried material was subsequently performed using a ball mixer for 2 min at 30 Hz (Retsch Mill MM400, fisher scientific, Bordeaux, France) after adding two metal beads (Beads inox AISI 400C 5 mm, CIMAP, Caen, France) to each tube (Micro-tube, 2 mL PP, Sarstedt, Germany). Ten milligrams of each replicated sample were weighed into 1.1 mL-micronic tubes (MP32033L, Micronic, Lelystad, Netherlands), randomised onto a 96-micronic rack (MPW51001BC6, Micronic, Lelystad, Netherlands) then capped using a robotised capper–decapper (Decapper 193000/00, Hamilton, Bienne, Switzerland). Each rack also contained an empty tube corresponding to the extraction blank. The resulting micronics were then stored at −80 °C. Extraction of metabolites was conducted on four biologically replicated pericarp samples (n = 4) using a robotised extraction method developed at *Bordeaux Metabolome Facility* (https://metabolome.cgfb.u-bordeaux.fr/en, Villenave d'Ornon, France). The robot was a bespoken piece of equipment that allowed for pipetting solvents, mixing, cooling and centrifuging racks of micronics. After decapping the micronics, the extraction began by adding 300 µL of solvent A containing 80% ethanol and 0.1% formic acid (v/v) with 250 µg/mL methyl vanillate as the internal standard. Racks were agitated on the robot (30 sec, 500 rpm) then placed for 15 min into a sonicator containing ice-cold water (Elmasonic S300, Elma, Singen, Germany). Racks were put back on the robot and centrifuged (5 min, 1350 g). The first round of extraction stopped by pipetting 300 µL of the resulting supernatant into new 1.1 mL-micronic tubes. A second round of extraction was performed with 300 µl of solvent A, and the resulting pellet was finally washed with solvent B (50% ethanol (v/v)). The micronics-containing supernatants were kept for filtration and the micronics with the pellets were kept for further starch and total protein analysis.

Filtration was also robotised (Microlab STARlet, Hamilton, Bienne, Switzerland) and allowed for the transfer of the supernatants onto a filtration 96-well sterile clear plate (MSGVS2210, 0.22 µM,

Hydrophil. Low Protein Binding Durapore, Millipore, Molsheim, France) according to the supplier's instruction. Filtrates were subsequently collected into a new micronic tube. Finally, quality control (QC) samples were prepared by robotically pipetting 15 μL of each sample into a single tube that was mixed afterwards (Microlab STARlet, Hamilton, Bienne, Switzerland). Each rack was supplemented with a micronic tube containing the QC mix. The QC sample was replicated six times along the project run.

4.6. Targeted Biochemical Phenotyping

Targeted analyses of sucrose, fructose, glucose, starch, fructose-6-P, glucose-6-P, glutamate, malate, fumarate, total soluble proteins and total polyphenols were conducted on the HiTMe plateau at *Bordeaux Metabolome Facility*. Measurements were based on coupled enzyme assays as described previously [25,26], except for total soluble proteins that were evaluated via Bradford assay [47], and total phenols that were measured colorimetrically using a redox reaction with Folin–Ciocalteu reagent and gallic acid as the standard [48].

4.7. Untargeted Metabolic Profiling

Untargeted metabolic profiling by UHPLC-LTQ-Orbitrap mass spectrometry (LCMS) was performed using an Ultimate 3000 ultra-high-pressure liquid chromatography (UHPLC) system coupled to an LTQ-Orbitrap Elite mass spectrometer interfaced with an electrospray (ESI) ionisation source (ThermoScientific, Bremen, Germany). The system was controlled by Thermo XCalibur v.3.0.63 software. Chromatographic separation was achieved at a flow rate of 350 μL/min using a GEMINI UHPLC C18 column (150 × 2 mm, 3 μm, Le Pecq, Phenomenex, France) coupled to a C18 SecurityGuard GEMINI pre-column (4 × 2 mm, 3 μm, Le Pecq, Phenomenex, France). The column was maintained at 35 °C and the injection volume was 5 μL. The mobile phase consisted of solvent A (0.05 % (v/v) formic acid in water) and solvent B (acetonitrile) with the following gradient: 0–0.5 min 3% B, 0.5–1 min 3% B, 1–9 min 50% B, 9–13 min 100% B, 13–14 min 100% B, 14–14.5 min 3% B, 14.5–18 min 3% B. Ionisation of samples was performed in both negative and positive mode with the following parameters: ESI^- (Heater temp: 300 °C, Sheath Gas Flow Rate: 45 (arb), Aux Gas Flow Rate: 15 (arb), Sweep Gas Flow Rate: 10 (arb), I Spray Voltage: 2.5 kV, Capillary Temp: 300 °C, S-Lens RF Level: 60%), and ESI^+ (Heater temp: 300 °C, Sheath Gas Flow Rate: 60 (arb), Aux Gas Flow Rate: 20 (arb), Sweep Gas Flow Rate: 10 (arb), I Spray Voltage: 3.2 kV, Capillary Temp: 300 °C, S-Lens RF Level: 55%). MS full scan detection of ions was operated by FTMS (50–1500 Da) at a resolution of 240,000. Prior to analyses, the LTQ-Orbitrap was calibrated by infusing a solution of the calibration dependent of the ionisation mode (Pierce© ESI Negative Ion Calibration Solution (ref: 88324); Pierce LTQ Velos ESI Positive Ion Calibration solution (ref: 88323). The injection sequence started with three blank extracts, then three QC samples, then one blank extract, and each group of samples was subsequently injected, followed by a blank extract. Another two QC samples were injected throughout the analysis. In total, six QC samples and 16 blank extracts were injected to correct for mass spectrometer signal drift, and to filter out variables detected in blanks, respectively.

4.8. Processing and Statistical Analysis of Metabolomic Datasets

Processing of raw LCMS data using XCMS in R (v 3.6.1) [49] yielded 10,875 detected RT-*m/z* pairs for ESI^+ and 5,796 for ESI^-. After data-cleaning (blank check, $\Delta_{RT} < 60$ s, $\Delta_{m/z} < 0.015$ Da, CV QC < 30%), 6887 variables were retained for further chemiometrics. Both untargeted and targeted metabolomic data were first normalised by median normalisation, cube-root transformation and Pareto scaling using MetaboAnalyst v.3 [50] before applying multivariate and univariate statistical analyses [51]. The normalised dataset is available as Supplemental Material 1. PCA and PLS-DA were performed with MetaboAnalyst v.3 providing satisfactory validation parameters of the multivariate models ($R^2 > 0.87$ and $Q^2 > 0.35$). PC coordinates for metabolomic features that are responsible for PC1 and PC2 are presented in Supplemental Material 2. Univariate statistical methods were performed using MeV v.4.9.0. [52] at $P < 0.05$ for two-factor ANOVA and $P < 0.01$ for binary comparisons by *t*-tests. In addition, MarVis v 1 was used to confirm the statistically significant variation om the priming markers through a Kruskal–Wallis test at $P < 0.05$ with correction for false discovery rate [53,54]. Putative annotation of such markers was performed by screening the detected exact *m/z* against multiple online databases, including METLIN chemical database (https://metlin.scripps.edu/) [55] and KNApSAcK (http://kanaya.naist.jp/KNApSAcK/) [56]. The resulting predicted pathways were checked using the PubChem database (https://pubchem.ncbi.nlm.nih.gov/).

4.9. Top-down Modelling Approach

Generalised linear models were constructed in R (v 3.6.1) using the *glmnet* package (v 3.0-2) [30] in order to identify potential links between detected metabolic markers and resistance to the fruit pathogens. Those models were used to predict resistance values based on the detected metabolic markers. Cross-validation was applied by randomly dividing the datasets into two parts: 80 % of the individuals were used to construct the models and 20 % to check for the quality of the prediction. The quality of the models was assessed based on the mean square error between real and predicted values. To cope with this randomisation, 500 models were constructed for each measurement. Generalised linear models contain a penalisation value, allowing less informative variables to be discarded as this value increases (1000 values were tested for each of the 500 models), hence variables occurring the most in the models can be seen as the most stable predictors of resistance to biotic challenges. Given the high number of metabolic variables and the relatively small set of plants, 500 randomly generated resistance datasets were created to estimate the chances of predicting random values. A Student's *t*-test was used to compare the quality of predictions of real and random resistances.

Supplementary Materials: The following are available online at http://www.mdpi.com/2218-1989/10/3/96/s1, Figure S1. Inoculation-responsive central metabolites. Major compounds involved in central metabolism statistically responded to the inoculation factor of a two-factor ANOVA (P given into brackets). Bar charts indicate means of normalised intensities of four independent bioreplicates (n = 4; ± SEM). Bot: *Botrytis cinerea*, Phy: *Phytophthora infestans*, Pst: *Pseudomonas syringae* pv. *tomato*, B: BABA-treated plants, W: water-treated plants, M: mock-inoculated fruit, P: pathogen-inoculated fruit. Figure S2. Partial least square discriminant analysis for each pathosystem. PLS-DA score plots (n = 4) of 6898 features (6887 LCMS variables + 11 major compounds) between the three different pathosystems. Validation parameters of the PLS model are given in red for each plot. Figure S3. Metabolic markers for BABA primed responses against fruit pathogens. LCMS significant markers that overlap between *Bot* and *Phy* (A), *Bot* and *Pst* (B) and *Phy* and *Pst* (C) in response to BABA priming and after infection (see Table 3). Markers are labelled according to their high-resolution detected *m/z*. Bar charts indicate means of normalised intensities (n = 4; ± SEM). See Figure S1 for sample labels. Table S1. Putative annotation of the top 15 predictors. Supplemental Material 1: metabolomic parameters detected y LCMS. 2: P indicating the statistical significance from a Kruskal–Wallis test followed by correction for false discovery rate using the Benjamini–Hochberg method. Supplemental Material 1. Normalised metabolomics dataset combining 6887 *m/z* features and 11 major compounds. Supplemental Material 2. Coordinates for Principal Component Analyses ranking the important metabolomic variables.

Author Contributions: Conceptualization, E.L. and P.P.; Data curation, E.L., A.F., C.C. (Cédric Cassan) and P.P.; Formal analysis, E.L., A.F., C.C. (Cédric Cassan), C.C. (Chloé Chevanne), C.F.K., S.P. and P.P.; Funding acquisition, E.L., Y.G. and P.P.; Investigation, E.L., A.F., C.C. (Cédric Cassan), C.C. (Chloé Chevanne), C.F.K., S.P. and P.P.; Methodology, E.L., A.F., C.C. (Cédric Cassan), S.P. and P.P.; Project administration, E.L. and P.P.; Resources, E.L., S.P., Y.G. and P.P.; Software, S.P.; Supervision, E.L. and P.P.; Validation, E.L., S.P., Y.G. and P.P.; Visualization, E.L., S.P. and P.P.; Writing—original draft, E.L. and P.P.; Writing—review & editing, E.L., S.P., Y.G. and P.P. All authors have read and agreed to the published version of the manuscript.

Funding: The authors are grateful for financial support from MetaboHUB (ANR-11-INBS-0010) and PHENOME (ANR-11-INBS-0012) projects to INRAE, and for the BBSRC Future Leader Fellowship BB/P00556X/1 and BB/P00556X/2 to EL.

Acknowledgments: The authors thank Sam Wilkinson for his comments on the project. *P. infestans* isolate was provided by Steve Whisson (The James Hutton Institute).

Conflicts of Interest: The authors declare no conflict of interest.

References

1. Donatelli, M.; Magarey, R.D.; Bregaglio, S.; Willocquet, L.; Whish, J.P.M.; Savary, S. Modelling the impacts of pests and diseases on agricultural systems. *Agric. Syst.* **2017**, *155*, 213–224. [CrossRef] [PubMed]

2. Sundström, J.F.; Albihn, A.; Boqvist, S.; Ljungvall, K.; Marstorp, H.; Martiin, C.; Nyberg, K.; Vågsholm, I.; Yuen, J.; Magnusson, U. Future threats to agricultural food production posed by environmental degradation, climate change, and animal and plant diseases – a risk analysis in three economic and climate settings. *Food Secur.* **2014**, *6*, 201–215. [CrossRef]
3. FAOSTAT. Available online: http://www.fao.org/faostat/en/#home (accessed on 7 January 2020).
4. Fry, W. Phytophthora infestans: The plant (and R gene) destroyer. *Mol. Plant Pathol.* **2008**, *9*, 385–402. [CrossRef] [PubMed]
5. Mansfield, J.; Genin, S.; Magori, S.; Citovsky, V.; Sriariyanum, M.; Ronald, P.; Dow, M.; Verdier, V.; Beer, S.V.; Machado, M.A.; et al. Top 10 plant pathogenic bacteria in molecular plant pathology. *Mol. Plant Pathol.* **2012**, *13*, 614–629. [CrossRef] [PubMed]
6. Williamson, B.; Tudzynski, B.; Tudzynski, P.; Van Kan, J.A.L. Botrytis cinerea: The cause of grey mould disease. *Mol. Plant Pathol.* **2007**, *8*, 561–580. [CrossRef] [PubMed]
7. Arah, I.K.; Ahorbo, G.K.; Anku, E.K.; Kumah, E.K.; Amaglo, H. Postharvest Handling Practices and Treatment Methods for Tomato Handlers in Developing Countries: A Mini Review. *Adv. Agric.* **2016**, *2016*, 1–8. [CrossRef]
8. Luna, E. Using Green Vaccination to Brighten the Agronomic Future. *Outlooks Pest Manag.* **2016**, *27*, 136–141. [CrossRef]
9. Pétriacq, P.; López, A.; Luna, E. Fruit Decay to Diseases: Can Induced Resistance and Priming Help? *Plants* **2018**, *7*, 77. [CrossRef]
10. Jones, J.D.G.; Dangl, J.L. The plant immune system. *Nature* **2006**, *444*, 323–329. [CrossRef]
11. Mauch-Mani, B.; Baccelli, I.; Luna, E.; Flors, V. Defense Priming: An Adaptive Part of Induced Resistance. *Annu. Rev. Plant Biol.* **2017**, *68*, 485–512. [CrossRef]
12. Martinez-Medina, A.; Flors, V.; Heil, M.; Mauch-Mani, B.; Pieterse, C.M.J.; Pozo, M.J.; Ton, J.; van Dam, N.M.; Conrath, U. Recognizing Plant Defense Priming. *Trends Plant Sci.* **2016**, *21*, 818–822. [CrossRef] [PubMed]
13. Cohen, Y.; Vaknin, M.; Mauch-Mani, B. BABA-induced resistance: Milestones along a 55-year journey. *Phytoparasitica* **2016**, *44*, 513–538. [CrossRef]
14. Thevenet, D.; Pastor, V.; Baccelli, I.; Balmer, A.; Vallat, A.; Neier, R.; Glauser, G.; Mauch-Mani, B. The priming molecule β-aminobutyric acid is naturally present in plants and is induced by stress. *New Phytol.* **2017**, *213*, 552–559. [CrossRef] [PubMed]
15. Ton, J.; Jakab, G.; Toquin, V.; Flors, V.; Iavicoli, A.; Maeder, M.N.; Métraux, J.-P.; Mauch-Mani, B. Dissecting the β-Aminobutyric Acid–Induced Priming Phenomenon in Arabidopsis. *Plant Cell* **2005**, *17*, 987–999. [CrossRef] [PubMed]
16. Zimmerli, L.; Jakab, G.; Métraux, J.-P.; Mauch-Mani, B. Potentiation of pathogen-specific defense mechanisms in *Arabidopsis* by β-aminobutyric acid. *Proc. Natl. Acad. Sci. USA* **2000**, *97*, 12920–12925. [CrossRef]
17. Koen, E.; Trapet, P.; Brulé, D.; Kulik, A.; Klinguer, A.; Spicher, L.; Meunier-Prest, R.; Boni, G.; Glauser, G.; Mauch-Mani, B.; et al. β-Aminobutyric Acid (BABA)-Induced Resistance in Arabidopsis thaliana : Link with Iron Homeostasis. *Mol. Plant-Microbe Interact.* **2014**. [CrossRef]
18. Luna, E.; Beardon, E.; Ravnskov, S.; Scholes, J.; Ton, J. Optimizing Chemically Induced Resistance in Tomato Against Botrytis cinerea. *Plant Dis.* **2016**, *100*, 704–710. [CrossRef]
19. van Hulten, M.; Pelser, M.; van Loon, L.C.; Pieterse, C.M.J.; Ton, J. Costs and benefits of priming for defense in *Arabidopsis*. *Proc. Natl. Acad. Sci. USA* **2006**, *103*, 5602–5607. [CrossRef]
20. Luna, E.; van Hulten, M.; Zhang, Y.; Berkowitz, O.; López, A.; Pétriacq, P.; Sellwood, M.A.; Chen, B.; Burrell, M.; van de Meene, A.; et al. Plant perception of β-aminobutyric acid is mediated by an aspartyl-tRNA synthetase. *Nat. Chem. Biol.* **2014**, *10*, 450. [CrossRef]
21. Bengtsson, T.; Holefors, A.; Witzell, J.; Andreasson, E.; Liljeroth, E. Activation of defence responses to Phytophthora infestans in potato by BABA. *Plant Pathol.* **2014**, *63*, 193–202. [CrossRef]
22. Worrall, D.; Holroyd, G.H.; Moore, J.P.; Glowacz, M.; Croft, P.; Taylor, J.E.; Paul, N.D.; Roberts, M.R. Treating seeds with activators of plant defence generates long-lasting priming of resistance to pests and pathogens. *New Phytol.* **2012**, *193*, 770–778. [CrossRef] [PubMed]
23. Wilkinson, S.W.; Pastor, V.; Paplauskas, S.; Pétriacq, P.; Luna, E. Long-lasting β-aminobutyric acid-induced resistance protects tomato fruit against Botrytis cinerea. *Plant Pathol.* **2018**, *67*, 30–41. [CrossRef]
24. Balmer, A.; Pastor, V.; Gamir, J.; Flors, V.; Mauch-Mani, B. The 'prime-ome': Towards a holistic approach to priming. *Trends Plant Sci.* **2015**, *20*, 443–452. [CrossRef] [PubMed]

25. Biais, B.; Bénard, C.; Beauvoit, B.; Colombié, S.; Prodhomme, D.; Ménard, G.; Bernillon, S.; Gehl, B.; Gautier, H.; Ballias, P.; et al. Remarkable Reproducibility of Enzyme Activity Profiles in Tomato Fruits Grown under Contrasting Environments Provides a Roadmap for Studies of Fruit Metabolism. *Plant Physiol.* **2014**, *164*, 1204–1221. [CrossRef]
26. Gibon, Y.; Vigeolas, H.; Tiessen, A.; Geigenberger, P.; Stitt, M. Sensitive and high throughput metabolite assays for inorganic pyrophosphate, ADPGlc, nucleotide phosphates, and glycolytic intermediates based on a novel enzymic cycling system. *Plant J.* **2002**, *30*, 221–235. [CrossRef]
27. Belouah, I.; Nazaret, C.; Pétriacq, P.; Prigent, S.; Bénard, C.; Mengin, V.; Blein-Nicolas, M.; Denton, A.K.; Balliau, T.; Augé, S.; et al. Modeling Protein Destiny in Developing Fruit. *Plant Physiol.* **2019**, *180*, 1709–1724. [CrossRef]
28. Colombié, S.; Beauvoit, B.; Nazaret, C.; Bénard, C.; Vercambre, G.; Le Gall, S.; Biais, B.; Cabasson, C.; Maucourt, M.; Bernillon, S.; et al. Respiration climacteric in tomato fruits elucidated by constraint-based modelling. *New Phytol.* **2017**, *213*, 1726–1739. [CrossRef]
29. Colombié, S.; Nazaret, C.; Bénard, C.; Biais, B.; Mengin, V.; Solé, M.; Fouillen, L.; Dieuaide-Noubhani, M.; Mazat, J.-P.; Beauvoit, B.; et al. Modelling central metabolic fluxes by constraint-based optimization reveals metabolic reprogramming of developing Solanum lycopersicum (tomato) fruit. *Plant J.* **2015**, *81*, 24–39. [CrossRef]
30. Friedman, J.; Hastie, T.; Tibshirani, R. Regularization Paths for Generalized Linear Models via Coordinate Descent. *J. Stat. Softw.* **2010**, *33*, 1–22. [CrossRef]
31. Beauvoit, B.P.; Colombié, S.; Monier, A.; Andrieu, M.-H.; Biais, B.; Bénard, C.; Chéniclet, C.; Dieuaide-Noubhani, M.; Nazaret, C.; Mazat, J.-P.; et al. Model-Assisted Analysis of Sugar Metabolism throughout Tomato Fruit Development Reveals Enzyme and Carrier Properties in Relation to Vacuole Expansion. *Plant Cell* **2014**, *26*, 3224–3242. [CrossRef]
32. Beauvoit, B.; Belouah, I.; Bertin, N.; Cakpo, C.B.; Colombié, S.; Dai, Z.; Gautier, H.; Génard, M.; Moing, A.; Roch, L.; et al. Putting primary metabolism into perspective to obtain better fruits. *Ann. Bot.* **2018**, *122*, 1–21. [CrossRef] [PubMed]
33. Decros, G.; Beauvoit, B.; Colombié, S.; Cabasson, C.; Bernillon, S.; Arrivault, S.; Guenther, M.; Belouah, I.; Prigent, S.; Baldet, P.; et al. Regulation of Pyridine Nucleotide Metabolism During Tomato Fruit Development Through Transcript and Protein Profiling. *Front. Plant Sci.* **2019**, 10. [CrossRef] [PubMed]
34. Gamir, J.; Cerezo, M.; Flors, V. The plasticity of priming phenomenon activates not only common metabolomic fingerprint but also specific responses against P. cucumerina. *Plant Signal Behav.* **2014**, *9*, e28916. [CrossRef] [PubMed]
35. Hilker, M.; Schwachtje, J.; Baier, M.; Balazadeh, S.; Bäurle, I.; Geiselhardt, S.; Hincha, D.K.; Kunze, R.; Mueller-Roeber, B.; Rillig, M.C.; et al. Priming and memory of stress responses in organisms lacking a nervous system. *Biol. Rev.* **2016**, *91*, 1118–1133. [CrossRef]
36. Tugizimana, F.; Mhlongo, M.I.; Piater, L.A.; Dubery, I.A. Metabolomics in Plant Priming Research: The Way Forward? *Int. J. Mol. Sci.* **2018**, *19*, 1759. [CrossRef]
37. Roch, L.; Dai, Z.; Gomès, E.; Bernillon, S.; Wang, J.; Gibon, Y.; Moing, A. Fruit Salad in the Lab: Comparing Botanical Species to Help Deciphering Fruit Primary Metabolism. *Front. Plant Sci.* **2019**, 10. [CrossRef]
38. Pinzón, A.; Barreto, E.; Bernal, A.; Achenie, L.; González Barrios, A.F.; Isea, R.; Restrepo, S. Computational models in plant-pathogen interactions: The case of Phytophthora infestans. *Theor. Biol. Med Model.* **2009**, *6*, 24. [CrossRef]
39. Kleessen, S.; Irgang, S.; Klie, S.; Giavalisco, P.; Nikoloski, Z. Integration of transcriptomics and metabolomics data specifies the metabolic response of Chlamydomonas to rapamycin treatment. *Plant J.* **2015**, *81*, 822–835. [CrossRef]
40. Botero, K.; Restrepo, S.; Pinzón, A. A genome-scale metabolic model of potato late blight suggests a photosynthesis suppression mechanism. *BMC Genom.* **2018**, *19*, 863. [CrossRef]
41. Rodenburg, S.Y.A.; Seidl, M.F.; Judelson, H.S.; Vu, A.L.; Govers, F.; de Ridder, D. Metabolic Model of the *Phytophthora infestans*-Tomato Interaction Reveals Metabolic Switches during Host Colonization. *mBio* **2019**, *10*, e00454-19. [CrossRef]
42. Lecompte, F.; Nicot, P.C.; Ripoll, J.; Abro, M.A.; Raimbault, A.K.; Lopez-Lauri, F.; Bertin, N. Reduced susceptibility of tomato stem to the necrotrophic fungus Botrytis cinerea is associated with a specific adjustment of fructose content in the host sugar pool. *Ann. Bot.* **2017**, *119*, 931–943. [CrossRef] [PubMed]

43. Faretra, F.; Pollastro, S. Genetic studies of the phytopathogenic fungus Botryotinia fuckeliana (Botrytis cinerea) by analysis of ordered tetrads. *Mycol. Res.* **1996**, *100*, 620–624. [CrossRef]
44. Whisson, S.C.; Boevink, P.C.; Moleleki, L.; Avrova, A.O.; Morales, J.G.; Gilroy, E.M.; Armstrong, M.R.; Grouffaud, S.; van West, P.; Chapman, S.; et al. A translocation signal for delivery of oomycete effector proteins into host plant cells. *Nature* **2007**, *450*, 115–118. [CrossRef] [PubMed]
45. Bais, H.P.; Fall, R.; Vivanco, J.M. Biocontrol of Bacillus subtilis against Infection of Arabidopsis Roots by Pseudomonas syringae Is Facilitated by Biofilm Formation and Surfactin Production. *Plant Physiol.* **2004**, *134*, 307–319. [CrossRef]
46. Bock, C.H.; Poole, G.H.; Parker, P.E.; Gottwald, T.R. Plant Disease Severity Estimated Visually, by Digital Photography and Image Analysis, and by Hyperspectral Imaging. *Crit. Rev. Plant Sci.* **2010**, *29*, 59–107. [CrossRef]
47. Bradford, M.M. A rapid and sensitive method for the quantitation of microgram quantities of protein utilizing the principle of protein-dye binding. *Anal. Biochem.* **1976**, *72*, 248–254. [CrossRef]
48. Vinson, J.A.; Su, X.; Zubik, L.; Bose, P. Phenol Antioxidant Quantity and Quality in Foods: Fruits. *J. Agric. Food Chem.* **2001**, *49*, 5315–5321. [CrossRef]
49. Smith, C.A.; Want, E.J.; O'Maille, G.; Abagyan, R.; Siuzdak, G. XCMS: Processing Mass Spectrometry Data for Metabolite Profiling Using Nonlinear Peak Alignment, Matching, and Identification. *Anal. Chem.* **2006**, *78*, 779–787. [CrossRef]
50. Xia, J.; Sinelnikov, I.V.; Han, B.; Wishart, D.S. MetaboAnalyst 3.0–making metabolomics more meaningful. *Nucleic Acids Res.* **2015**, *43*, W251–W257. [CrossRef]
51. Pétriacq, P.; Williams, A.; Cotton, A.; McFarlane, A.E.; Rolfe, S.A.; Ton, J. Metabolite profiling of non-sterile rhizosphere soil. *Plant J.* **2017**, *92*, 147–162. [CrossRef]
52. Saeed, A.I.; Sharov, V.; White, J.; Li, J.; Liang, W.; Bhagabati, N.; Braisted, J.; Klapa, M.; Currier, T.; Thiagarajan, M.; et al. TM4: A Free, Open-Source System for Microarray Data Management and Analysis. *BioTechniques* **2003**, *34*, 374–378. [CrossRef] [PubMed]
53. Hochberg, Y.; Benjamini, Y. More powerful procedures for multiple significance testing. *Stat. Med.* **1990**, *9*, 811–818. [CrossRef] [PubMed]
54. Kaever, A.; Landesfeind, M.; Possienke, M.; Feussner, K.; Feussner, I.; Meinicke, P. MarVis-Filter: Ranking, Filtering, Adduct and Isotope Correction of Mass Spectrometry Data. *J. Biomed. Biotechnol.* **2012**, *2012*, 263910. [CrossRef] [PubMed]
55. Smith, C.; O'Maille, G.; Want, E.; Qin, C.; Trauger, S.; Brandon, T.; Custodio, D.; Abagyan, R.; Siuzdak, G. METLIN: A metabolite mass spectral database. *Ther. Drug Monit.* **2006**, *27*, 747–751. [CrossRef] [PubMed]
56. Afendi, F.M.; Okada, T.; Yamazaki, M.; Hirai-Morita, A.; Nakamura, Y.; Nakamura, K.; Ikeda, S.; Takahashi, H.; Altaf-Ul-Amin, M.; Darusman, L.K.; et al. KNApSAcK Family Databases: Integrated Metabolite–Plant Species Databases for Multifaceted Plant Research. *Plant Cell Physiol.* **2011**, *53*, e1. [CrossRef] [PubMed]

metabolites

Article

Identification of Bioactive Phytochemicals in Mulberries

Gilda D'Urso [1], Jurriaan J. Mes [2], Paola Montoro [1], Robert D. Hall [3,4] and Ric C.H. de Vos [3,*]

1 Department of Pharmacy, University of Salerno, 84084 Fisciano SA, Italy; gidurso@unisa.it (G.D.); pmontoro@unisa.it (P.M.)
2 Business Unit Fresh Food and Chains, Wageningen Food & Biobased Research, Wageningen University and Research, 6708 WG Wageningen, The Netherlands; jurriaan.mes@wur.nl
3 Business Unit Bioscience, Wageningen Plant Research, Wageningen University and Research, 6708 PB Wageningen, The Netherlands; robert.hall@wur.nl
4 Laboratory of Plant Physiology, Wageningen University and Research, 6708 PB Wageningen, The Netherlands
* Correspondence: ric.devos@wur.nl; Tel.: +31-317480841

Received: 29 November 2019; Accepted: 18 December 2019; Published: 20 December 2019

Abstract: Mulberries are consumed either freshly or as processed fruits and are traditionally used to tackle several diseases, especially type II diabetes. Here, we investigated the metabolite compositions of ripe fruits of both white (*Morus alba*) and black (*Morus nigra*) mulberries, using reversed-phase HPLC coupled to high resolution mass spectrometry (LC-MS), and related these to their in vitro antioxidant and α-glucosidase inhibitory activities. Based on accurate masses, fragmentation data, UV/Vis light absorbance spectra and retention times, 35 metabolites, mainly comprising phenolic compounds and amino sugar acids, were identified. While the antioxidant activity was highest in *M. nigra*, the α-glucosidase inhibitory activities were similar between species. Both bioactivities were mostly resistant to in vitro gastrointestinal digestion. To identify the bioactive compounds, we combined LC-MS with 96-well-format fractionation followed by testing the individual fractions for α-glucosidase inhibition, while compounds responsible for the antioxidant activity were identified using HPLC with an online antioxidant detection system. We thus determined iminosugars and phenolic compounds in both *M. alba* and *M. nigra*, and anthocyanins in *M. nigra* as being the key α-glucosidase inhibitors, while anthocyanins in *M. nigra* and both phenylpropanoids and flavonols in *M. alba* were identified as key antioxidants in their ripe berries.

Keywords: mulberry; high resolution mass spectrometry; antioxidant activity; in vitro gastrointestinal digestion; α-glucosidase inhibitory activity

1. Introduction

Mulberry belongs to the genus *Morus*, plant family Moraceae, which comprises 24 different species and one subspecies, with at least 100 varieties [1]. The most commonly known *Morus* species are *Morus alba* (white mulberry), *Morus nigra* (black mulberry) and *Morus rubra* (red mulberry.) [2]. They are deciduous trees originating from China and Japan, and have spread into America and Europe for silkworm breeding. Globally, the main use of mulberry trees is to produce leaves as feed for cultivating silkworms, but in various regions, they are also much appreciated for their fruit, which can be consumed both fresh and as an ingredient in processed food products [3].

Mulberry species have been used in Traditional Chinese Medicines (TCM) for the treatment of several diseases, especially diabetes mellitus type II; they contain specific molecules, i.e., iminosugars or iminocyclitols, which are low molecular weight carbohydrates in which the endocyclic oxygen atom has been replaced by a nitrogen atom. These compounds are known to be able to inhibit the enzyme α-glucosidase which is present in the brush border of the human intestine [4–6]. Inhibition of α-glucosidase leads to a decreased rate of glucose absorption thus resulting in a lower postprandial blood glucose level. Inhibitors of α-glucosidase can prevent the development of diabetes in individuals with impaired glucose tolerance and/or impaired fasting of blood glucose [7].

Mulberries are also a good nutritional source of a variety of phenolic compounds, like flavonols and phenolic acids, as well as coloured anthocyanins in the case of black and red mulberry fruits [8–10]. Phenolic compounds are the subject of increasing scientific interest; they are natural antioxidants in plant-derived foods and food products and their intake is frequently related to human health. Many of the bioactivities ascribed to mulberries, such as antioxidant action, hypolipidemic effect and macrophage activating effect, have also been linked to their phenolic compound composition [11–14].

The present study aimed to determine differences in chemical composition between the ripe fruits of *M. alba* and *M. nigra*, which are the major species growing in Italy, by metabolomic analysis and to identify the bioactive compounds responsible for their antioxidant and α-glucosidase inhibitory bioactivities. To this purpose, we used I) HPLC-PDA with an online, post-column antioxidant detection system, and II) HPLC-PDA- HR Orbitrap FTMS with on-line fractionation into 96-wells plates followed by off line in vitro α-glucosidase inhibitory activity testing of the contents of the individual wells. A series of compounds present in mulberry fruits as well as major bioactives were further characterized using MS^n fragmentation. In addition, we subjected mulberries to an in vitro gastrointestinal digestion system in order to investigate potential effects of digestion on the observed bioactivities upon consumption of these berries.

2. Results

2.1. *Identification of Phytochemicals in Morus using LC-PDA-Orbitrap FTMS*

To identify the metabolites present in white and black mulberry fruits, HPLC-PDA-Orbitrap FTMS analysis was performed on their aqueous methanol extracts, thus generating both an LC-PDA and an LC-MS profile per extract (Figures S1 and S2, respectively). Based on the exact mass of their molecular $[M+H]^+$ ion masses, their MS^n fragmentation patterns and their UV/Vis absorbance spectra, we putatively identified 35 compounds in the fruits (Table 1).

Table 1. Metabolites manually identified in *Morus alba* and *Morus nigra* using accurate mass LC-MS^n in positive ESI mode. L. I.: level of identification according to the Metabolomics Society Initiative [15].

N°	RT	Accurate Mass	Molecular Ion $[M+H]^+$	Molecular Formula	Putative ID	Fragment Ions $[M+H]^+$	L. I.	References
1	1.58	163.0844	164.0923	$C_6H_{13}NO_4$	1-deoxynojirimycin	-	3	[16]
2	1.70	289.2253	290.2332	$C_{15}H_{31}NO_4$	n-nonil-deoxynojirimycin	206.8857/122.9243	3	[16]
3	2.14	147.0895	148.0974	$C_6H_{13}NO_3$	fagomine	-	3	[16]
4	2.19	181.0738	182.0817	$C_9H_{11}NO_3$	2-formyl-1H-pyrrole-1-butanoic acid	165.0544/136.0755	3	[11]
5	8.16	354.0951	355.1024	$C_{16}H_{18}O_9$	caffeoylquinic acid isomer I	163.0386	2	[17]
6	9.30	449.1084	449.1084	$C_{21}H_{21}O_{11}^+$	cyanidin hexoside	287.0546	2	[17]
7	9.35	507.3043	508.3122	$C_{24}H_{45}NO_{10}$	morusimic acid E	346.2587/284.2579	3	[18]
8	10.21	595.1662	595.1662	$C_{27}H_{30}O_{15}^+$	cyanidin hexose deoxyhexose	449.1058/287.0546	2	[17]
9	10.93	433.1135	433.1135	$C_{21}H_{21}O_{10}^+$	pelargonidin hexoside	271.0596	2	[17]
10	11.52	354.0951	355.1024	$C_{16}H_{18}O_9$	caffeoylquinic acid isomer II	163.0386	2	[17]
11	11.92	579.1714	579.1714	$C_{27}H_{31}O_{14}^+$	pelargonidin hexose deoxyhexose	433.1115/271.0596	2	[17]
12	12.24	626.1483	627.1542	$C_{27}H_{30}O_{17}$	quercetin hexose hexose	465.1023/303.0489	2	[17]
13	12.38	354.0951	355.1024	$C_{16}H_{18}O_9$	caffeoylquinic acid isomer III	163.0386	2	[17]
14	12.42	772.2062	773.2135	$C_{33}H_{40}O_{21}$	quercetin-3-*O*-rutinoside-7-*O*-glucoside	303.0496/465.0995/611.1576	1	[19]
15	13.25	466.1111	467.1190	$C_{21}H_{22}O_{12}$	dihydroquercetin hexoside/taxifolin hexoside	449.1069/305.0650	3	[20]
16	14.41	712.1487	713.1544	$C_{30}H_{32}O_{20}$	quercetin hexoside malonyl hexoside	551.1015/463.1021/303.0496	2	[21]
17	14.71	756.2112	757.2192	$C_{33}H_{40}O_{21}$	kaempferol-3-*O*-rutinoside-7-*O*-glucoside	611.1576/449.1065/287.547	1	[19]
18	15.35	386.1940	387.2020	$C_{19}H_{30}O_8$	roseoside	370.1118/208.0599	3	[22]
19	16.37	772.2062	773.2135	$C_{33}H_{40}O_{21}$	quercetin-hexose-hexose-deoxyhexose	303.0496/465.0995/611.1576	3	[19]
20	16.52	450.1162	451.1235	$C_{21}H_{22}O_{11}$	dihydrokaempferol-hexoside	289.0703	3	[23]
21	17.02	696.1517	697.1597	$C_{30}H_{32}O_{19}$	kaempferol hexoside malonyl hexoside	287.0545/449.1065/535.1076	2	[24]
22	17.22	772.2062	773.2135	$C_{34}H_{40}O_{31}$	quercetin-hexose-hexose-deoxyhexose	303.0496/465.0995/611.1576	2	[19]
23	18.28	756.2112	757.2192	$C_{33}H_{40}O_{20}$	kaempferol hexose-hexose deoxyhexose	611.1576/449.4065/287.0547	2	[19]
24	18.63	756.2112	757.2192	$C_{33}H_{40}O_{20}$	kaempferol hexose hexose deoxyhexose	611.1576/449.4065/287.0547	2	[19]
25	19.59	491.3094	492.3173	$C_{24}H_{46}O_9N$	morusimic acid C	330.2640	3	[18]
26	21.39	610.1534	611.1606	$C_{27}H_{30}O_{16}$	quercetin-3-*O*-rutinoside	303.0495/465.1019	1	[25]
27	22.06	329.2566	330.2645	$C_{18}H_{35}NO_4$	morusimic acid B	312.2529/268.2630/250.2525	3	[18]
28	22.42	464.0954	465.1027	$C_{21}H_{20}O_{12}$	quercetin-hexoside	303.0498	2	[17]
29	24.39	594.1584	595.1664	$C_{27}H_{30}O_{15}$	kaempferol-3-*O*-rutinoside	449.1066/287.0546	1	[17]
30	24.88	550.0958	551.1038	$C_{24}H_{22}O_{15}$	quercetin-malonylhexoside	303.0499	2	[26]
31	25.10	516.1268	517.1341	$C_{25}H_{24}O_{12}$	dicaffeoylquinic acid I	163.0387	2	[27]
32	25.59	448.1006	449.1078	$C_{21}H_{20}O_{11}$	kaempferol-hexoside	287.0546	2	[17]
33	25.66	516.1268	517.1341	$C_{25}H_{24}O_{12}$	dicaffeoylquinic acid II	325.0913/163.0387	2	[27]
34	27.83	516.1268	517.1341	$C_{25}H_{24}O_{12}$	dicaffeoylquinic acid III	325.0913/163.0387	2	[27]
35	28.53	534.1009	535.1088	$C_{24}H_{22}O_{14}$	kaempferol malonyl hexoside	287.0546	2	[28]

In the fruits of *M. nigra*, four anthocyanins with a characteristic absorbance maximum at around 500–520 nm were identified. Compound **6**, cyanidin hexoside, showed a pseudomolecular ion at *m*/*z* 449.1084, corresponding to the molecular formula $C_{21}H_{21}O_{11}^{+}$ that, upon fragmentation, gave one principal product ion at *m*/*z* 287.0546; Compound **8**, cyanidin hexose-deoxyhexose, showed a pseudomolecular ion at *m*/*z* 595.1662, corresponding to the molecular formula $C_{27}H_{31}O_{15}^{+}$, that fragmented into two principal product ions at 449.1058 ($C_{21}H_{21}O_{11}^{+}$: loss of deoxyhexose) and 287.0546 (loss of hexose+deoxyhexose); Compound **9**, pelargonidin hexoside, showed a pseudomolecular ion at *m*/*z* 433.1135, corresponding to the molecular formula $C_{21}H_{21}O_{10}^{+}$, and gave one principal product ion at 271.0596; Compound **11**, pelargonidin hexose-deoxyhexose showed a pseudomolecular ion at *m*/*z* 579.1714 corresponding to molecular formula $C_{27}H_{31}O_{14}^{+}$ that, upon fragmentation, gave two principal product ions of *m*/*z* 433.1115 and 271.0596. Thus, anthocyanins detected represented both pelargonidin and cyanidin conjugated with one, two or three C_6-sugars. These anthocyanin compounds have previously been identified in *Morus alba* fruits [17] and are responsible for the dark color of black mulberry fruits. These anthocyanins were present in berries of *M. nigra* and were not detectable in those of *M. alba* (Supplemental Table S1), as was expected from their differential colors.

A series of flavonols with different substituents were present in both white and black mulberry fruits (**12, 14–17, 19–24, 26, 28–30, 32, 35**). All these compounds showed the characteristic flavonol absorbance peaks at around 260 nm (resulting from the A-ring) and at around 350 nm (due to B-ring) and producing daughter ions of *m*/*z* 303.0496 (quercetin) or 287.0547 (kaempferol); many of the compounds detected have not previously been described for mulberry fruits. For instance, compounds **14, 19, 22** show the same pseudomolecular ion $[M+H]^+$ at *m*/*z* 773.2135 with the same fragments of *m*/*z* 303.0496, 465.0995 and 611.1565, corresponding to the loss of two hexose and one deoxyhexose moiety, but with different RTs. These compounds were thus identified as different isomers of quercetin-hexose-hexose-deoxyhexose. A similar fragmentation pattern has been reported for a quercetin-trisaccharide in tomato fruit [19]. Specifically, compound **14** was confirmed as quercetin-3-O-rutinoside-7-O-glucoside based on the retention time of the reference compound reported in tomato fruit database [19].

Compounds **17**, **23**, and **24** likewise showed a similar pseudomolecular ion $[M+H]^+$ at *m*/*z* 757.2192 with the same fragments of *m*/*z* 287.0547, 449.1065 and 611.1576, corresponding to the loss of two hexoses and one deoxyhexose moiety, but with different RTs. These were thus identified as kaempferol-hexose-hexose-deoxyhexose isomers. The fragmentation pattern of these compounds agrees with known kaempferol glycosides in tomato [19]. Moreover, compound **17** was confirmed as kaempferol-3-O-rutinoside-7-O-glucoside based on the retention time of the reference compound reported in tomato fruit database [19].

Compound **16** showed a pseudomolecular ion at *m*/*z* 713.1544 that upon fragmentation, gave three principal product ions at *m*/*z* 303.0496, 463.1021 and 551.1015 corresponding to the loss of two hexose moieties and one malonyl moiety; this compound was thus tentatively identified as quercetin hexose-malonyl-hexoside. This fragmentation pattern is consistent with that reported for a quercetin malonyl glucoside in lettuce [21]. Compound **21** showed a pseudomolecular ion at *m*/*z* 697.1597 which gave three principal product ions at *m*/*z* 287.0545, 449.1065 and 535.1076, corresponding to the loss of two hexose and one malonyl moiety; this compound was identified as kaempferol hexose-malonyl-hexoside. This compound showed a similar fragmentation pattern as reported in Cycorium intibus [24]. Thus, similar flavonol conjugates consisting of both quercetin and kaempferol esterified with one to three C_6-sugars, or one or two sugars with one malonyl group, were present in both white and black mulberries.

Compound **20** showed a pseudomolecular ion at *m*/*z* 451.1235 that gave one principal product ion at *m*/*z* 289.0703, corresponding to the loss of a hexose moiety; this compound was tentatively identified as dihydrokaempferol-hexoside. A similar fragmentation pattern was reported for dihydrokaempferol-hexoside in raspberry [23].

Three N-containing sugars, i.e., compounds (**1**) 1-deoxynojirimicin, (**2**) N-nonil deoxynojirimicin and (**3**) fagomine, were found in fruits of both mulberry species. These compounds have previously been reported for leaves of *M. alba* [16] and are well known for inhibiting the enzyme α-glucosidase and consequently, might contribute to an antihyperglycemic effect [4–6].

Four piperidine alkaloids, morusimic acids B, C and E (compounds **7**, **25**, **27**) were also identified based on their exact molecular mass and fragmentation; these compounds have previously been reported in fruits of *M. alba* from Turkey [18].

Both *M. alba* and *M. nigra* fruits also contained caffeoylquinic acids monomers (**5**, **10**, **13**) as well as dimers (**31**, **33**, **34**). All three caffeoylquinic acid isomers (**5**, **10**, **13**) showed a pseudomolecular ion at 355.1024 that, upon fragmentation, gave the same daughter ion at *m/z* 163.0386, corresponding to the loss of their quinic acid moiety. These compounds have also been reported in *M. alba* fruits from Serbia [17]. The three dicaffeoylquinic acid isomers (**31**, **33**, **34**) showed a pseudomolecular ion at 517.1341 that produced the same MS/MS base peak at 163.0387. These compounds have previously been reported in leaves of *M. alba* [27].

Compound **4** showed a pseudomolecular ion at *m/z* 182.0817, that fragmented into two principal product ions at 165.0544 and 136.0755. It was putatively identified as the alkaloid 2-formyl-1H-pyrrole-1-butanoic acid, previously reported in *M. alba* fruits by Kim et al. [11].

2.2. Global Metabolome Differences between Morus Alba and Morus Nigra Fruits

Ripe fruits of *Morus alba* and *Morus nigra* were collected from trees growing at various locations in the Campania Region (Italy) and subjected to untargeted LCMS-based metabolite profiling. Subsequent unbiased data processing generated a dataset with the relative intensities and in-source mass spectra of 361 putative metabolites in the samples (Supplemental Table S1; note that this metabolite list misses some of the manually identified compounds described in Table 1, indicating that one or more parameter settings in the untargeted data processing workflow appears suboptimal for these specific compounds). An unsupervised multivariate statistical method, Principal Components Analysis (PCA), was subsequently applied to the entire metabolite dataset resulting in a clear differentiation of *M. alba* and *M. nigra* fruit samples (Figure S3). Among the most significantly ($p < 0.05$) differing metabolites were anthocyanins (Supplemental Table S1), as was expected from the differential fruit colours of both species. In addition, it was possible to identify two other flavonoids only detectable in *M. nigra*, namely dihydroquercetin hexoside and dihydrokaempferol hexoside. In fact, an important step for the biosynthesis of anthocyanidins is the reduction of dihydroflavonols catalysed by the enzyme DFR (dihydroflavonol 4-reductase) converting dihydroquercetin and dihydrokaempferol into colorless leucoanthocyanidins, which are further converted by the enzyme anthocyanin synthase (ANS) into cyanidin and pelargonidin, respectively [29], thereby providing the fruit colour in *M. nigra*. Several flavonol conjugates, including quercetin glycosides **19** and **26** and kaempferol glycoside **29** (Table 1), were also significantly ($p < 0.05$) higher in *M. nigra* fruit, while the mono- and di-caffeoyl quinic acids (phenylpropanoids) were not differential between both fruit species (Supplemental Table S1). These data suggest that *M. nigra* fruits exhibit a higher activity of the general flavonoid pathway than *M. alba* fruit. The alkaloids identified did not significantly differ between the *M. nigra* and *M. alba* fruit samples analyzed (Supplemental Table S1).

2.3. α-Glucosidase Inhibitory Activity and Effect of In Vitro Gastrointestinal Digestion

The α-glucosidase inhibitory activity of the black and white mulberries was firstly evaluated using the crude aqueous-methanol extracts of the fresh fruits. The extraction solvent was evaporated by freeze-drying and the metabolites re-dissolved in MQ-water. These water extracts were then tested for inhibiting α-glucosidase enzyme activity, monitored through the increase in the pNP product, detected at 412 nm, using 96-wells plates kept at 30 °C; the α-glucosidase inhibitor acarbose was used as a positive control (Figure 1b). Both mulberry extracts showed a marked and similar α-glucosidase inhibitory activity as compared to the water blank (Figure 1a).

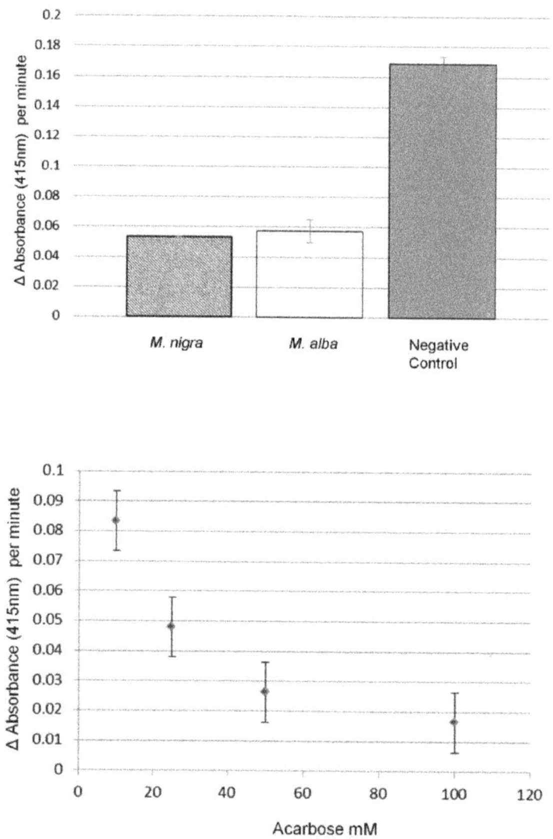

(a)

(b)

Figure 1. α-glucosidase inhibitory activity of mulberry methanol extracts. The Y axis represents the α-glucosidase activity (increase in 415 nm absorbance per minute) and the X axis the sample type tested. (**a**) inhibitory activity of water extracts of *Morus alba* and *Morus nigra* compared to the negative control (water). (**b**) enzyme activity inhibition by acarbose (positive control) at increasing concentrations (mM) in the assay. Data represent means and standard deviations (n = 3 assays).

The *M. nigra* fruit showed an IC_{50} value of 0.75 ± 0.004 mg/g DW (n = 3), while that for *M. alba* fruit was 0.93 ± 0.003 mg/g DW (n = 3). In comparison, the IC_{50} value of acarbose was 13.83 ± 0.02 mg/g.

A simulated gastrointestinal digestion was then applied to estimate the effect of consumption and digestion on the α-glucosidase inhibitory activity of mulberry fruits (Figure 2). For both fruit types, the bioactivity measured in the original fruit extracts was partially lost upon this in vitro digestion (GI samples compared to MN and MA samples). Gastric digestion (PG) resulted in a slight decrease in bioactivity in *M. nigra* only. The control incubation consisting of water instead of fruit extract in the digestion test (DC samples) showed a slight inhibition of the α-glucosidase activity as compared to the negative control (NC of undigested fruits: water instead of both fruit extract and digestion enzymes and buffer): a decrease of 0.04 enzyme units. Taking this inhibiting effect of the digestion conditions on α-glucosidase into account, it was calculated that the simulated gastrointestinal digestion resulted in an overall reduction in α-glucosidase inhibitory activity of about 50% (a decrease of about 0.055 units from 0.13 in DC to 0.075–0.08 in GI samples, compared to a decrease of about 0.115, i.e., from 0.17 units in NC to about 0.055 in original MN and MA extracts).

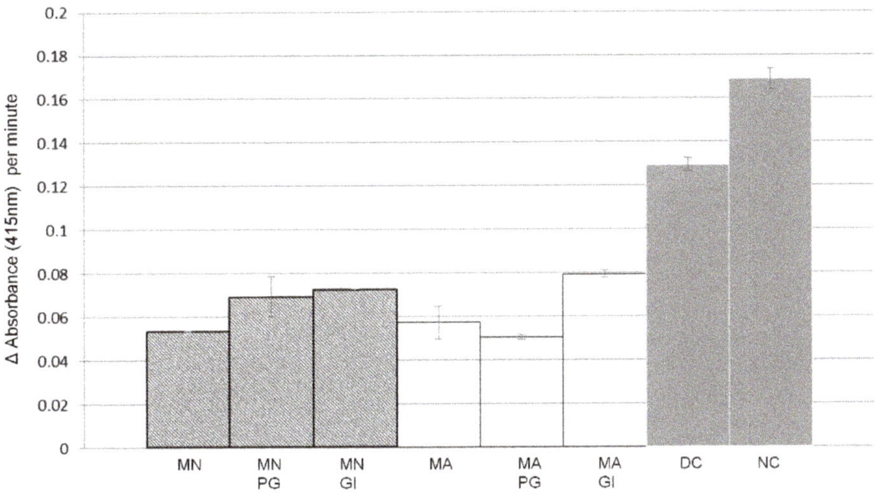

Figure 2. α-glucosidase inhibitory activity after in vitro gastrointestinal digestion. Inhibition activity of original *Morus nigra* (MN) and *Morus alba* (MA) fruit extracts, and after their in vitro stomach (Post Gastric, PG) digestion and in vitro gastrointestinal (GI) digestion. DC: digestion control, representing the digestion process, including all enzymes, without plant material; NC: negative control (NC), representing only water. Data represent means values and standard deviations ($n = 3$ measurements).

2.4. LCMS Combined with 96-Well Format Fractionation

In order to pinpoint those compounds in *Morus* fruits that are responsible for the observed α-glucosidase inhibitory activity, we subsequently used HPLC separation combined with both 96-well plate fractionation and Orbitrap FTMS detection. Injection, fractionation and FTMS analyses of the *M. alba* and *M. nigra* crude extracts, as used in the α-glucosidase inhibition assay, were performed in triplicate; a water blank was injected as a control. The fractionation plates were subsequently dried under a gentle N_2 flow at 30 °C, the dried well contents re-dissolved in water and tested for α-glucosidase inhibitory activity. Compounds present in bioactive wells were then further characterized from their corresponding UV/Vis spectra and FT-MS data.

The results of the α-glucosidase inhibitory activity of individual wells are shown in Figure 3A,B for *M. alba* and *M. nigra*, respectively. Based on the average enzyme activity measured in the wells of the water control sample, we set a threshold value at 0.17 absorbance units per minute, below which we considered a sample well to possess α-glucosidase inhibitory bioactivity.

The compounds identified in the active wells of both fruit types were the amino sugar acids **1–3** and the flavonoids **15–17**, **19–20** and **32**. Two anthocyanins, **6** and **9**, only present in *M. nigra* (see Table 1), also showed bioactivity. In addition, other fractions clearly showing α-glucosidase inhibitory were detected e.g., between 34.7 and 35.2 min in *M. alba*, although we have yet been unable to pinpoint and identify the specific bioactive compound(s) (Table 2).

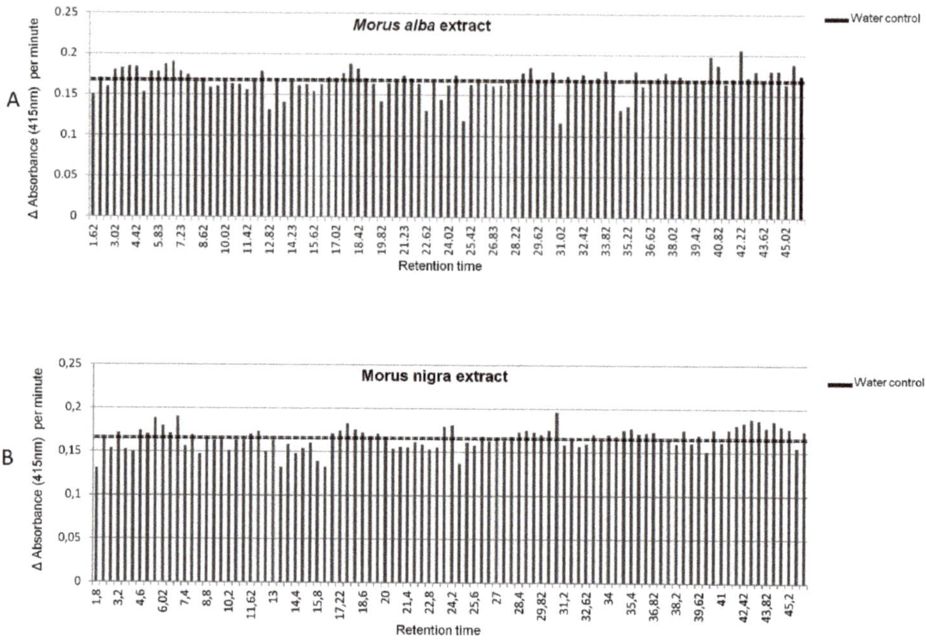

Figure 3. α-glucosidase inhibitory activity of 96-well LC-MS fractions of (**A**) *M. alba* and (**B**) *M. nigra* extracts. The Y axis shows the enzyme activity and the X axis the retention time corresponding to the LC-MS fraction. The vertical line at an enzyme activity of 0.17 indicates the average value in the water control. The wells considered bioactive are the ones below an enzyme activity value of 0.15.

Table 2. Retention time window of bioactive 96-well fractions and putatively corresponding compounds (numbers refer to Table 1) in *M. alba* and *M. nigra*. n.i. = not identified.

M. alba Retention Time (min)	Bioactive Metabolite	*M. nigra* Retention Time (min)	Bioactive Metabolite
1.6–2.08	**1-2-3**	1.8–2.27	**1-2-3**
4.88–5.35	n.i.	4.13–4.6	n.i.
12.82–13.28	**15**	8.33–8.8	**6**
13.75–14.23	**16/17**	10.2–10.67	**9**
19.82–20.3	**25**	10.2–10.67	**9**
22.62–23.08	**28**	12.53–13	**15**
23.55–24.02	**29**	13.4–13.93	**16**
24.95–25.42	**32**	14.4–14.87	**17**
31.02–31.5	n.i.	15.8–16.27	**19**
34.7–35.2	n.i.	16.27–16.73	**20**
35.22–35.68	n.i.	24.67–25.12	**31/32**
-		40.07–40.53	n.i.

2.5. Total Antioxidant Activity and HPLC with Online Antioxidant Detection

The total antioxidant activity of the mulberry fruits was compared between other fruits well known for their antioxidant activity: cultivated strawberry (*Fragaria* × *ananassa*) and wild strawberry (*Fragaria vesca*). This antioxidant assay (Table 3) indicated that the aqueous-methanol extract of *M. nigra* is slightly more active than that of *M. alba*; in fact the mulberry fruits showed about the same activity as strawberry, which is among the fruit species with the highest antioxidant capacity [30].

Table 3. Antioxidant capacity of *Morus alba* and *Morus nigra* fruits compared with strawberry fruits. Data represent average values ± standard deviation of three independent extractions. All antioxidant values are expressed as mg Trolox per g of fresh weight. TEAC: Trolox-equivalent antioxidant capacity.

Extracts	TEAC mg/g FW
Morus alba (White mulberry)	39.40 ± 0.02
Morus nigra (Black mulberry)	49.42 ± 0.01
Fragaria vesca (Wild strawberry)	50.61 ± 0.01
Fragaria ananassa (Strawberry)	51.31 ± 0.01

Subsequently, a HPLC-PDA system coupled to online $ABTS^{\cdot+}$ cation radical reaction and detection [31] was used to determine the relative contribution of each individual component to the total antioxidant activity (Figure 4). Several antioxidant components could be identified by comparison of their retention times and absorption spectra with those of the LC-PDA-FTMS/MS analysis using the same chromatographic conditions. According to this online antioxidant assay, the key antioxidants in *M. nigra* corresponded to anthocyanins, in particular, compounds **6** and **9**. The other compounds responsible for antioxidant activity in both *M. alba* and *M. nigra* were caffeoylquinic acids, like compounds **10** and **13**, and flavonols like compounds **15, 26, 28, 32** (see Table 1).

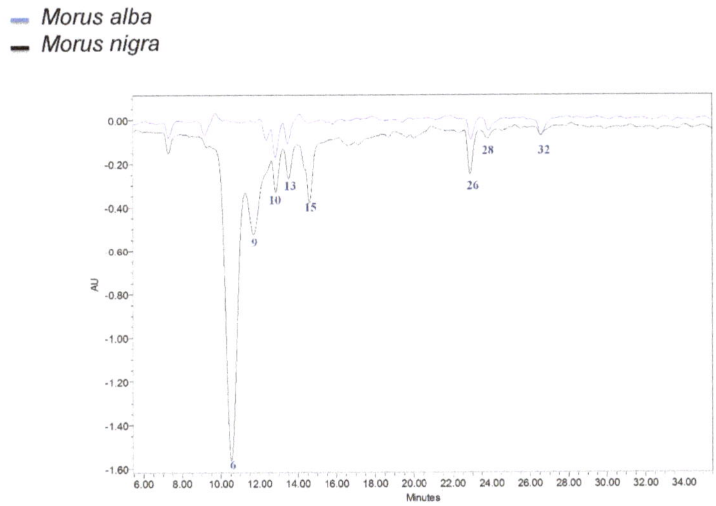

Figure 4. Antioxidant activity. Overlay of representative antioxidant chromatograms of fruit of *Morus alba* (in blue) and *Morus nigra* (in black). Antioxidant profiles of fruit extracts were determined online, by a post column reaction with $ABTS^{\cdot+}$ cation radicals after HPLC separation and PDA detection of compounds. The ABTS-radicals remaining after post-column reaction were recorded at 600 nm: negative peaks thus indicate antioxidant activity. The numbers refer to the main peaks identified (see Table 1): **6** cyanidin hexoside, **9** pelargonidin hexoside, **10** and **13** caffeoylquinic acid isomers, **15** dihydroquercetin hexoside, **26** quercetin hexose deoxyhexose, **28** quercetin hexoside, and **32** kaempferol hexoside.

3. Discussion

In the present study, we compared ripe fruits of *Morus alba* and *Morus nigra* for their metabolite composition in relation to their potential relevant bioactivities upon consumption, i.e., α-glucosidase inhibiting and antioxidative activities. Using HPLC-PDA-Orbitrap FTMS analysis of aqueous-methanol extracts, we were able to detect a large series of compounds and identified a number of metabolites, previously reported for mulberry or other fruit species, as well as new compounds being present in either or both *M. alba* and *M. nigra*. Fruit of both species exhibited a marked α-glucosidase inhibiting activity in vitro, an indication of their potential beneficial effect with regard to type II diabetes. Moreover, we showed that this α-glucosidase inhibiting activity was partially resistant to simulated gastric and intestinal digestion. Anthocyanins appear among the potential bioactive compounds in *M. nigra* fruit (Figure 3) and the general instability of anthocyanins at the alkaline conditions of gastrointestinal digestion [32] may at least partly explain the loss of α-glucosidase inhibitory activity in *M. nigra* fruits. When calculating the α-glucosidase inhibiting activity of mulberries in units of acarbose, a well known type II diabetes drug based on its α-glucosidase inhibiting activity (https://www.drugs.com/pro/precose.html), our data suggest that consumption of about 20–25 g of fresh mulberry fruit corresponds to 50 mg of acarbose, taking into account a 50% bioactivity loss upon digestion. It has been shown that an intake of 100 mg acarbose 3 times a day can significantly reduce type II diabetes risk factors [33]. Thus, a daily consumption of 100–150 g fresh mulberries may exert relevant pharmacological effects with regard to type II diabetes. Using analytical LC-based extract fractionation, it was possible to pinpoint three known iminosugar acids, i.e., [1-deoxynojirimycin (**1**), N-nonil-deoxynojirimycin (**2**) and fagomine (**3**)], and 7 phenolic compounds, including five flavonols [dihydroquercetin hexoside, (**15**) quercetin hexoside malonyl hexoside (**16**), kaempferol-3-O-rutinoside-7-O-glucoside (**17**), quercetin hexose (**28**) and kaempferol hexoside (**32**)] present in both *M. alba* and *M. nigra*, and 2 anthocyanins [cyanidin hexoside (**6**), pelargonidin hexoside (**9**)] only present in *M. nigra*, as the key α-glucosidase inhibitors in mulberry fruits. While compounds **1**, **2**, **3**, **6**, **28** and **32** have already been reported to exert this bioactivity [5,16,34,35], in our study, we were able to detect novel α-glucosidase inhibitory compounds in mulberries. A similar approach, using accurate mass LCMS coupled to 96-well fractionation and bioactivity testing, has recently been used to identify novel compounds in pepper fruits interacting with the human hot-taste receptor [36].

Although it was not yet possible to identify the novel α-glucosidase inhibitory compounds in mulberry fruits on the basis of the observed accurate mass only, this method can well be optimized and adapted for further structural characterization of these bioactives, e.g., by using so-called multistage mass spectrometry at high mass resolution [19], if needed, combined with NMR experiments. For the latter approach, the same bioactive wells from replicate plates may be pooled to get sufficient NMR signals for the de novo identification. Alternatively, bioactive extracts can be re-injected in a LC-MS-SPE set up to collect and concentrate individual LC-MS peaks upon repeated injections; the SPE cartridges containing the active compounds (based on their known accurate mass and LC-retention time) can then be subjected to NMR for structural elucidation [37].

In addition to the α-glucosidase inhibitory activity, the $ABTS^{+}$-radical based total antioxidant assay indicated significant antioxidant activity present in the same mulberries, comparable to that of strawberries, which are among the fruit species with the highest antioxidant capacity [32]. The higher activity in *M. nigra* compared to *M. alba* fruits is likely due to the presence of anthocyanins, which both provide fruits with their dark color and contribute to antioxidant activity [38]. Indeed, using HPLC with online antioxidant detection [32], we were able to pinpoint anthocyanins as the main phenolic antioxidants in *M. nigra*, while both phenylpropanoids and flavonols were the key phenolic antioxidants in *M. alba*.

This work shows that it is well possible, using analytical scale techniques, to pinpoint the compounds that are key to the well described bioactivities of mulberry fruits, and to validate the value of metabolomics technologies in the phytochemical and bioactivity evaluation of functional foods. However, further studies towards, for example growth conditions, genotypic variation, fruit development and ripening are needed to obtain the best material for preparation of such functional foods with optimal composition of bioactive ingredients or for purification of the bioactive compounds.

4. Materials and Methods

4.1. Mulberry Materials

Fruits of *M. alba* and *M. nigra* were manually picked at ripe stage in May 2014 in different areas of the Campania region in Italy, in particular, the geographical locations Solofra (GPS coordinates latitude: 40.8291: longitude: 14.8456), Roccadaspide (GPS coordinates latitude: 40.4253: longitude: 15.1917), Fisciano (GPS coordinates latitude: 40.7728: longitude: 14.7994), San Sossio Baronia (GPS coordinates latitude: 410712: longitude: 15.2005). *Morus alba* fruits (MAF) were collected from 4 locations and coded as MAF-S (collected in Solofra) MAF-big (collected in San Sossio Baronia), MAF-wt (collected in San Sossio Baronia), MAF-R (collected in Roccadaspide); *M. nigra* fruits (MNF) were collected at five locations and coded as MNF-R (collected in Roccadaspide), MNF-U14 (collected in Fisciano), MNF-U13 (collected in Fisciano), MNF-S (collected in Solofra) and MNF-U13 (collected in Fisciano). All samples were botanically identified by Prof. V. De Feo (Department of Pharmacy University of Salerno) and compared with reference materials, then were freeze-dried before being transported to The Netherlands. They were then ground to a fine powder and stored at −80 °C until analysis.

For comparing antioxidant activities, fresh frozen mulberries were compared with fresh fruit of strawberry and wild strawberry collected in June 2014 in Campania (Italy).

4.2. Extract Preparation

The sample extracts used for LC-MS analysis were prepared essentially as described in De Vos et al. [31]: 30 mg of freeze dried samples were extracted with 1200 µL of 75% methanol in MQ water containing 0.1% of formic acid. Mixtures were then sonicated for 15 min, centrifuged at 12,500 g for 10 min and filtered over a 0.45 µm filter (Minisart SRP4, Biotech GmbH, Germany).

For α-glucosidase inhibitory activity testing, for both NanoMate fractionation and HPLC with online antioxidant analysis, 1 mL of supernatant was dried in a speedvac and taken up in 250 µL of water, sonicated and filtered through a 0.45 µm filter (Minisart SRP4, Biotech GmbH, Germany). These concentrated extracts were prepared in 3 independent replicates.

4.3. LC-PDA-Orbitrap FTMS Analysis

A metabolite analysis was performed using an HPLC (Waters Aquity) coupled to both a photodiode array detector (PDA; Waters) and an LTQ Ion trap-Orbitrap Fourier transformed Mass spectrometer (FTMS; Thermo) hybrid system. A Luna 3 µm C18 150 × 2 mm column (Phenomenex, USA) at 40 °C was used to separate the extracted metabolites, with MQ water with 0.1% formic acid (A) and acetonitrile with 0.1% formic acid (B) as solvents. A linear gradient from 5% to 35% B in 45 min, at a flow rate of 0.19 mL/min, was used [33]. In order to prevent possible order and batch effects in the LCMS analysis, all samples were analysed in a single series and in random order. Three quality control samples (QCs) prepared from a mix of all samples were included and equally distributed over the study samples, in order to check system stability and estimate overall technical variation.

Electrospray ionization (ESI) in positive mode was used to generate ions from eluting compounds. ESI source parameters were as follows: capillary voltage 43 V; tube lens voltage 120 V; capillary temperature 295 °C; Sheath and Auxiliary Gas flow at 40 and 3 (arbitrary units), respectively, Sweep gas 0 n, Spray voltage 5 V. MS spectra were acquired by full range acquisition covering *m/z* 104–1350 at a resolution of 60,000 FWHM.

4.4. LCMS data Processing and Multivariate Analysis

The raw LCMS data files were processed using Metalign software [34] for baseline correction, noise estimation, and ion-wise mass spectral alignment. MSClust software [35] was then used to assemble redundant mass signals derived from the same metabolite, including natural isotopes, adducts and in-source fragments, based on their corresponding retention times and relative abundance patterns across samples. This resulted in the relative intensities of 361 mass peak clusters, each representing a (reconstructed) putative metabolite, present in at least two samples. These metabolite intensity data were then subjected to a multivariate analysis using GeneMaths XT software version 2.12 (Applied Maths, Belgium). Metabolite intensities were firstly log2-transformed and then mean-centred across samples.

Using multiple online databases, including KNApSAcK (http://kanaya.naist.jp/KNApSAcK/), Dictionary of Natural Products (http://dnp.chemnetbase.com), Metlin (https://metlin.scripps.edu/), HMD (http://www.hmdb.ca), in-house libraries based on standards, as well as the mass spectra information within the clustered mass peaks and from additional LC-MS^n runs generating accurate mass spectral trees from the top 3 intensity ions every 30 s [19]. Selected metabolites (Table 1) were manually annotated as far as was possible using the mass data and the UV/Vis-absorbance spectral data available.

4.5. *In Vitro Simulated Gastrointestinal Digestion*

In vitro digestion was carried out on freeze-dried fruit samples of both *M. alba* and *M. nigra*, following the protocol described by McDougall et al. [36] with slight modification. Release of phytochemicals from fruit was checked by LC-MS at different stages of digestion, i.e., after gastric digestion (post gastric, PG) and gastrointestinal digestion (GI). Both PG and GI samples were stored at −80 °C until further analysis. During the process, three different controls were used: (1) plant material without digestion solutions and enzymes, diluted in water using the same ratio used for the samples coming from the digestion process (2) the solutions with all the ingredients for digestion but without plant material (DC: Digestion control), (3) plant material with all the ingredients for digestion but without active enzymes (the enzymes where added at the end of the digestion process, to the cold extract). Both α-glucosidase inhibitory activity and antioxidant activity were investigated for each of these PG and GI samples using the methods described below.

4.6. *α-Glucosidase Inhibition Assay*

The α-glucosidase assay uses the synthetic substrate p-nitrophenyl-α-D-glucopyranoside (pNPG), which is hydrolyzed by α-glucosidase to release p-nitrophenol (pNP), a color agent that can be monitored at 415 nm. Briefly, 10 μL of extract was combined with 40 μL of 100 mM phosphate buffer (pH 6.8) and 20 μL of α-glucosidase (0.6 units per mL buffer). After mixing and incubation for 5 min at 37 °C, 20 μL of a 20 mM pNPG solution in buffer was added to start the reaction. The reaction was monitored in time at 415 nm by a TECAN SpectraFluor microplate reader. Acarbose was used as a positive control, while water was used as a negative control for enzyme inhibition. The enzyme activities were evaluated as increase in the absorbance at 415 nm per minute and the percentage of enzyme inhibition was calculated. Three dilution series of extracts were used for IC_{50} determination. Dose−response curves and IC_{50} values were obtained by use of GraphPad Prism (version 6.00.283). The assay was performed with 3 replicates.

4.7. *NanoMate LC-Fractionation of Extracts*

The HPLC–PDA–FTMS system was adapted with a chip-based nano-electrospray ionization source/fractionation robot (NanoMate Triversa, Advion BioSciences) coupled between the PDA and the inlet of the Ion Trap/Orbitrap hybrid instrument [19]. In this system, the compounds separated and eluting from the analytical column firstly passed the PDA detector for determining their UV/Vis absorbance spectra and then the eluent was automatically split by a NanoMate LC-fraction collector/injection robot (Advion) into a nanoflow for chip-based ESI nanospray Orbitrap FTMS analysis and the rest for fractionation into microwells with a collection time of 28 sec per well. The sample injection volume was 5 μL. The gradient and flow conditions were the same as described above, with an additional 30 μL/min 100% isopropanol added into the LC flow via a T-junction between the PDA and the NanoMate, in order to improve the solvent composition for generating a stable nano-electrospray. The eluent flow was split by the NanoMate at a ratio of 219.5 μL/min to the fraction collector and 0.5 μL/min to the nano-electrospray source. LC-fractions were collected every 28.2 s (i.e., 100 μL solvent) into 96-well plates (Twin tec, Eppendorf). After collection, the plates were dried at 30 °C under a gentle N_2 flow, and then tested for α-glucosidase inhibitory activity as described above (performed in 3 replicates).

4.8. Antioxidant Activity and HPLC Analysis with Online Antioxidant Detection

The total antioxidant capacity of fruits was analyzed using the $ABTS^{\cdot+}$ radical scavenging method, essentially according to Capanoglu et al. [37] with slight modifications. The fruits were collected in June 2014 and the antioxidant activity was tested on basis of fresh weight. 0.5 g of samples (fresh fruit) were extracted in 1.5 mL of methanol (final MeOH concentration about 77%, taking into account a fruit water content of 95%) containing 0.05% of formic acid, sonicated for 15 min and centrifuged at 12,500 g for 15 min, filtered through 0.45 μm (Minisart SRP4, Biotech GmbH, Germany) and then 10 μL of extract was used to test the antioxidant activity. Trolox was used as a reference.

To determine total antioxidant capacity, 10 μL of sample extracts or standard solution was mixed with 90 μL of ABTS-radical working solution (pH 7.4) and after 40 s, the remaining $ABTS^{\cdot+}$ radicals were measured at 415 nm using 96-well microplates (Nunc, Roskilde, Denmark) and an Infinite® M200 micro plate reader (Tecan, Gröding, Austria). The analyses were done using 3 replicates and the results were expressed in terms of mg Trolox Equivalent Antioxidant Capacity (TEAC) per g fruit FW. In addition, the contribution of individual antioxidants to the total antioxidant capacity of the crude mulberry extracts was determined using an HPLC-PDA system coupled to post-column on-line antioxidant detection [37,38]. For this, the extracts of *M. alba* and *M. nigra* fruits, also used for the LC-MS analysis, were analyzed using a W600 Waters HPLC system coupled to a Waters 996 PDA detector (240–600 nm) [37,38]. Eluted compounds were allowed to react online for 30 s at 40 °C in a buffered solution of $ABTS^{\cdot+}$ cation radicals (pH 7.4). Then, the absorption of the remaining $ABTS^{\cdot+}$ radicals was monitored at 412 nm by a second detector (Waters 2487, dual-wavelength UV–vis detector). Peak identification was done by comparing PDA-absorbance spectra and retention times of eluting peaks with data taken from the literature and annotations were confirmed by HPLC-FTMS and MS/MS analyses, as described above.

Supplementary Materials: The following are available online at http://www.mdpi.com/2218-1989/10/1/7/s1, Figure S1: Representative LC-MS chromatograms, recorded in ESI positive ionization mode, of *M. nigra* (upper trace) and *M. alba* (lower trace) fruit extracts; Y-axes are on the same scale (1.5*E7; base peak intensity in ion counts/sec). Values above peaks indicate retention times (in minutes) and detected *m/z* value, Figure S2: Representative LC-PDA chromatograms of aqueous-methanol extracts from ripe fruits of *M. nigra* (A and B) and *M. alba* (C and D). Figures show absorbance at 520 nm (A and C) representing elution profile of anthocyanins, and at 355 nm (B and D) representing mainly flavonoids and phenylpropanoids, Values above peaks indicate retention times (in minutes). Note: intensity scales (Y-axes) are similar for all traces, Figure S3: 3 dimensional PCA plot of 5 *Morus alba* and 4 *M. nigra* fruit samples, harvested from trees spread over region Campania, Italy, based on their variation in 371 metabolites detected by the untargeted LCMS approach. The 3 quality control samples are technical replicates of a mix of samples. The X-axis (PC1) explains 33.2% of the total metabolites variation, the Y-axis (PC2) 18.6% and the Z-axis (PC3) 14.2%, Table S1: Relative intensity of all putative metabolite features (clusterID's) for each of the analyzed mulberry samples, Table S2: Description of column heads, Table S3: MSI Identification level.

Author Contributions: Conceptualization, G.D., P.M. and R.C.H.d.V.; methodology, G.D., J.J.M. and R.C.H.d.V.; software, G.D. and R.C.H.d.V.; formal analysis, G.D.; investigation, G.D., P.M. and R.C.H.d.V.; resources, P.M., J.J.M. and R.D.H.; data curation, G.D. and R.C.H.d.V.; writing—original draft preparation, G.D.; writing—review and editing, G.D., J.J.M., P.M., R.D.H. and R.C.H.d.V.; visualization G.D'U and R.C.H.d.V.; supervision, P.M., J.J.M., and R.C.H.d.V.; project administration, G.D., P.M. and R.C.H.d.V. All authors have read and agreed to the published version of the manuscript.

Funding: This research received no external funding.

Acknowledgments: This research was carried out at the Bioscience department of Wageningen Plant Research of Wageningen UR. Special thanks to Harry Jonker and Bert Schipper for their technical help in HPLC-antioxidant and LC-PDA-FTMS analysis, and to Monic Tomassen (Fresh Food and Chains, Food & Biobased Research, Wageningen UR) for helping with the in vitro gastrointestinal digestions.

Conflicts of Interest: The authors declare no conflict of interest.

References

1. Ercisli, S.; Orhan, E. Chemical composition of white (*Morus alba*), red (*Morus rubra*) and black (*Morus nigra*) mulberry fruits. *Food Chem.* **2007**, *103*, 1380–1384. [CrossRef]
2. Gundogdu, M.; Muradoglu, F.; Sensoy, R.I.G.; Yilmaz, H. Determination of fruit chemical properties of *Morus nigra* L., *Morus alba* L. and *Morus rubra* L. by HPLC. *Sci. Hortic Amst.* **2011**, *132*, 37–41. [CrossRef]
3. Liang, L.H.; Wu, X.Y.; Zhao, T.; Zhao, J.L.; Li, F.; Zou, Y.; Mao, G.H.; Yang, L.Q. In vitro bioaccessibility and antioxidant activity of anthocyanins from mulberry (*Morus atropurpurea* Roxb.) following simulated gastro-intestinal digestion. *Food Res. Int.* **2012**, *46*, 76–82. [CrossRef]
4. Banu, S.; Jabir, N.R.; Manjunath, N.C.; Khan, M.S.; Ashraf, G.M.; Kamal, M.A.; Tabrez, S. Reduction of post-prandial hyperglycemia by mulberry tea in type-2 diabetes patients. *Saudi J. Biol. Sci.* **2015**, *22*, 32–36. [CrossRef] [PubMed]
5. He, H.; Lu, Y.H. Comparison of Inhibitory Activities and Mechanisms of Five Mulberry Plant Bioactive Components against alpha-Glucosidase. *J. Agric. Food Chem.* **2013**, *61*, 8110–8119. [CrossRef]
6. Rodriguez-Sanchez, S.; Quintanilla-Lopez, J.E.; Soria, A.C.; Sanz, M.L. Evaluation of different hydrophilic stationary phases for the simultaneous determination of iminosugars and other low molecular weight carbohydrates in vegetable extracts by liquid chromatography tandem mass spectrometry. *J. Chromatogr. A* **2014**, *1372*, 81–90. [CrossRef] [PubMed]
7. Kongstad, K.T.; Ozdemir, C.; Barzak, A.; Wubshet, S.G.; Staerk, D. Combined Use of High-Resolution a-Glucosidase inhibition Profiling and High-Performance Liquid Chromatography High-Resolution. Mass Spectrometry Solid-Phase Extraction Nuclear Magnetic Resonance Spectroscopy for Investigation of Antidiabetic Principles in Crude Plant Extracts. *J. Agric. Food Chem.* **2015**, *63*, 2257–2263. [CrossRef] [PubMed]
8. Sanchez-Salcedo, E.M.; Mena, P.; Garcia-Viguera, C.; Hernandez, F.; Martinez, J.J. (Poly)phenolic compounds and antioxidant activity of white (*Morus alba*) and black (*Morus nigra*) mulberry leaves: Their potential for new products rich in phytochemicals. *J. Funct. Foods* **2015**, *18*, 1039–1046. [CrossRef]
9. Sanchez-Salcedo, E.M.; Mena, P.; Garcia-Viguera, C.; Martinez, J.J.; Hernandez, F. Phytochemical evaluation of white (*Morus alba* L.) and black (*Morus nigra* L.) mulberry fruits, a starting point for the assessment of their beneficial properties. *J. Funct. Foods* **2015**, *12*, 399–408. [CrossRef]
10. Bao, T.; Xu, Y.; Gowd, V.; Zhao, J.C.; Xie, J.H.; Liang, W.K.; Chen, W. Systematic study on phytochemicals and antioxidant activity of some new and common mulberry cultivars in China. *J. Funct. Foods* **2016**, *25*, 537–547. [CrossRef]
11. Kim, S.B.; Chang, B.Y.; Jo, Y.H.; Lee, S.H.; Han, S.B.; Hwang, B.Y.; Kim, S.Y.; Lee, M.K. Macrophage activating activity of pyrrole alkaloids from *Morus alba* fruits. *J. Ethnopharmacol.* **2013**, *145*, 393–396. [CrossRef] [PubMed]
12. Liu, L.K.; Chou, F.P.; Chen, Y.C.; Chyau, C.C.; Ho, H.H.; Wang, C.J. Effects of Mulberry (*Morus alba* L.) Extracts on Lipid Homeostasis In Vitro and In Vivo. *J. Agric. Food Chem.* **2009**, *57*, 7605–7611. [CrossRef] [PubMed]
13. Yang, X.L.; Yang, L.; Zheng, H.Y. Hypolipidemic and antioxidant effects of mulberry (*Morus alba* L.) fruit in hyperlipidaemia rats. *Food Chem. Toxicol.* **2010**, *48*, 2374–2379. [CrossRef] [PubMed]
14. Krishna, P.G.A.; Sivakumar, T.R.; Jin, C.; Li, S.H.; Weng, Y.J.; Yin, J.; Jia, J.Q.; Wang, C.Y.; Gui, Z.Z. Antioxidant and Hemolysis Protective Effects of Polyphenol-Rich Extract from Mulberry Fruits. *Pharmacogn. Mag.* **2018**, *14*, 103–109. [CrossRef]
15. Sumner, L.W.; Amberg, A.; Barrett, D.; Beale, M.H.; Beger, R.; Daykin, C.A.; Fan, T.W.; Fiehn, O.; Goodacre, R.; Griffin, J.L.; et al. Proposed minimum reporting standards for chemical analysis Chemical Analysis Working Group (CAWG) Metabolomics Standards Initiative (MSI). *Metabolomics* **2007**, *3*, 211–221. [CrossRef]
16. Asano, N.; Oseki, K.; Tomioka, E.; Kizu, H.; Matsui, K. N-Containing Sugars from *Morus alba* and Their Glycosidase Inhibitory Activities. *Carbohydr. Res.* **1994**, *259*, 243–255. [CrossRef]
17. Natic, M.M.; Dabic, D.C.; Papetti, A.; Aksic, M.M.F.; Ognjanov, V.; Ljubojevic, M.; Tesic, Z.L. Analysis and characterisation of phytochemicals in mulberry (*Morus alba* L.) fruits grown in Vojvodina, North Serbia. *Food Chem.* **2015**, *171*, 128–136. [CrossRef]
18. Kusano, G.; Orihara, S.; Tsukamoto, D.; Shibano, M.; Coskun, M.; Guvenc, A.; Erdurak, C.S. Five new nortropane alkaloids and six new amino acids from the fruit of *Morus alba* LINNE growing in Turkey. *Chem. Pharm. Bull.* **2002**, *50*, 185–192. [CrossRef]

19. Van Der Hooft, J.J.J.; Vervoort, J.; Bino, R.J.; De Vos, R.C.H. Spectral trees as a robust annotation tool in LC-MS based metabolomics. *Metabolomics* **2012**, *8*, 691–703. [CrossRef]
20. Yang, J.F.; Wen, H.C.; Zhang, L.; Zhang, X.X.; Fu, Z.; Li, J.M. The influence of ripening stage and region on the chemical compounds in mulberry fruits (Morus atropurpurea Roxb.) based on UPLC-QTOF-MS. *Food Res. Int.* **2017**, *100*, 159–165. [CrossRef]
21. Llorach, R.; Martinez-Sanchez, A.; Tomas-Barberan, F.A.; Gil, M.I.; Ferreres, F. Characterisation of polyphenols and antioxidant properties of five lettuce varieties and escarole. *Food Chem.* **2008**, *108*, 1028–1038. [CrossRef] [PubMed]
22. Li, H.X.; Jo, E.; Myung, C.S.; Kim, Y.H.; Yang, S.Y. Lipolytic effect of compounds isolated from leaves of mulberry (*Morus alba* L.) in 3T3-L1 adipocytes. *Nat. Prod. Res.* **2018**, *32*, 1963–1966. [CrossRef] [PubMed]
23. Paudel, L.; Wyzgoski, F.J.; Scheerens, J.C.; Chanon, A.M.; Reese, R.N.; Smiljanic, D.; Wesdemiotis, C.; Blakeslee, J.J.; Riedl, K.M.; Rinaldi, P.L. Nonanthocyanin Secondary Metabolites of Black Raspberry (*Rubus occidentalis* L.) Fruits: Identification by HPLC-DAD, NMR, HPLC-ESI-MS, and ESI-MS/MS Analyses. *J. Agric. Food Chem.* **2013**, *61*, 12032–12043. [CrossRef] [PubMed]
24. Carazzone, C.; Mascherpa, D.; Gazzani, G.; Papetti, A. Identification of phenolic constituents in red chicory salads (*Cichorium intybus*) by high-performance liquid chromatography with diode array detection and electrospray ionisation tandem mass spectrometry. *Food Chem.* **2013**, *138*, 1062–1071. [CrossRef] [PubMed]
25. Zhao, S.C.; Park, C.H.; Yang, J.L.; Yeo, H.J.; Kim, T.J.; Kim, J.K.; Park, S.U. Molecular characterization of anthocyanin and betulinic acid biosynthesis in red and white mulberry fruits using high-throughput sequencing. *Food Chem.* **2019**, *279*, 364–372. [CrossRef] [PubMed]
26. Mena, P.; Sanchez-Salcedo, E.M.; Tassotti, M.; Martinez, J.J.; Hernandez, F.; Del Rio, D. Phytochemical evaluation of eight white (*Morus alba* L.) and black (*Morus nigra* L.) mulberry clones grown in Spain based on UHPLC-ESI-MSn metabolomic profiles. *Food Res. Int.* **2016**, *89*, 1116–1122. [CrossRef]
27. Dugo, P.; Donato, P.; Cacciola, F.; Germano, M.P.; Rapisarda, A.; Mondello, L. Characterization of the polyphenolic fraction of *Morus alba* leaves extracts by HPLC coupled to a hybrid IT-TOF MS system. *J. Sep. Sci.* **2009**, *32*, 3627–3634. [CrossRef]
28. Thabti, I.; Elfalleh, W.; Hannachi, H.; Ferchichi, A.; Campos, M.D. Identification and quantification of phenolic acids and flavonol glycosides in Tunisian Morus species by HPLC-DAD and HPLC-MS. *J. Funct. Foods* **2012**, *4*, 367–374. [CrossRef]
29. Qi, X.W.; Shuai, Q.; Chen, H.; Fan, L.; Zeng, Q.W.; He, N.J. Cloning and expression analyses of the anthocyanin biosynthetic genes in mulberry plants. *Mol. Genet. Genom.* **2014**, *289*, 783–793. [CrossRef]
30. Wang, H.; Cao, G.H.; Prior, R.L. Total antioxidant capacity of fruits. *J. Agric. Food Chem.* **1996**, *44*, 701–705. [CrossRef]
31. De Vos, R.C.H.; Moco, S.; Lommen, A.; Keurentjes, J.J.B.; Bino, R.J.; Hall, R.D. Untargeted large-scale plant metabolomics using liquid chromatography coupled to mass spectrometry. *Nat. Protoc.* **2007**, *2*, 778–791. [CrossRef] [PubMed]
32. Hu, D.W.; Xu, Y.; Xie, J.H.; Sun, C.D.; Zheng, X.D.; Chen, W. Systematic evaluation of phenolic compounds and protective capacity of a new mulberry cultivar J33 against palmitic acid-induced lipotoxicity using a simulated digestion method. *Food Chem.* **2018**, *258*, 43–50. [CrossRef] [PubMed]
33. Van Der Hooft, J.J.J.; De Vos, R.C.H.; Ridder, L.; Vervoort, J.; Bino, R.J. Structural elucidation of low abundant metabolites in complex sample matrices. *Metabolomics* **2013**, *9*, 1009–1018. [CrossRef]
34. Lommen, A. MetAlign: Interface-Driven, Versatile Metabolomics Tool for Hyphenated Full-Scan Mass Spectrometry Data Preprocessing. *Anal. Chem.* **2009**, *81*, 3079–3086. [CrossRef] [PubMed]
35. Tikunov, Y.M.; Laptenok, S.; Hall, R.D.; Bovy, A.; De Vos, R.C.H. MSClust: A tool for unsupervised mass spectra extraction of chromatography-mass spectrometry ion-wise aligned data. *Metabolomics* **2012**, *8*, 714–718. [CrossRef] [PubMed]
36. McDougall, G.J.; Dobson, P.; Smith, P.; Blake, A.; Stewart, D. Assessing potential bioavallability of raspberry anthocyanins using an in vitro digestion system. *J. Agric. Food Chem.* **2005**, *53*, 5896–5904. [CrossRef] [PubMed]
37. Capanoglu, E.; Beekwilder, J.; Boyacioglu, D.; Hall, R.; De Vos, R. Changes in antioxidant and metabolite profiles during production of tomato paste. *J. Agric. Food Chem.* **2008**, *56*, 964–973. [CrossRef]

38. Beekwilder, J.; Jonker, H.; Meesters, P.; Hall, R.D.; Van Der Meer, I.M.; De Vos, C.H.R. Antioxidants in raspberry: On-line analysis links antioxidant activity to a diversity of individual metabolites. *J. Agric. Food Chem.* **2005**, *53*, 3313–3320. [CrossRef]

metabolites

MDPI

Article

Untargeted and Targeted Metabolomic Profiling of Australian Indigenous Fruits

Vuanghao Lim [1,2,*], Sara Ghorbani Gorji [1], Venea Dara Daygon [1] and Melissa Fitzgerald [1,*]

1 School of Agriculture and Food Sciences, The University of Queensland, Brisbane, QLD 4072, Australia; s.ghorbanigorji@uq.edu.au (S.G.G.); v.daygon@uq.edu.au (V.D.D.)
2 Integrative Medicine Cluster, Advanced Medical and Dental Institute, Universiti Sains Malaysia, Bertam, Kepala Batas 13200, Penang, Malaysia
* Correspondence: vlim@usm.my (V.L.); m.fitzgerald2@uq.edu.au (M.F.)

Received: 18 February 2020; Accepted: 11 March 2020; Published: 19 March 2020

Abstract: Selected Australian native fruits such as Davidson's plum, finger lime and native pepperberry have been reported to demonstrate potent antioxidant activity. However, comprehensive metabolite profiling of these fruits is limited, therefore the compounds responsible are unknown, and further, the compounds of nutritional value in these native fruits are yet to be described. In this study, untargeted and targeted metabolomics were conducted using the three fruits, together with assays to determine their antioxidant activities. The results demonstrate that targeted free and hydrolysed protein amino acids exhibited high amounts of essential amino acids. Similarly, important minerals like potassium were detected in the fruit samples. In antioxidant activity, Davidson's plum reported the highest activity in ferric reducing power (FRAP), finger lime in antioxidant capacity (ABTS), and native pepperberry in free radical scavenging (DPPH) and phosphomolybdenum assay. The compounds responsible for the antioxidant activity were tentatively identified using untargeted GC×GC-TOFMS and UHPLC-QqQ-TOF-MS/MS metabolomics. A clear discrimination into three clusters of fruits was observed using principal component analysis (PCA) and partial least squares (PLS) analysis. The correlation study identified a number of compounds that provide the antioxidant activities. GC×GC-TOFMS detected potent aroma compounds of limonene, furfural, and 1-R-α-pinene. Based on the untargeted and targeted metabolomics, and antioxidant assays, the nutritional potential of these Australian bush fruits is considerable and supports these indigenous fruits in the nutraceutical industry as well as functional ingredients for the food industry, with such outcomes benefiting Indigenous Australian communities.

Keywords: Davidson's plum; finger lime; native pepperberry; antioxidant; amino acids; metabolomics; GC×GC-TOFMS; UHPLC-QqQ-TOF-MS/MS; bush fruit

1. Introduction

Australia is famous for its rich diversity of native plant foods, which are also known as bush tucker, or bush food. There are about 6500 types of bush foods, and only a handful have been commercialised, and are considered to be worth about $18–25 million to the Australian economy [1,2]. Among these, there are about 2400 native fruits found in Queensland alone. In the stocktake published by the Australian native food industry, [joint collaboration of the Australian Native Food Industry Limited (ANFIL) and Rural Industries Research and Development Corporation (RIRDC) now AgriFutures Australia], emphasis has focussed on twelve key crops for further development. In the list, several native fruits have been identified, such as Davidson's plum, desert limes, quandong, lemon aspen, riberry, muntries, finger lime, kakadu plum, and native pepperberry [3]. These native edible fruits possess health benefits and can be used in applications such as functional foods and nutraceuticals, contributing to the emerging commercialisation in pharmaceutical industries. In the present work, three fruits were selected for in depth metabolomic profiling based on future potential: Davidson's plum, native pepperberry, and finger lime.

Davidson's plum (*Davidsonia pruriens* F. Muell) belongs to the Davidsoniaceae family, and it grows mostly in north-east Queensland in areas like coastal and upland rainforests (Djirbalngan, Yidinjdji, Djabuganjdji, Kuku-yalanji Nations). There are two other varieties; *D. johnsonii*, which is found in the south-east Queensland and New South Wales (Bundjalung Nation) while *D. jerseyana*, cultivates in the northern New South Wales (Bundjalung and Gumbainggir Nations) [4]. The tree can reach up to 20 m high [5]. The fruit is a purple plum described as tasting intensely sour [6], due to a high amount of acid with very little sugar to counteract it [6]. The fruit is rich in flavonoids, vitamins, minerals and several other important secondary metabolites, such as anthocyanins as well as proanthocyanidins [4]. Few preliminary studies have proclaimed that the extracts of Davidson's plum fruit inhibited in vitro cancer cells, metabolic syndrome enzymes, and contained high antioxidant activity [7]. Belonging to the Rutaceae family, finger lime, *Citrus australasica* var. *sanguinea* is also called Rainforest Pearl, found in the rainforests of Queensland (Bundjalung Nation) and northern part of New South Wales (Gumbainggir Nation) [8,9]. The fruits come in various shapes and sizes, and a range of colours including purple, green, yellow and pink. The native finger lime cultivated in Australia is one of the seven citrus species with 'caviar like' appearance of the fruit pulp. In general, finger lime is rich in vitamins, minerals and terpenes, such as limonene [10]. The bioactivities and phytoconstituents of finger lime are not well-established with reported properties limited to antioxidant and anti-inflammatory activity [9,11]. Native pepperberry [Tasmannia lanceolata (Poir.) A.C. Smith], which belongs to the family of Winteraceae grows at highland areas in Tasmania (Palawa Nation) and southeastern Australia (Boonwurrung, Woiworung, Jaitmatang, Bidwell, Yuin, Ngarigo Nations). As a native shrub tree, the fruit is black (dark purple) in colour and contains many tiny black seeds [12]. Native pepperberry has been ethnopharmacologically used as an ailment to treat stomach discomfort, and as an antifungal for skin diseases by Indigenous people, and it has scientifically proven to inhibit in vitro platelet aggregation, microbial activity, as well as antioxidant capacity [13–15]. Sesquiterpene polygodial, a major phytoconstituent which is mostly found in the oil of native pepperberry makes the spicy and pungent flavour [14]. Additionally, other major secondary metabolites like guaiol, calamenene, hexacosanal, drimenol and linalool were also reported to possess antimicrobial properties [12,16].

In order to support the medical, nutritional and food significance of these bushfoods, it is important to identify the functional compounds and understand their activities. In this regard, the identification of phytoconstituents from various analyses is emphasised, considering the synergistic or antagonistic activity of the metabolite-metabolite interactions for certain bioactivities. Plant metabolomics provides the tools necessary to analyse and potentially identify all the metabolites that possess bioactive properties. Analytical platforms, such as ultra-high performance liquid chromatography mass spectrometry (UHPLC-MS), have been widely used in plant science for metabolomics applications to identify and quantify compounds [17,18]. The comprehensive profiling and metabolomics studies of bush fruits are important in reaping better insight into commercial viability of these fruits. Antioxidant activity has been reported for the fruits [7,11,14,15,19], however the correlation of the activity with bioactive compounds through metabolomics approaches has not been reported. Therefore, this study was conducted to bridge the gap for the discovery of antioxidant-based active metabolites together with comprehensive profiling of Davidson's plum, finger lime and native pepperberry. In the current study, profiling of the fruits was conducted using targeted applications such as mineral analysis and amino acid analysis, and untargeted applications for semi-polar and aromatic compounds. Multivariate data analysis (MVDA), both supervised and un-supervised was applied to assess the association and discrimination of the compounds in the fruit samples. Data comparison of antioxidant scavenging activities was carried out together with total phenolic, flavonoids and flavonols content. The correlation between identified compounds and antioxidant activity was then conducted to identify potential bioactive markers in the fruit samples, as a mean of comprehensive findings for the fruits to be used in the nutraceutical industry.

2. Results

2.1. Untargeted Metabolic Profiling Using GC×GC-TOFMS and UHPLC-QqQ-TOF-MS/MS

The aromatic compounds in the powdered fruits were analysed by two-dimensional gas chromatography time of fight mass spectrometry (GC × GC-TOFMS), and approximately 616 peaks were found from each sample. The samples were screened using ChromaTOF software for the presence of common components. A total of 604 compounds were tentatively identified in the samples of fruits based on the library match searching data in NIST 11 v 2.0 and our in-house library. The metabolite profiles were compared in the PCA scores plots by submitting the combined Davidson's plum, finger lime and native pepperberry triplicate samples as shown in Figure 1a. Clustering of the scores was observed in three groups based on the three types of fruits. Two PCs were found with the greatest eigenvalues recorded at 55.7% and 25.7% of the total variance. The discrimination of the samples in three clusters indicates the aroma components in these fruits were different.

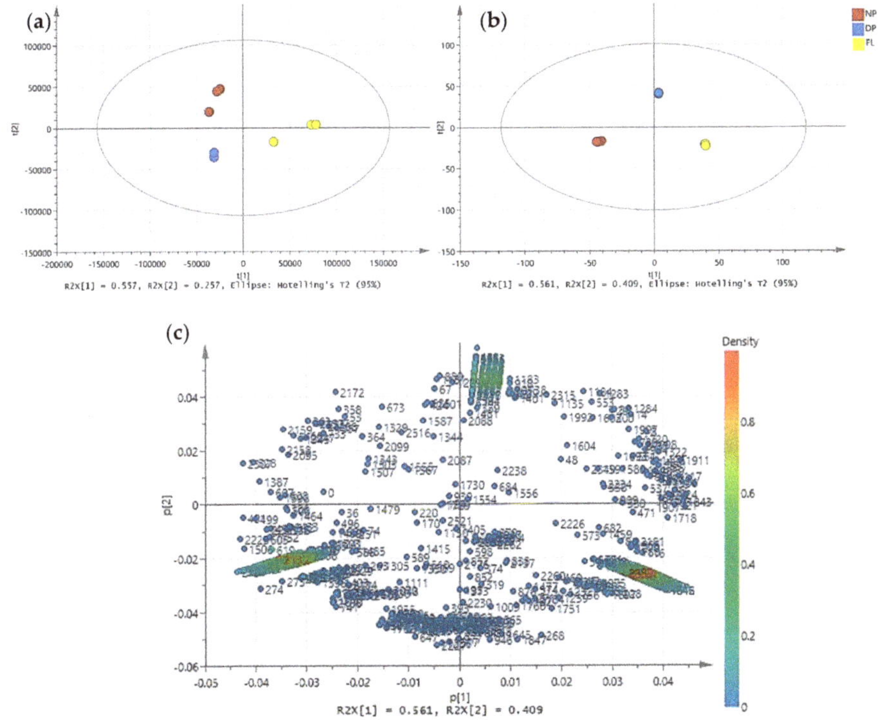

Figure 1. The principal component analysis of Davidson's plum (DP), finger lime (FL) and native pepperberry (NP) using (**a**) GC×GX-TOFMS; (**b**) UHPLC-QqQ-TOF-MS/MS; and (**c**) loading score plot representing compounds using UHPLC-QqQ-TOF-MS/MS. Compounds are coloured to indicate the relative density of peak areas.

The untargeted metabolic profile of the fruit samples obtained from the negative mode of ultrahigh-performance liquid chromatography triple quadrupole-time-of-flight mass spectrometry/mass spectrometry (UHPLC-QqQ-TOF-MS/MS) provided 1166 compounds (after data processing), with 542 annotated compounds tentatively identified by MS1 and MS2 matchings; 478 by MS1 and 64 by MS2 matching only. The PCA analysis dataset of the 3 types of fruits provided distinct differences of clustering patterns as shown in Figure 1b. Noticeably, PC1 indicated 56.1% deviation between the 3 fruits with finger lime clustering on the right side, whereas native pepperberry was observed towards the negative quadrant of PC1. The PCA was further analysed in loading plot to observe the discrimination of the compounds (Figure 1c).

2.2. *Antioxidant Activity Using DPPH, ABTS, FRAP and Phosphomolybdenum Assays*

The fruit samples were subjected to in vitro antioxidant activity using different spectrophotometric assays as shown in Table 1. From the table, finger lime exhibited the strongest capability in ABTS, however the lowest in DPPH assay despite the same radical scavenging activity. In reducing power ability, Davidson's plum reported the highest, while finger lime and native pepperberry did not show any significant difference. Phosphomolybdate method was later investigated for total antioxidant activity and expressed as gallic acid equivalent (μmol/gDW). The activity was found to decrease in the order of native pepperberry > finger lime > Davidson's plum.

2.3. *Total Phenolic Content (TPC), Total Flavonoid Content (TFC) and Total Flavonol Content (TFlC)*

Further investigation was conducted for TPC, TFC and TFlC. Native pepperberry exhibited a significantly higher amount of TPC, TFC and TFlC compared to finger lime, as shown in Table 1. Finger lime showed the lowest gallic acid and quercetin equivalent for all the assays. These assays showed a similar trend for the samples, with native pepperberry the highest, followed by Davidson's plum, and then finger lime.

2.4. *Correlation Between Antioxidant Activity And Compounds in The GCMS Dataset*

A supervised multivariate data, partial least squares (PLS) was applied to fathom the relationship between antioxidant activity and the fruit extracts. Based on Figure S1 in Supplementary Materials, the model showed cumulative $R^2X = 0.864$ and $Q^2 = 0.886$, indicating good fitness and high predictability (>0.5). In this analysis, the X-variables denote the aroma compounds and Y-variables are the antioxdant activity. Both PC1 and PC2 explained 86.4% of the variation in the aroma compounds showing discrimination in the compounds for antioxidant activity, with Y-variables recorded at 98%. In order to determine the aroma compounds that may contribute to the antioxidant activity, variable importance in projection (VIP) was carried out. The potential bioactive aroma compounds were chosen from the variables with VIP values of greater than 2.5. A total of 21 aroma compounds were sorted and the details are tabulated in Table 2. The identified aroma compounds were categorised into 6 different groups, namely, terpenes, aldehydes, terpenoids, furans, isoprenoids, and alkanes.

Table 1. In vitro antioxidant activity of the 3 fruit samples. Data is expressed in dry weight, DW and presented as mean values ± standard deviation ($n = 3$). Different superscript letters within each column indicate significant ($p < 0.01$) difference between samples.

Samples	ABTS (GAEμmol/gDW)	DPPH (GAEμmol/gDW)	FRAP (μmol Fe^{2+}/gDW)	Phosphomolybdenum (GAEμmol/gDW)	TPC (GAEμmol/gDW)	TFC (QTE μmol/gDW)	TFlC (QTE (μmol/gDW)
DP	21.92 ± 1.60^{a}	97.38 ± 3.93^{b}	500.38 ± 64.32^{c}	52.71 ± 5.40^{a}	113.58 ± 14.20^{b}	11.31 ± 0.52^{a}	5.55 ± 0.08^{b}
FL	62.73 ± 0.55^{c}	17.23 ± 2.04^{a}	46.16 ± 3.74^{a}	114.89 ± 3.25^{b}	63.46 ± 1.10^{a}	10.59 ± 0.97^{a}	0.56 ± 0.18^{a}
NP	22.06 ± 3.43^{a}	119.49 ± 2.72^{c}	48.48 ± 3.23^{a}	291.94 ± 2.23^{c}	134.82 ± 11.08^{b}	35.86 ± 3.99^{b}	9.99 ± 0.21^{c}

ABTS: 2,2′-azino-bis(3-ethylbenzothiazoline-6-sulfonic acid); DPPH: 2,2-diphenyl-1-picrylhydrazyl; FRAP: Ferric Reducing of Antioxidant Power Assay; TPC: Total Phenolic Content; TFC: Total Flavonoid Content; TFlC: Total Flavonol Content; DP: Davidson's plum; FL: Finger lime; NP: Native pepperberry; GA: Gallic acid equivalent; QTE: Quercetin equivalent.

Table 2. Aroma compounds identified by variable importance in projection (VIP) > 2.5 selection method for Davidson's plum, finger lime and native pepperberry.

Peak No.	Compounds	Group (VIP Score)	Retention Time ($1t_R$, $2t_R$) [a]	CAS No.	Mf [b]	Exact Mass	Level [c]
526	1-methyl-4-(prop-1-en-2-yl) cyclohex-1-ene [Limonene]	Terpene (7.01)	769.55, 0.72	138-86-3	$C_{10}H_{16}$	136.23	L2
329	1-R-α-Pinene	Terpene (6.34)	716.15, 637.29	7785-70-8	$C_{10}H_{16}$	136.23	L2
498	Furfural	Aldehyde (5.58)	577.36, 1.28	98-01-1	$C_5H_4O_2$	96.08	L1
508	Hexanal	Aldehyde (5.12)	503.75, 0.69	66-25-1	$C_6H_{12}O$	100.15	L1
437	1-methyl-4-(1-methylethylidene) -cyclohexene [Terpinolene]	Terpenoid (4.74)	972.00, 0.98	586-62-9	$C_{10}H_{16}$	136.23	L2
410	γ-Terpinene	Terpene (4.70)	1072.61, 1.18	99-85-4	$C_{10}H_{16}$	136.23	L2
599	Terpinen-4-ol	Terpenoid (4.49)	1038.69, 1.00	562-74-3	$C_{10}H_{18}O$	154.24	L2
397	2,2-dimethyl-3-methylidenebicy clo[2.2.1]heptane [Camphene]	Terpenoid (4.28)	953.92, 1.06	79-92-5	$C_{10}H_{16}$	136.23	L2
326	α-Phellandrene	Terpene (3.78)	778.35, 0.70	99-83-2	$C_{10}H_{16}$	136.23	L2
342	1-methyl-4-(1-methylethenyl) -benzene [p-Cymenene]	Terpene (3.63)	960.45, 0.88	1195-32-0	$C_{10}H_{12}$	132.20	L2
493	2-ethylfuran	Furan (3.55)	367.27, 0.88	3208-16-0	C_6H_8O	96.12	L2
281	1-methyl-4-(1-methylethyl)-7 -Oxabicyclo [2.2.1]heptane	Alkane (3.38)	757.76, 0.68	470-67-7	$C_{10}H_{18}O$	154.24	L2
1	(-)-Carvone	Terpenoid (3.11)	1170.77, 1.10	6485-40-1	$C_{10}H_{14}O$	150.21	L2
331	α-Terpineol	Terpenoid (3.05)	1072.61, 1.18	98-55-5	$C_{10}H_{18}O$	154.24	L2
527	l-menthone	Terpenoid (2.90)	1018.87, 0.87	10458-14-7	$C_{10}H_{18}O$	154.24	L2
364	4-methylene-1-(1-methylethyl)-Bicyclo[3.1.0]hexane, [Sabinene]	Terpene (2.89)	799.78, 0.70	3387-41-5	$C_{10}H_{16}$	136.23	L2
399	Caryophyllene	Isoprenoid (2.82)	1399.12, 0.71	87-44-5	$C_{15}H_{24}$	204.35	L2
569	Pentanal	Aldehyde (2.81)	382.87, 0.67	110-62-3	$C_5H_{10}O$	86.13	L2
272	5-isopropenyl-2-methylcyclo pent-1-enecarboxaldehyde	Terpenoid (2.78)	1275.00, 0.95	3865-09-6	$C_{10}H_{14}O$	150.21	L2
271	4-methylene-5-hexenal	Aldehyde (2.52)	629.52, 0.90	17844-21-2	$C_7H_{10}O$	110.15	L2
469	Dodecane	Alkanes (2.51)	984.85, 0.61	112-40-3	$C_{12}H_{26}$	170.33	L1

[a] $1t_R$: First dimension retention time, $2t_R$: Second dimension retention time; [b] Mf: Molecular formula; [c] Level: Level of identification based on the guidelines [20]. L1—level 1 identified through authentic chemical standards; L2—putatively identified compounds through library matching.

2.5. Correlation Between Antioxidant Activity And Compounds in The LCMS Dataset

With regard to the clear variance defined by the PCA (Figure 1b) and antioxidant activity (Table 2), a supervised multivariate data analysis, i.e., partial least squares (PLS) was utilised to organise and distinguish the samples according to their MS dataset. The correlation of the antioxidant activity and identified compounds was conducted by setting ABTS, DPPH, FRAP and phosphomolybdenum assays as Y variables, while the identified compounds were assigned as the X variables (Figure 2). In this experiment, a clear discrimination was achieved between Davidson's plum, finger lime and native pepperberry with good discriminant model indicators, R^2Y at 0.968 with the goodness-of-prediction value, Q^2 at 0.965. The robustness of the PLS model on the antioxidant activity for discriminating the identified compounds was confirmed by the results of permutation tests (Figure S2 in Supplementary Materials). The correlation of identified compounds with antioxidant acitivty showed discrimination of the fruits with scattered Y variables as shown in Figure 2b. Y variables of DPPH and FRAP are seen in the right quadrant of PC1 (X = 47.5%; Y = 55.1%), whereas the ABTS and phosphomolybdenum assays are towards the left. PC2 accounts for 49.5% of X variables and 43.3% in Y variables.

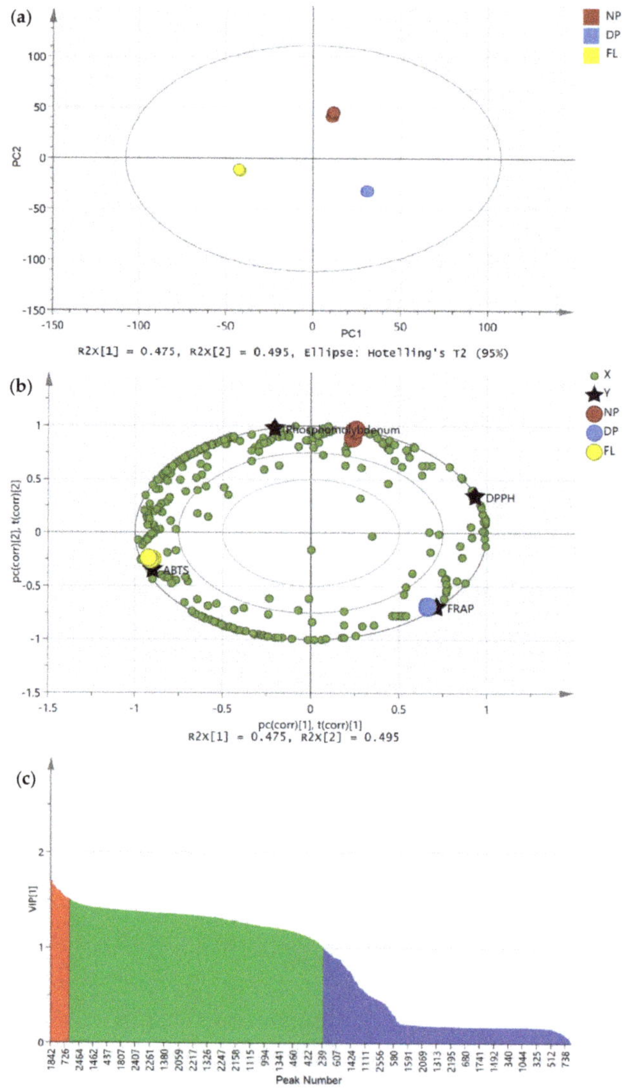

Figure 2. (**a**) Partial least square (PLS) score plot derived from UHPLC-QqQ-TOF-MS/MS on 3 fruit sample. (**b**) PLS biplot plots showing correlation between identified compounds with antioxidant activity. X = compounds, Y = antioxidant activity. (**c**) The variable importance in the projection (VIP) values (>1.5) represented by red coloured bars, (1.0–1.5) in green while blue coloured bars are VIP values (<1.0). The coloured compounds in (**b**) are VIP values >1.5.

Compounds that are responsible for the separation in Figure 2b were observed through variable importance in projection (VIP) selection approach. In this model, we set the VIP values at greater than 1.5 and loadings correlation coefficient [p(corr)] values above 0.5 to be significant for sample separation in the PLS model. From the analysis, a total of 44 compounds displayed good VIP values as listed in Table 3, with 7 unknown compounds. From the VIP scores, sugars, flavonoids and terpenes were the most important classes of compounds for antioxidant activity, followed by other phenolic compounds. Next, in order to better understand the distinctive incidence in antioxidant activity, we provide a fold-change analysis on the compounds (LogFC, Table 3). This analysis aims to correlate the significant value changes between the two group means. In this analysis, the fold-change was set at two, and any numbers that surpassed the threshold were considered significant. In particular, finger lime and native pepperberry showed the highest series of fold change with the fold-change distributions ranging from 3 to 10 times higher compared to Davidson's plum and finger lime. Nevertheless, Davidson's plum and native pepperberry only exhibited 5 significant fold-change of putatively identified compounds, ie. isovitexin, quercetagetin, racemosic acid, quercetin 3-[rhamnosyl-(1->2)-alpha-L-arabinopyranoside], and {3-[2-(3-hydroxy-5-methoxyphenyl)ethyl]phenyl}oxidanesulfonic acid, with the last compound also exhibited the highest fold-change (17.55).

2.6. Targeted Free And Protein Amino Acid Profilling Using UHPLC-MS

In total, 11 free amino acids were found in Davidson's plum, 19 in finger lime and 14 in native pepperberry (Table 4). Interestingly, essential amino acid, lysine, exhibited the highest free amino acid for all the three fruit samples with Davidson's plum (74.11%), finger lime (53.74%), and native pepperberry (67.88%). Other amino acids such as isoleucine, cystine and histidine are relatively high among the samples. Most of the free amino acids were detected in finger lime, and included aromatic amino acids, phenylalanine, tyrosine and tryptophan; sulfur-containing amino acids, cysteic acid, taurine and cystine; and non-essential amino acids, arginine, aspartic acid, glutamic acid, glycine, proline, and serine, but not alanine. Trace amounts of some free amino acids (<0.7%) were detected in native pepperberry compared to lysine, and isoleucine.

Analysis of the protein amino acids showed that Davidson's plum exhibited 10 amino acids, and both finger lime and native pepperberry reported all the protein amino acids (Table 4). Similar to the profile of free amino acids, the three fruit samples demonstrated that lysine was the highest amount of hydrolysed protein amino acid, followed by isoleucine indicating a significant difference between the samples for isoleucine. lysine is significantly higher in Davidson's plum compared to both finger lime and native pepperberry, in which the content is insignificant. Different trends of amino acids were observed for the three samples, where the percentage of amino acids in Davidson's plum ranged from 55.53% (lysine) to 0.08% (valine), finger lime, from 45.185% (lysine) to 0.235% (norleucine), and native pepperberry, from 37.07% (lysine) to 0.18% (cysteic acid).

Un-supervised, PCA of the free and hydrolyed protein amino acids was later conducted to observe the association of the amino acids with the fruits (Figure S3 in Supplementary Materials). From the biplots, finger lime (positive PC1) is discriminated from Davidson's plum and native pepperberry (negative PC1) and most of the free amino acids are associated with finger lime. In contrast, hydrolysed protein amino acids showed clustering of finger lime with native pepperberry and separating from Davidson's plum with negative PC1. Almost all the amino acids were associated with finger lime and native pepperberry except lysine, cysteine and isoleucine.

Table 3. PLS modelling of discriminant putatively identified compounds by variable importance in the projection (VIP) selection method based on antioxidant activity. The list of VIP scores are provided together with Log fold-change (LogFC) values.

No	Var ID	VIP Compounds	Class	VIP Score	p(corr)	LogFC [DP vs. FL]	LogFC [DP vs. NP]	LogFC [FL vs. NP]
1	1842	β-D-Glucuronopy ranosyl-(1->3)-a-D-galacturonopyranosyl-(1->2)-L-rhamnose	Oligosaccharides	1.70	-9.67×10	−5.83	-	16.85
2	1845	Dicaffeoylquinic acid	Quinic acids and derivatives	1.70	-9.67×10^3	−5.79	-	16.81
3	1846	Formononetin 7-(6″-malonylglucoside)	Isoflavonoid O-glycosides	1.69	-9.67×10^3	−5.73	-	16.75
4	2010	Octotiamine	Aminopyrimidines and derivatives	1.66	-9.67×10^3	−4.83	-	15.85
5	268	Quercetagetin	Flavonoid-7-O-glycosides	1.66	-9.39×10^3	−4.94	−7.61	8.36
6	2447	2″-O-Acetylrutin	Flavonoid-3-O-glycosides	1.65	-9.67×10^3	−4.69	-	15.71
7	2172	Quercetin 3-[rhamnosyl-(1->2)-alpha-L-arabinopyranoside]	Flavonoid-3-O-glycosides	1.64	9.92×10^3	3.69	6.88	-7.83
8	820	Suaveolenine	3-alkylindoles	1.64	-9.67×10^3	−4.43	-	15.44
9	1797	Kurzichalcolactone B	2′-Hydroxychalcones	1.63	-9.67×10^3	−4.23	-	15.25
10	1789	Syringetin-3-O-glucoside	Flavonoid-3-O-glycosides	1.63	-9.67×10^3	−4.16	-	15.18
11	310	3-(5-methoxy-2,2-dimethyl-1-benzopyran-8-yl)-3 -oxopropanoic acid	2,2-dimethyl-1-benzopyrans	1.62	-9.67×10^3	−4.07	-	15.09
12	1085	Bakkenolide D	Terpene lactones	1.62	-9.67×10^3	−3.87	-	14.89
13	712	3-Feruloyl-1,5-quinolactone	Coumarins and derivatives	1.61	-9.67×10^3	−3.81	-	14.83
14	1857	6-{7-Acetoxy-5-chloro-3-[(1E,3E)-3,5-dimethyl-1, 3-heptadien-1-yl]-7-methyl-6,8-dioxo-7,8-dihydro-2(6H) -isoquinolinyl}norleucine	Amino acid derivatives	1.61	-9.67×10^3	−3.64	-	14.66
15	1592	5,7-dihydroxy-2-(4-hydroxy-3-methoxyphenyl)-3 -{[3,4,5-trihydroxy-6-(hydroxymethyl)oxan-2-yl]oxy} -4H-chromen-4-one	Flavonoid-3-O-glycosides	1.60	-9.67×10^3	−3.49	-	14.51
16	1189	Unknown	-	1.60	-9.67×10^3	−3.47	-	14.49
17	819	Unknown	-	1.60	-9.67×10^3	−3.44	-	14.46
18	1188	farnochrol	Terpene lactones	1.59	-9.67×10^3	−3.40	-	14.42
19	581	Perilloside B	Terpene glycosides	1.59	-9.66×10^3	−3.37	-	14.39
20	1847	Racemosic acid	Phenolic glycosides	1.59	-8.95×10^3	−4.28	−9.18	6.12
21	2341	Kaempferol 3-[2‴-acetyl-alpha-L-arabinopyranosyl-(1->6)-galactoside]	Flavonoid-3-O-glycosides	1.58	-9.67×10^3	−3.16	-	14.18

Table 3. *Cont.*

No	Var ID	VIP Compounds	Class	VIP Score	p(corr)	LogFC [DP vs. FL]	LogFC [DP vs. NP]	LogFC [FL vs. NP]
22	1851	Dracunculifoside G	Coumaric acids and derivatives	1.58	-9.66×10^3	−3.12	-	14.14
23	583	Betuloside	Fatty acyl glycosides of mono- and disaccharides	1.57	-9.66×10^3	−2.83	-	13.85
24	2404	Isosakuranetin-7-O-neohesperidoside	Flavanone	1.56	-9.66×10^3	−2.76	-	13.78
25	2000	6-(benzoyloxy)-1-(hexopyranosyloxy)-1,4a,5,6,7,7a-hexahydro -5-hydroxy-7-methyl-cyclopenta[c]pyran-4-carboxylic acid	Carboxyl	1.56	-9.67×10^3	−2.68	-	13.69
26	413	Unknown	-	1.55	-9.66×10^3	−2.52	-	13.54
27	1304	Unknown	-	1.55	-9.67×10^3	−2.35	-	13.37
28	1802	Pungiolide A	Xanthanolides	1.54	-9.67×10^3	−2.28	-	13.30
29	817	Speradine A	Isoindolones	1.54	-9.67×10^3	−2.25	-	13.27
30	1228	Isovitexin	Flavone	1.53	-9.49×10^3	−1.85	−3.22	9.65
31	2016	6-methoxy-7-[3,4,5-trihydroxy-6-[(3,4,5-trihydroxy-6-methyloxan-2-yl)oxymethyl]oxan-2-yl]oxychromen-2-one	Ketone	1.53	-9.66×10^3	−1.94	-	12.96
32	2388	Unknown	-	1.53	-9.66×10^3	−1.90	-	12.92
33	726	4-Methylumbelliferyl glucuronide	Coumarins and derivatives	1.52	-9.67×10^3	−1.77	-	12.78
34	1697	3,4,5-trihydroxy-6-{3,4,5-trihydroxy-2-[3-(4-methoxyphenyl)-2-oxopropanoyl]phenoxy}oxane-2-carboxylic acid	Flavonoid O-glycosides	1.52	-9.81×10^3	−1.12	-	11.36
35	1949	Unknown	-	1.52	-9.67×10^3	−1.70	-	12.72
36	582	3-Hydroxy-4-isopropylbenzyl alcohol 3-glucoside	Terpene glycosides	1.52	-9.66×10^3	−1.70	-	12.72
37	2256	Unknown	-	1.52	-9.65×10^3	−1.80	-	12.81
38	552	{3-[2-(3-hydroxy-5-methoxyphenyl)ethyl]phenyl}oxidanesulfonic acid	Stilbenes	1.52	7.05×10^3	6.53	17.55	
39	1020	Picraquassioside A	Phenolic glycosides	1.52	-9.66×10^3	−1.65	-	12.67
40	834	(2S)-2-Butanol O-[b-D-Apiofuranosyl-(1->6)-b-D-glucopyranoside]	O-glycosyl compounds	1.51	-9.67×10^3	−1.62	-	12.64
41	1939	Longipedunin A	Hydrolyzable tannins	1.50	-9.66×10^3	−1.43	-	12.45
42	441	10,11-epoxycurvularin	Aryl alkyl ketones	1.50	-9.67×10^3	−1.36	-	12.38
43	727	Moschamine	N-acylserotonins	1.50	-9.67×10^3	−1.34	-	12.36
44	2131	Unknown	-	1.50	-9.66×10^3	−1.32	-	12.34

Var ID: Peak number; p(corr): correlation coefficient; DP: Davidson's plum; FL: Finger lime; NP: Native pepperberry.

Table 4. Amino acid composition of Davidson's plum (DP), finger lime (FL) and native pepperberry (NP) for (**a**) free amino acids; and (**b**) hydrolysed protein amino acids. Results are expressed as mean values (µmol/g) ± SD ($n = 3$). Superscript letters within each column indicate statistically significant ($p < 0.05$; Tukey test). ND = not detected.

(a)											
Sample	**Alanine**	**Arginine**	**Aspartic Acid**	**Cysteic Acid**	**Cystine**	**Glutamic Acid**	**Glycine**	**Histidine**	**Isoleucine**	**Leucine**	**Lysine**
DP	ND	0.16 ± 0.05^{a}	0.04 ± 0.04^{a}	ND	0.63 ± 0.15^{a}	0.18 ± 0.16^{a}	0.30 ± 0.03^{b}	0.57 ± 0.05^{a}	7.48 ± 0.71^{a}	0.03 ± 0.01^{a}	27.43 ± 0.96^{a}
FL	ND	5.60 ± 0.35^{b}	0.77 ± 0.09^{b}	1.80 ± 0.12^{a}	0.59 ± 0.12^{a}	0.16 ± 0.14^{a}	0.25 ± 0.04^{b}	2.91 ± 0.10^{b}	7.58 ± 1.13^{a}	0.04 ± 0.01^{a}	26.52 ± 1.62^{a}
NP	ND	0.06 ± 0.007^{a}	0.04 ± 0.003^{a}	1.79 ± 0.11^{a}	0.53 ± 0.21^{a}	0.56 ± 0.28^{a}	0.05 ± 0.02^{a}	ND	8.20 ± 1.62^{a}	0.04 ± 0.005^{a}	25.18 ± 1.59^{a}
Sample	**Methionine**	**Norleucine**	**Phenylalanine**	**Proline**	**Serine**	**Taurine**	**Threonine**	**Tryptophan**	**Tyrosine**	**Valine**	
DP	ND	0.11 ± 0.01^{a}	0.08 ± 0.007^{a}	ND	ND	ND	ND	ND	ND	ND	
FL	ND	0.11 ± 0.02^{a}	0.14 ± 0.003^{a}	0.50 ± 0.04^{b}	0.45 ± 0.09	1.07 ± 0.01	0.19 ± 0.06	0.03 ± 0.02	0.36 ± 0.05	0.27 ± 0.07^{a}	
NP	0.04 ± 0.01	0.10 ± 0.01^{a}	0.08 ± 0.007^{a}	0.22 ± 0.03^{a}	ND	ND	ND	ND	ND	0.20 ± 0.18^{a}	
(b)											
Sample	**Alanine**	**Arginine**	**Aspartic Acid**	**Cysteic Acid**	**Cystine**	**Glutamic Acid**	**Glycine**	**Histidine**	**Isoleucine**	**Leucine**	**Lysine**
DP	ND	ND	0.00 ± 0.006^{a}	ND	0.39 ± 0.31^{a}	0.46 ± 0.42^{a}	ND	ND	17.62 ± 0.43^{a}	0.34 ± 0.10^{a}	66.22 ± 2.92^{a}
FL	1.71 ± 0.78^{a}	5.21 ± 1.75^{a}	28.71 ± 2.17^{c}	0.64 ± 0.56^{a}	0.62 ± 0.04^{a}	6.99 ± 2.32^{b}	6.42 ± 2.88^{a}	2.76 ± 1.03^{a}	30.01 ± 2.19^{b}	5.05 ± 1.41^{b}	114.68 ± 16.62^{b}
NP	9.72 ± 0.91^{b}	4.25 ± 1.10^{a}	13.38 ± 0.93^{b}	0.60 ± 0.52^{a}	0.61 ± 0.02^{a}	36.58 ± 1.94^{c}	11.93 ± 0.90^{b}	3.70 ± 0.50^{a}	28.45 ± 5.25^{b}	14.70 ± 2.56^{c}	117.28 ± 14.24^{b}
Sample	**Methionine**	**Phenylalanine**	**Proline**	**Serine**	**Threonine**	**Tryptophan**	**Tyrosine**	**Valine**			
DP	ND	2.33 ± 0.27^{b}	ND	ND	ND	16.18 ± 1.18^{a}	15.24 ± 1.16^{b}	0.10 ± 0.01^{a}			
FL	1.07 ± 0.68^{a}	1.48 ± 0.54^{a}	8.45 ± 1.50^{b}	3.66 ± 0.27^{a}	6.80 ± 1.52^{a}	15.29 ± 0.43^{a}	12.90 ± 0.07^{a}	0.74 ± 0.10^{a}			
NP	5.10 ± 0.28^{b}	4.11 ± 0.06^{c}	1.83 ± 0.67^{a}	12.61 ± 1.38^{b}	6.52 ± 1.12^{a}	15.48 ± 0.05^{a}	13.97 ± 0.82^{a}	14.13 ± 0.80^{b}			

2.7. *Targeted Minerals and Heavy Metals Profilling using ICP-OES*

Here, we targeted 18 minerals and heavy metals for the fruit samples and the analysed data are shown in Figure S4 in Supplementary Materials. The heat map shows the relative concentration of all the minerals and heavy metals in each sample. Minerals (Mn, Na, Fe, Zn) are abundantly distributed in native pepperberry compared to Davidson's plum and finger lime. However, Davidson's plum contained elevated levels of Mg (816.1 ± 7.4 mg/kg) and Al (114.2 ± 1.2 mg/kg) while finger lime with higher levels of Ca (1390.1 ± 35.8 mg/kg) and P (870.63 ± 24.2 mg/kg) than any other elements analysed in the study (Table S1 in Supplementary Materials). From the statistical analysis, 9 elements varied significantly ($p < 0.05$), i.e., Al, Ca, Fe, K, Mg, Mn, Na, P, and Zn as shown in Figure 3 and represented in box and whiskers plot. Interestingly, the relative amount of Al, K, and Mg were highest in Davidson's plum compared to native pepperberry and finger lime. Al, Ca, Mg and P were moderately high in native pepperberry except for potassium. Finger lime exhibited a relatively low amount of most elements, but did show an amount of Ca and P compared to the other fruits.

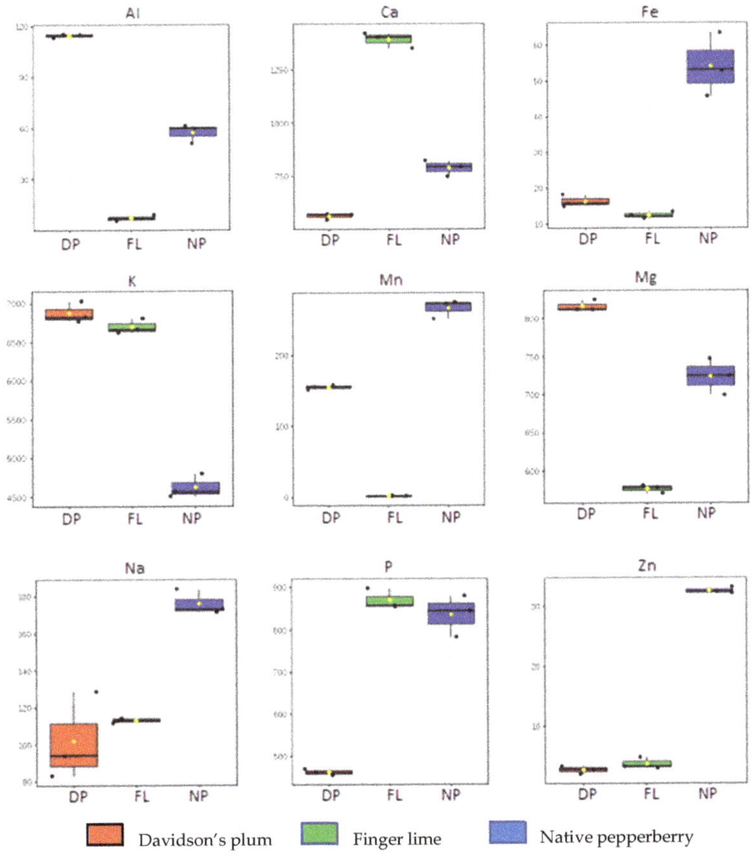

Figure 3. Box and whisker plots of nine elements varied significantly. DP: Davidson's plum, FL: Finger lime, NP: Native pepperberry. The Y-axis of box and whisker plots indicates the amount in mg/kg.

3. Discussion

Davidson's plum, finger lime and native pepperberry are among Australian native foods that are important to be classified as functional foods due to vast biologically active primary and secondary metabolites, especially terpenes, and flavonoids. Recently, with the inclusion of Indigenous foods in the food industry, natural antioxidants from bush fruits have gained expanding attention due to the growing demand for novel flavours, new functional compounds, and clean labelling. These fruits are rich in polyphenolic compounds, which are prominent natural antioxidants. The identification of aroma compounds using GC×GC-TOFMS provides an important factor in discriminating the clusters of the fruit samples. Correlation between antioxidant activity and aroma compounds exhibited abundance in terpenoid and terpene groups such as limonene, which generally contributes to the fruity smell due to its low odour threshold. Similar with previous studies, limonene was relatively high in finger lime, showing one of the major volatile compounds compared to other terpene groups [2,9,10]. The abundance of terpenes and terpenoids in the fruit samples suggest that they are suitable for processing of the fruits into jam and chutneys or even other food products with pleasant aroma and colour [9].

In the current study, we also conducted a correlation between bioactive compounds with antioxidant activity of the fruits using LCMS metabolomics. Through the comprehensive metabolite profiling, the results of various scavenging activities along with TPC, TFC and TFlC are presented in Table 2. In relation to the reported studies, Davidson's plum exhibited proportionate amount of total phenolic content with damson plum, *Prunus domestica* subsp. *Insititia* L. (124.32 GAE μmol/gDW) cultivated in France [21], but three times higher (38.20 GAE μmol/gDW) compared with the same plum planted in Serbia [22]. In general, bush fruits are rich in phenolics, terpenes and flavonoids, forming the vital class of compounds for scavenging activities in human body [14]. Each antioxidant assay provides different data, therefore four types of antioxidant assays were conducted in order to completely evaluate the efficacy of the powdered fruit extracts. Generally, different fruits exhibited different strengths in their respective antioxidant assays when compared with gallic acid. The antioxidant capacity changes with in vitro assays depending on the affinities of the active compounds present in the fruits [23]. The outcome of this study is also synergistic with the TPC, TFC and TFlC activities with their respective antioxidant activity. Thus, these findings support the antioxidant function of the fruits, which is in good agreement with previous results [2,11,14,24,25].

The relative variability of the compounds in the fruits was measured using MVDA to characterise the tentative identification of peaks that are bio-actively related to antioxidant activity. PCA and PLS were employed with clear discrimination of the score plots among the fruits. The correlation of the compounds with antioxidant activity (DPPH, ABTS, FRAP and phosphomolybdenum) was conducted using a PLS biplot to reveal the dominant compounds that may be responsible for the activities. The search of the compounds that attributed to antioxidant activity was conducted through the VIP values greater than 1.5 (Table 3). Therefore, from the findings of both GC and LC chromatographic analyses, the identified notable compounds were highly possible for the antioxidant activity. Some aroma compounds such as limonene, furfural, α-pinene, terpinen-4-ol, and γ-terpinene and polar compounds like dicaffeoylquinic acid, quercetagetin, and 2′′-O-acetylrutin have been reported to possess potent antioxidant activity [22–32].

In addition to the secondary compounds identified, the fruits, especially finger lime and native pepperberry, are rich in amino acids. Nevertheless, most of the essential amino acids were present in the samples and can therefore act as a functional food in dietary supplements [33]. Though lysine is well represented in some vegetable species [34,35], it is absent in many plants compared to other amino acids and this is a disadvantage for vegans. According to the Academy of Nutrition and Dietetics (AND), vegetarians are recommended to consume an array of plant-based food that are rich in protein in order to meet nutritional and health requirements [36]. Interestingly, lysine exhibited the highest amount in the fruits, which is suitable for vegetarians as well as those who are allergic to beans and legumes [37].

Minerals and trace elements in fruits are acquired and transported naturally for biological processes in the plants. However, some of these minerals this could be toxic to humans when they occur at particular levels. Instead of posing a threat, mineral nutrients under the normal level assist in mitigating toxicity caused by heavy metals. In our mineral study, the ANOVA statistical analysis highlighted nine important minerals from the heatmap (Figure S4 in Supplementary Materials) of the fruit samples. As the major mineral in human dietary intake, potassium contributed the highest percentage for all the samples ranging from 52–68% showing a good source of potassium, especially for Davidson's plum (68%). The requirement of potassium by humans is greater than 100 mg/day and with this fact, it is anticipated that the contribution of these fruits to dietary intake will grow in future. The potassium levels of Davidson's plum (6877 ± 138 mg/kg), and finger lime (6697 ± 98 mg/kg) found in this study are higher than potassium levels in any Colombian fruits [38], and subtropical fruits grown in Spain, such as custard apple, avocado, mango, banana, papaya, persimmon and starfruit [39]. The amount of phosphorus (835 ± 49 mg/kg), calcium (788 ± 37 mg/kg) and magnesium (723 ± 23 mg/kg) were relatively high in native pepperberry, compared to jujube fruits grown in China [40]. Bush fruits have been reported for rich sources of calcium and magnesium with an added advantage because both minerals are important in forming DNA and are especially important in repairing damaged DNA. They are equally crucial in segregating the chromosomes during DNA synthesis [19].

Taken together, all the data indicate that these three native fruits are excellent sources of important compounds, that offer nutritional value to consumers. For the food industry, these fruits offer potential to extend shelf-life naturally, and increase the functional value of the food, enabling the industry to make health claims about the food. This information could also assist with the development of Indigenous businesses to supply these high value foods to the food industry. In addition, this study illustrates that the reason for different antioxidants values from the different methods is based on the nature and classes of the secondary compounds much more than any of the other compounds.

4. Materials and Methods

4.1. Chemicals and Reagents

All organic solvents were purchased from Fisher Scientific (Pittsburgh, PA, USA) unless stated otherwise. Ammonium ferrous sulfate, sulfuric acid, aluminium chloride, sodium nitrate, sodium hydroxide, sodium acetate, aluminium hexahydrate, ammonium molybdate, sodium phosphate, sodium nitrite, quercetin, gallic acid, glacial acetic acid, 2,4,6-Tris(2-pyridyl)-s-triazine (TPTZ), 2,2-Di(4-tert-octylphenyl)-1-picrylhydrazyl (DPPH), potassium persulfate, formic acid, 2,2''-azinobis (3-ethylbenzothiazoline-6-sulfonic acid) diammonium salt (ABTS), were purchased from Sigma-Aldrich (St. Louis, MO, USA).

4.2. Plant Materials

Davidson's plum, native pepperberry, and finger lime and were purchased from Taste Australia Bush Food Shop (Queensland, Australia). The samples were pulverised into fine powder using TissueLyser II (Qiagen, Tokyo, Japan). The pulverised samples were extracted with solvent or water (specified in respective method) in triplicates for chemical assays.

4.3. GC×GC-TOFMS Analysis

The aroma in fruits was comprehensively analysed using static headspace extraction coupled with separation by two-dimensional gas chromatography time-of-flight mass spectrometry (GC × GC-TOFMS, LECO Pegasus 4D, Castle Hill, Australia). Briefly, sample (500 mg) was weighed into 20 mL silicon capped GC headspace vials (Restek, Germany) and kept at −80 °C until further analysis. 2.5 mL headspace syringe was used to collect 1.5 mL of sample headspace after sample agitation of 10 min at 80 °C. An empty vial was used as blank and quality assurance (QA) standard was prepared by mixing all the samples. The program settings, conditions and parameters are provided in Table S2 in Supplementary Materials [41]. LECO ChromaTOF 4.50 software was used to process the GC×GC-TOFMS data for pre-processing baseline correction and identification was conducted by library matching (NIST 11 v2.0) and from authentic reference standards created in an in-house library. The similarity of ≥80% with the NIST library was defined as putative identification (when standards were not available) [42–44].

4.4. UHPLC-QqQ-TOF-MS/MS Analysis

The pulverised fruits powder (10 mg) from 3 plant species were dissolved in methanol (600 µL) and vortexed for 15 min at 30 °C. The mixture was then centrifuged at 28,500× *g* for 15 min and an aliquot of each supernatant (100 µL) was transferred into new Eppendorf tubes. Millipore water was added in ratio 3:1 to the supernatant. The mixture was vortexed and filtered through a 0.2 µm PTFE filter into an autosampler vial. For pooled biological quality control (PBQC) samples, 5 µL of each sample was taken and vortexed for 1 min. For the blank, 100 µL of Millipore water was used. The UHPLC-QqQ-TOF-MS/MS analysis was performed using Shimadzu Nexera UHPLC system (Kyoto, Japan; LC-30AD pump, SIL-30AC autosampler and CTO-30A column oven) equipped with Shimadzu Q-TOFMS-9030 detector. Separation of the sample analytes was conducted on a Shimadzu Velox C18 (2.1 × 100 mm, 1.8 µm, part number 227-32007-03, Shimadzu, Kyoto, Japan). The mobile phase consisted of A (0.1% [*v*/*v*] formic acid in acetonitrile) and B (0.1% [*v*/*v*] formic acid in water). The flow of the solvent gradient flow was set as follows: 97% A for 0–0.75 min, 5% A for 0.75–13 min, 97% A for 13–16 min. The sample injection volume was 1 µL with consistent flow rate at 0.4 mL/min. The column temperature and auto-sampler were set at 40 °C. Negative ionisation mode was operated for mass spectrometry analysis equipped with electrospray ionisation (ESI) source with collision energy set at 70 eV. The MS data were collected from *m*/*z* 70–700 Da; nebulisation gas at 3 L/min; source temperature, 120 °C; and desolvation temperature, 200 °C. The MS data were centroided and acquired with Lab Solutions software version 5.80. The raw data files were exported to Lab Solutions Insight software in LCD (*.lcd) format for pre-processing, correction of retention time, and baseline. The raw files were also converted to mzML format for peak discrimination, filtering and alignment using MS-DIAL [45,46].

4.5. Targeted Analysis of Free Amino Acids

4.5.1. Extraction of Free Amino Acids from Fruit Samples

The samples were weighed (20 mg) in Eppendorf tube. Methanol (500 µL) was added and vortexed. The mixture was centrifuged at 30,000× *g* for 5 min, then the supernatant was transferred into a 2 mL tube. The precipitate was resuspended in 500 µL of Millipore water and vortexed. Next, the mixture was centrifuged at the same speed for 5 min. The methanol and water extract were combined and filtered through a 0.2 µm PTFE filter.

4.5.2. Free Amino Acid Derivatisation

The free amino acids were derivatised using AccQ.Fluor^{TM} reagent according to the Waters AccQ.Tag^{TM} pre-column derivatisation procedure [47]. Briefly, 35 µL of borate buffer was put into a tube (2 mL). 5 µL of sample was then mixed and vortexed for several seconds. Ten µL of reconstituted derivatisation reagent were then admixed to the buffered samples and immediately vortexed. Next, the mixture was left at room temperature for 1 min. After that, the samples were heated in a heating block for 10 min at 55 °C to finalise the derivatisation. The derivatised free amino acids were then transferred to autosampler vial for analysis. PBQC samples were prepared by pooling 5 µL from each sample, vortexed for 1 min and derivatised using the same method as mentioned previously.

4.5.3. Standards

Standard amino acid mix solution (Batch. No SLBS2232V; Sigma-Aldrich, St. Louis, MO, USA) was used for the identification and quantification of amino acids. The standards were prepared in serial dilutions of 40% stock solutions to 20%, 10%, 5%, 2.5% and 1.25%. Derivatisation of the standard amino acids were conducted according to the methods explained in Section 4.5.2. As for blank, 5 μL of Millipore water was used to replace sample at the beginning step, then followed by the derivatisation steps mentioned in Section 4.5.2.

4.5.4. UHPLC-MS Conditions

The analysis of amino acid derivatives was analysed on a Shimadzu Nexera UHPLC system (Kyoto, Japan; LC-30AD pump, SIL-30AC autosampler and CTO-30A column oven) equipped with Shimadzu MS-2020 detector. Waters Acquity UPLC^{TM} BEH C_{18} column (2.1 × 100 mm, 1.7 μm, part number 186003837, Waters, Milford, MA, USA) was used for the chromatographic separation at consistent temperature of 55 °C. The MS parameters were set as follows: Acquisition mode, SIM (refer Table S3 in Supplementary Materials); detector voltage, 0.1 V; interface voltage, 2.5 V; ionisation mode, positive; heat block temperature, 500 °C; interface temperature, 350 °C; nebulising gas flow, 1.5 L/min; injection volume, 10 μL. The mobile phase consists of A: 0.1% formic acid (*v*/*v*) in Millipore water, and B: 0.1% formic acid (*v*/*v*) in acetonitrile. The flow rate was set at 0.7 mL/min based on the gradient profile: initial-0.54 min (0–0.1% B); 0.54–5.74 min (0.1–15% B); 5.74–8.74 min (15–21.2% B); 8.74–10.50 min (21.2–59.6% B); 10.50–11.50 min (59.6% B); 11.50–12.00 min (59.6–0.1% B) and finally at 0.1% B until 13 min. The interconnected cleaning purge was set within 1 min (rinsing speed 35 μL/sec), and equilibrium was repeated for 5 min at initial conditions. The whole cycle time took 13 min to complete before the next injection.

4.6. Targeted Analysis of Protein Amino Acids

4.6.1. Sample Digestion

Fruit samples (20 mg each pulverised) were digested in a glass vessel containing 2 mL of 6 N HCl and 0.1% of phenol. The glass vessels were flushed with N_2 before sealing. Samples were hydrolysed at 110 °C for 20 h. Next, the samples were filtered using 0.2-μm filter, and the filtrate was neutralised with freshly prepared 6 N NaOH solution [48].

4.6.2. Protein Amino Acid Derivatisation

The protein amino acid derivatisation, PBQC and standards were prepared according to the methods mentioned in Sections 4.5.2 and 4.5.3.

4.6.3. UHPLC-MS Analysis

The standards and derivatised protein amino acid were injected and analysed according to the UHPLC-MS conditions mentioned in Section 4.5.4.

4.7. Targeted Analysis of Minerals and Heavy Metals

4.7.1. Sample Preparation

The fruits samples were weighed (100 mg) into new and clean 15 mL digestion tubes. To each tube, concentrated nitric acid (2.0 mL) was added and samples were pre-digested overnight. They were then placed into a digestion rack of Hotblock® Digestor SC100 Digestion System (Environmental Express, Vernon Hills, IL, USA). The samples were digested for 1 h at 100 °C. Upon completion, the samples were cooled, and then water added to bring the volume to 15 mL. The samples were shaken and centrifuged for 5 min [49].

4.7.2. ICP-OES Analysis

The operating conditions for inductively coupled plasma optical emission spectrometer (ICP-OES) were performed according to the parameters: RF incident power, 1000 W; plasma argon flow rate, 15.0 mL/min; auxillary argon flow rate, 1.50 mL/min; nebulizer argon flow rate, 0.75 mL/min; mist chamber, tracey and nebulizer SeaSpray, 2 mL/min flow rate. The wavelength measured for each element is as follows: Al 396.152 nm, As 188.980 nm, Ca 422.673 nm, Cd 214.439 nm, Co 238.892 nm, Cr 267.716 nm, Cu 324.754 nm, Fe 238.204 nm, K 766.491 nm, Mg 279.553 nm, Mn 257.610 nm, Mo 202.032 nm, Na 588.995 nm, Ni 231.604 nm, P 213.618 nm, Pb 220.353 nm, S 181.972 nm, and Zn 213.857 nm.

4.8. Determination of in Vitro Antioxidant Activity

The DPPH, FRAP, ABTS and phosphomolybdenum activities of the fruit samples were evaluated by the methods reported by [44,50,51], respectively. The ABTS, DPPH and phosphomolybdenum assays were reported as gallic acid equivalents (GAE μmol/gDW), whereas FRAP as Fe^{2+} equivalents (Fe^{2+} μmol/gDW).

4.9. Total Phenolic, Flavonoid and Flavonol Contents

Spectrophotometric technique was used for total phenolic, flavonoid and flavonol contents and were quantified following the protocol of [52,53]. Total phenolic contents were expressed as gallic acid equivalents (GAE μmol/gDW), whereas total flavonoid and flavonol contents were presented as quercetin equivalents (QTE μmol/gDW).

4.10. Data Processing and Analysis

Data for bioactivities were presented as mean ± SD for all triplicate analysis and a one-way analysis of the variance (ANOVA) was carried out using SPSS Statistics version 25 (IBM Corp, Armonk, NY, USA). The mean comparisons were conducted using the post-hoc Tukey's (HSD) multiple comparison test. Values with $p < 0.05$ were considered statistically significant.

The processed data from GC×GC-TOFMS were analysed using SIMCA-P software version 15 (Umeå, Sweden) for multivariate data analysis (MVDA). In order to access the clustering and trends of the comprehensive depiction of the fruit samples, principal component analysis (PCA) was used. Partial least squares (PLS) chemometric method was conducted to further analyse the correlation of antioxidant activity with volatile compounds in the fruit samples [41,44]. The variables were pareto scaled for PCA and PLS analyses. Afterwards, the variables selection namely VIP was used to select those notable volatile compounds (VIP > 2.5).

For UHPLC-QqQ-TOF-MS/MS data, the variables were pareto scaled and log transformed for PCA and PLS to lower heteroscedasticity and asymmetry in the statistical distribution [44]. In order to tentatively identify the compounds, the acquired raw data in mzML format were processed using MS-DIAL by comparing the MS/MS spectra with those in the spectral library. VIP was applied to identify the most significant compounds contributing to the antioxidant activity [54]. Compounds with VIP score > 1.5 were maintained for further elaboration using methods from MetaboAnalyst 4.0, an open source web-based tools for metabolomics data analysis [55].

For targeted MVDA of amino acids, both targeted free and hydrolysed amino acids were pareto scaled and log transformed. The heatmap and box and whiskers were generated using MetaboAnalyst 4.0 [55].

5. Conclusions

We have successfully reported that the Australian bush fruits, Davidson's plum, finger lime and native pepperberry are rich in terpenes, phenolics, flavonoids, flavonols, minerals, essential and non-essential free and hydrolysed protein amino acids, and functional bioactive compounds, which are promising fruits to be commercialised in the nutraceutical industry and food industry. We found that the fruits are an abundant source of antioxidant compounds (sugars, terpenes, flavonoids etc.) and that may serve as the source of natural antioxidants in food products or as new medicines. The use of metabolomics in correlation to the primary and secondary metabolites as well as the minerals would render an opportunity to obtain more potential bioactivities of the Australian bush fruits as functional food in nutraceutical industry. We also show that when reporting antioxidant activity, it is important to use more than one method to obtain a true understanding of the activity.

Supplementary Materials: The following are available online at http://www.mdpi.com/2218-1989/10/3/114/s1, Table S1: Targeted analysis of 18 minerals and heavy metals from Davidson's plum (DP), finger lime (FL) and native pepperberry (NP). Results are expressed as mean values (mg/DWkg) ± SD (n = 3). Superscript letters within each column indicate statistically significant ($p < 0.05$; Tukey test), Table S2: GC×GC-TOF-MS parameters for comprehensive profiling of fruit samples, Table S3: Amino acids' single ion monitoring for mass spectrometry detector, Figure S1: (**a**) Partial least square (PLS) score plot based on GC×GX-TOFMS data. (**b**) PLS biplot plots showing correlation between identified aroma compounds with antioxidant activity. X = compounds, Y = antioxidant activity, Figure S2: Permutation test of the PLS model based on UHPLC-QqQ-TOF-MS/MS data (**a**) ABTS; (**b**) DPPH; (**c**) FRAP; (**d**) Phosphomolybdenum assays, Figure S3: Biplot association of (**a**) free amino acid; (**b**) hydrolysed protein amino acid in Davidson's plum (DP), finger lime (FL) and native pepperberry (NP), Figure S4: Heat map of 18 mineral nutrients and heavy metals found in the fruit samples.

Author Contributions: Conceptualization, M.F. and V.L.; methodology, V.L. and S.G.G.; software, V.L. and S.G.G.; validation, V.L., S.G.G., V.D.D. and M.F.; formal analysis, V.L., S.G.G.; investigation, V.L., S.G.G., V.D.D.; resources, M.F.; data curation, V.L.; writing—original draft preparation, V.L.; writing—review and editing, V.L., S.G.G., V.D.D., M.F.; visualization, M.F.; supervision, M.F.; project administration, M.F.; funding acquisition, M.F. All authors have read and agreed to the published version of the manuscript.

Funding: This research received no external funding.

Acknowledgments: V.L. would like to thank Universiti Sains Malaysia and Ministry of Higher Education, Malaysia for the SLAB Scholarship under the Post-Doctoral program. We also thank Shimadzu Australia for access to some equipment and assistance with data analysis.

Conflicts of Interest: The authors declare no conflict of interest. The funders had no role in the design of the study; in the collection, analyses, or interpretation of data; in the writing of the manuscript, or in the decision to publish the results.

References

1. Australian Bush Foods Are All around Us, But the Industry Is Just Waking up. Available online: https://www.abc.net.au/news/2019-04-06/australian-bush-foods-under-threat-just-as-industry-develops/10950522 (accessed on 25 September 2019).
2. Richmond, R.; Bowyer, M.; Vuong, Q. Australian native fruits: Potential uses as functional food ingredients. *J. Funct. Foods* **2019**, *62*, 103547. [CrossRef]
3. Clarke, M. *Australian Native Food Industry Stocktake*; Rural Industries Research and Development Corporation: Wagga, Australia, 2012; Publication No. 12/066.
4. John, O.D.; Mouatt, P.; Prasadam, I.; Xiao, Y.; Panchal, S.K.; Brown, L. The edible native Australian fruit, Davidson's plum (*Davidsonia pruriens*), reduces symptoms in rats with diet-induced metabolic syndrome. *J. Funct. Foods* **2019**, *56*, 204–215. [CrossRef]
5. New South Wales. *Department of Environment and Conservation 2004, Approved NSW Recovery Plan: Davidson's Plum (Davidsonia jerseyana)*; Dept. of Environment and Conservation (NSW): Hurstville, Sydney, Australia, 2004.
6. Australian Native Food and Botanicals. Available online: https://anfab.org.au/main.asp?_=DavidsonPlum (accessed on 10 October 2019).
7. Chuen, T.L.K.; Vuong, Q.V.; Hirun, S.; Bowyer, M.C.; Predebon, M.J.; Goldsmith, C.D.; Sakoff, J.A.; Scarlett, C.J. Antioxidant and anti-proliferative properties of Davidson's plum (*Davidsonia pruriens* F. Muell) phenolic-enriched extracts as affected by different extraction solvents. *J. Herb. Med.* **2016**, *6*, 187–192. [CrossRef]
8. Department of Industry and Investment (Industry & Investment NSW). Growing Australian native finger limes. *PrimeFacts* **2010**, *979*, 1–11.
9. Delort, E.; Yuan, Y.-M. Finger lime/The Australian Caviar—*Citrus australasica*. In *Exotic Fruits*; Rodrigues, S., de Oliveira Silva, E., de Brito, E.S., Eds.; Academic Press: Cambridge, MA, USA, 2018; pp. 203–210.

10. Delort, E.; Jaquier, A.; Decorzant, E.; Chapuis, C.; Casilli, A.; Frérot, E. Comparative analysis of three Australian finger lime (*Citrus australasica*) cultivars: Identification of unique citrus chemotypes and new volatile molecules. *Phytochemistry* **2015**, *109*, 111–124. [CrossRef] [PubMed]
11. Wang, Y.; Ji, S.; Zang, W.; Wang, N.; Cao, J.; Li, X.; Sun, C. Identification of phenolic compounds from a unique citrus species, finger lime (*Citrus australasica*) and their inhibition of LPS-induced NO-releasing in BV-2 cell line. *Food Chem. Toxicol.* **2019**, *129*, 54–63. [CrossRef] [PubMed]
12. Sultanbawa, Y. *Tasmanian Pepper Leaf (Tasmannia Lanceolata) Oils*; Elsevier Inc.: Amsterdam, The Netherlands, 2016.
13. Read, C.; Menary, R. Analysis of the contents of oil cells in *Tasmannia lanceolata* (Poir.) A. C. Smith (Winteraceae). *Ann. Bot.* **2000**, *86*, 1193–1197. [CrossRef]
14. Konczak, I.; Zabaras, D.; Dunstan, M.; Aguas, P. Antioxidant capacity and phenolic compounds in commercially grown native Australian herbs and spices. *Food Chem.* **2010**, *122*, 260–266. [CrossRef]
15. Beckett, J.M.; Ahuja, K.D.K.; Booth, C.; Ovenden, S.; Ball, M.J. In vitro anti-platelet and antioxidant effect of *Tasmannia lanceolata* (native pepperberry). *J. Nutr. Intermed. Metab.* **2014**, *1*, 47. [CrossRef]
16. Wilson, M.D.; Menary, R.C.; Close, D.C. Effects of tree guards and mulching on plantation establishment of Tasmanian Native Pepper (*Tasmannia lanceolata* (Poir.) A.C. Smith). *J. Appl. Res. Med. Aromat. Plants* **2015**, *2*, 154–159. [CrossRef]
17. Wang, X.; Zhang, A.; Yan, G.; Han, Y.; Sun, H. UHPLC-MS for the analytical characterization of traditional Chinese medicines. *TrAC Trends Anal. Chem.* **2014**, *63*, 180–187. [CrossRef]
18. Aszyk, J.; Byliński, H.; Namieśnik, J.; Kot-Wasik, A. Main strategies, analytical trends and challenges in LC-MS and ambient mass spectrometry–based metabolomics. *TrAC Trends Anal. Chem.* **2018**, *108*, 278–295. [CrossRef]
19. Konczak, I.; Roulle, P. Nutritional properties of commercially grown native Australian fruits: Lipophilic antioxidants and minerals. *Food Res. Int.* **2011**, *44*, 2339–2344. [CrossRef]
20. Salek, R.M.; Steinbeck, C.; Viant, M.R.; Goodacre, R.; Dunn, W.B. The role of reporting standards for metabolite annotation and identification in metabolomic studies. *Gigascience* **2013**, *2*, 1. [CrossRef] [PubMed]
21. Mousavi, M.; Zaiter, A.; Modarressi, A.; Baudelaire, E.; Dicko, A. The positive impact of a new parting process on antioxidant activity, malic acid and phenolic content of *Prunus avium* L., *Prunus persica* L. and *Prunus domestica* subsp. *Insititia* L. powders. *Microchem. J.* **2019**, *149*, 103962. [CrossRef]
22. Miletić, N.; Mitrović, O.; Popović, B.; Nedović, V.; Zlatković, B.; Kandić, M. Polyphenolic Content and Antioxidant Capacity in Fruits of Plum (*Prunus domestica* L.) Cultivars Valjevka and Mildora as Influenced by Air Drying. *J. Food Qual.* **2013**, *36*, 229–237. [CrossRef]
23. Floegel, A.; Kim, D.O.; Chung, S.J.; Koo, S.I.; Chun, O.K. Comparison of ABTS/DPPH assays to measure antioxidant capacity in popular antioxidant-rich US foods. *J. Food Compos. Anal.* **2011**, *24*, 1043–1048. [CrossRef]
24. Rayan, P.; Matthews, B.; Mc Donnell, P.; Cock, I.E. Phytochemical analysis of *Tasmannia lanceolata* extracts and inhibition of giardia duodenalis proliferation. *Pharmacogn. J.* **2016**, *8*, 291–299. [CrossRef]
25. Science, F. *Potential Health Properties of Selected Commercially Grown Native Australian Herbs and Fruits*; The University of New South Wales: Sydney, Australia, 2012.
26. Kang, Y.; Li, P.; Zeng, X.; Chen, X.; Xie, Y.; Zeng, Y.; Zhang, Y.; Xie, T. Biosynthesis, structure and antioxidant activities of xanthan gum from *Xanthomonas campestris* with additional furfural. *Carbohydr. Polym.* **2019**, *216*, 369–375. [CrossRef]
27. Shahriari, M.; Zibaee, A.; Sahebzadeh, N.; Shamakhi, L. Effects of α-pinene, trans-anethole, and thymol as the essential oil constituents on antioxidant system and acetylcholine esterase of *Ephestia kuehniella* Zeller (Lepidoptera: Pyralidae). *Pestic. Biochem. Physiol.* **2018**, *150*, 40–47. [CrossRef]
28. Souza, C.F.; Baldissera, M.D.; Silva, L.; de Silva, L.L.; Geihs, M.A.; Baldisserotto, B. Is monoterpene terpinen-4-ol the compound responsible for the anesthetic and antioxidant activity of *Melaleuca alternifolia* essential oil (tea tree oil) in silver catfish? *Aquaculture* **2018**, *486*, 217–223. [CrossRef]
29. Milde, J.; Elstner, E.F.; Graßmann, J. Synergistic inhibition of low-density lipoprotein oxidation by rutin, γ-terpinene, and ascorbic acid. *Phytomedicine* **2004**, *11*, 105–113. [CrossRef]
30. de Christo Scherer, M.M.; Marques, F.M.; Figueira, M.M.; Peisino, M.C.O.; Schmitt, E.F.P.; Kondratyuk, T.P.; Endringer, D.C.; Scherer, R.; Fronza, M. Wound healing activity of terpinolene and α-phellandrene by attenuating inflammation and oxidative stress in vitro. *J. Tissue Viability* **2019**, *28*, 94–99. [CrossRef] [PubMed]

31. Singh, P.; Shukla, R.; Prakash, B.; Kumar, A.; Singh, S.; Mishra, P.K.; Dubey, N.K. Chemical profile, antifungal, antiaflatoxigenic and antioxidant activity of *Citrus maxima* Burm. and *Citrus sinensis* (L.) Osbeck essential oils and their cyclic monoterpene, DL-limonene. *Food Chem. Toxicol.* **2010**, *48*, 1734–1740. [CrossRef] [PubMed]
32. Ahmad, S.; Beg, Z.H. Hypolipidemic and antioxidant activities of thymoquinone and limonene in atherogenic suspension fed rats. *Food Chem.* **2013**, *138*, 1116–1124. [CrossRef]
33. Patel, M.K.; Mishra, A.; Jha, B. Non-targeted metabolite profiling and scavenging activity unveil the nutraceutical potential of psyllium (*Plantago ovata* forsk). *Front. Plant Sci.* **2016**, *7*, 1–17. [CrossRef]
34. Caruso, G.; Conti, S.; Villari, G.; Borrelli, C.; Melchionna, G.; Minutolo, M.; Russo, G.; Amalfitano, C. Effects of transplanting time and plant density on yield, quality and antioxidant content of onion (*Allium cepa* L.) in southern Italy. *Sci. Hortic. (Amst.)* **2014**, *166*, 111–120. [CrossRef]
35. Caruso, G.; Villari, G.; Borrelli, C.; Russo, G. Effects of crop method and harvest seasons on yield and quality of green asparagus under tunnel in southern italy. *Adv. Hortic. Sci.* **2012**, *26*, 51–58.
36. Craig, W.J.; Mangels, A.R. Position of the American dietetic association: Vegetarian diets. *J. Am. Diet. Assoc.* **2009**, *109*, 1266–1282.
37. Rogerson, D. Vegan diets: Practical advice for athletes and exercisers. *J. Int. Soc. Sports Nutr.* **2017**, *14*, 1–15. [CrossRef]
38. Leterme, P.; Buldgen, A.; Estrada, F.; Londoño, A.M. Mineral content of tropical fruits and unconventional foods of the Andes and the rain forest of Colombia. *Food Chem.* **2006**, *95*, 644–652. [CrossRef]
39. Barea-Álvarez, M.; Delgado-Andrade, C.; Haro, A.; Olalla, M.; Seiquer, I.; Rufián-Henares, J.Á. Subtropical fruits grown in Spain and elsewhere: A comparison of mineral profiles. *J. Food Compos. Anal.* **2016**, *48*, 34–40. [CrossRef]
40. Wang, L.; Fu, H.; Wang, W.; Wang, Y.; Zheng, F.; Ni, H.; Chen, F. Analysis of reducing sugars, organic acids and minerals in 15 cultivars of jujube (*Ziziphus jujuba* mill.) fruits in China. *J. Food Compos. Anal.* **2018**, *73*, 10–16. [CrossRef]
41. Ghorbani Gorji, S.; Calingacion, M.; Smyth, H.E.; Fitzgerald, M. Effect of natural antioxidants on lipid oxidation in mayonnaise compared with BHA, the industry standard. *Metabolomics* **2019**, *15*, 1–14. [CrossRef] [PubMed]
42. Daygon, V.D.; Prakash, S.; Calingacion, M.; Riedel, A.; Ovenden, B.; Snell, P.; Mitchell, J.; Fitzgerald, M. Understanding the Jasmine phenotype of rice through metabolite profiling and sensory evaluation. *Metabolomics* **2016**, *12*, 1–15. [CrossRef]
43. Daygon, V.D.; Calingacion, M.; Forster, L.C.; de Voss, J.J.; Schwartz, B.D.; Ovenden, B.; Alonso, D.E.; Mccouch, S.R.; Garson, M.J.; Fitzgerald, M.A. Metabolomics and genomics combine to unravel the pathway for the presence of fragrance in rice. *Sci. Rep.* **2017**, *7*, 8767. [CrossRef]
44. Ghorbani Gorji, S.; Calingacion, M.; Smyth, H.E.; Fitzgerald, M. Comprehensive profiling of lipid oxidation volatile compounds during storage of mayonnaise. *J. Food Sci. Technol.* **2019**, *56*, 4076–4090. [CrossRef]
45. Tsugawa, H.; Cajka, T.; Kind, T.; Ma, Y.; Higgins, B.; Ikeda, K.; Kanazawa, M.; Vandergheynst, J.; Fiehn, O.; Arita, M. MS-DIAL: Data-independent MS/MS deconvolution for comprehensive metabolome analysis. *Nat. Methods* **2015**, *12*, 523–526. [CrossRef]
46. Tsugawa, H.; Nakabayashi, R.; Mori, T.; Yamada, Y.; Takahashi, M.; Rai, A.; Sugiyama, R.; Yamamoto, H.; Nakaya, T.; Yamazaki, M.; et al. A cheminformatics approach to characterize metabolomes in stable-isotope-labeled organisms. *Nat. Methods* **2019**, *16*, 295–298. [CrossRef]
47. Fiechter, G.; Mayer, H.K. Characterization of amino acid profiles of culture media via pre-column 6-aminoquinolyl-N-hydroxysuccinimidyl carbamate derivatization and ultra performance liquid chromatography. *J. Chromatogr. B Anal. Technol. Biomed. Life Sci.* **2011**, *879*, 1353–1360. [CrossRef]
48. Ali, N.; Paul, S.; Gayen, D.; Sarkar, S.N.; Datta, K.; Datta, S.K. Development of Low Phytate Rice by RNAi Mediated Seed-Specific Silencing of Inositol 1,3,4,5,6-Pentakisphosphate 2-Kinase Gene (IPK1). *PLoS ONE* **2013**, *8*, e68161. [CrossRef] [PubMed]
49. Dolan, S.P.; Nortrup, D.A.; Bolger, P.M.; Capar, S.G. Analysis of dietary supplements for arsenic, cadmium, mercury, and lead using inductively coupled plasma mass spectrometry. *J. Agric. Food Chem.* **2003**, *51*, 1307–1312. [CrossRef] [PubMed]
50. Ong, M.G.; Mat Yusuf, S.N.A.; Lim, V. Pharmacognostic and antioxidant properties of *Dracaena sanderiana* leaves. *Antioxidants* **2016**, *5*, 28. [CrossRef] [PubMed]

51. Ibrahim, N.M.; Mat, I.; Lim, V.; Ahmad, R. Antioxidant activity and phenolic content of *Streblus asper* leaves from various drying methods. *Antioxidants* **2013**, *2*, 156–166. [CrossRef] [PubMed]
52. Shan, S.; Huang, X.; Shah, M.H.; Abbasi, A.M. Evaluation of polyphenolics content and antioxidant activity in edible wild fruits. *BioMed Res. Int.* **2019**, *2019*, 1381989. [CrossRef]
53. Batool, R.; Khan, M.R.; Sajid, M.; Ali, S.; Zahra, Z. Estimation of phytochemical constituents and in vitro antioxidant potencies of *Brachychiton populneus* (Schott & Endl.) R.Br. *BMC Chem.* **2019**, *13*, 1–15.
54. Chong, H.W.; Rezaei, K.; Chew, B.L.; Lim, V. Chemometric Profiling of *Clinacanthus nutans* Leaves Possessing Antioxidant Activities using. *Chiang Mai J. Sci.* **2018**, *45*, 1377–1388.
55. Chong, J.; Wishart, D.S.; Xia, J. Using MetaboAnalyst 4.0 for Comprehensive and Integrative Metabolomics Data Analysis. *Curr. Protoc. Bioinform.* **2019**, *68*, 1–128. [CrossRef]

metabolites

MDPI

Article

Non-Targeted Metabolite Profiles and Sensory Properties Elucidate Commonalities and Differences of Wines Made with the Same Variety but Different Cultivar Clones

Álvaro Cuadros-Inostroza [1], Claudio Verdugo-Alegría [2], Lothar Willmitzer [3], Yerko Moreno-Simunovic [2,*] and José G. Vallarino [2,*]

1 metaSysX GmbH, 14476 Potsdam-Golm, Germany; Inostroza@metasysx.com
2 Centro Tecnológico de la Vid y el Vino, Universidad de Talca, Av. Lircay s/n, 3460000 Talca, Maule, Chile; cverdugo@utalca.cl
3 Max-Planck-Institut für Molekulare Pflanzenphysiologie, 14476 Potsdam-Golm, Germany; willmitzer@mpimp-golm.mpg.de
* Correspondence: ymoreno@utalca.cl (Y.M.-S.); vallarino@uma.es (J.G.V.)

Received: 28 April 2020; Accepted: 22 May 2020; Published: 28 May 2020

Abstract: Grapes, one of the oldest agricultural crops, are cultivated to produce table fruits, dried fruits, juice, and wine. Grapevine variety is composed of clones that share common morphological traits. However, they can differ in minor genetic mutations which often result in not only notorious morphological changes but also in other non-visible sensorial distinctive attributes. In the present work, we identified three *Vitis vinifera* cv. Pinot noir clones grown under identical field conditions that showed different grape cluster types. Here, sensorial analysis together with non-targeted metabolite profiles by Ultra High performance Liquid Chromatography (UPLC) couples to Ultra High Resolution Mass Spectrometry (FT-ICR-MS) of wines elaborated from the three different grape cluster types was studied with the aim of (i) finding sensorial differences among these three types of wines, and, if there were, (ii) determining the molecular features (metabolites) associated with these sensorial attributes by a multivariate statistical approach.

Keywords: metabolomics; wine; clones; *Vitis vinifera*; sensory analysis

1. Introduction

With several thousand varieties, the *Vitis* genus is characterized by high levels of genetic diversity. The *Vitis* international variety catalogue identifies more than 21 thousand names of varieties, including 12.250 for *Vitis vinifera*. However, we have to consider that this number includes also synonyms and homonyms [1]. Moreover, the actual number of vine varieties for *V. vinifera* species is estimated at 6000 [2]. Additionally, vines are one of the most important crops in the world with a total surface area planted reaching 7.4 million hectares (Mha and world-wine market reaching 31.3 bn EUR in 2018. The variety Pinot noir is the 6th most widely planted red grape variety worldwide [3]. This cultivar is particularly well adapted to cool growing areas and can be grown at higher latitudes in comparison to other varieties. In addition, the Pinot noir variety is appreciated by the quality of its fresh and fruity monovarietal wine.

Wine can be considered a complex product whose quality is attributable to its chemical composition, which is highly influenced by grape variety [4], viticultural practices [5,6], fermentation conditions [7], vineyard geographical location that is related to soil and climate characteristics [8–10], technical conditions of wine-making [11], and the quality of the grapes used for its production [12,13]. Determination of grape quality is important as grower payment per ton of fruit. Aroma is one of the most important factors in determining wine character and quality [14]. Another equally important wine quality attribute is mouthfeel. Aroma descriptors include red berries, smoky, herbal, and woody, which represent what a wine smells like; while taste is described by attributes such as astringency, bitter, acidity, and sweetness. These characteristics are the result of complex interactions among volatile and non-volatile compounds. So far, the contributions of these attributes to wine are determined by a tasting panel, who must be rigorous and well trained using an appropriate and standardized vocabulary.

Metabolomic approaches can take a snapshot of the current biochemical status and are useful to compare varieties and evaluate changes in metabolic pools. In this regard, metabolomics-based approaches can reveal several thousand signals of candidate biomarkers providing high sensitivity and good resolution for wine authentication purposes. Recently, there have been a number of reports utilizing high-throughput techniques focusing on wine or grape metabolomes to discriminate wines from distinct varieties, fermentation conditions, geographical origin, as well as different qualities [15–26]. Moreover, most of the current analysis of wine with high-throughput analytical technologies contributes to a volatile profiling using different multivariate techniques [17,18,21,27]. However, very little is reported on non-targeted and non-volatile metabolite profiles as a basis [19,20,22,26,28,29]. Furthermore, to the best of our knowledge, none of the non-volatile biomarkers associated with wine quality (sensory properties) have been targeted to assist or confirm the rating score provided by wine panelists.

Due to the unstudied phenotypic cluster variation observed in an established clonal-commercial vineyard, the scope of this work is to use the combination of sensorial analysis and non-targeted metabolite profiling by ultra-high-performance liquid chromatography coupled to Ultra-High-Resolution Mass Spectrometry (UPLC-FT-ICR-MS) to determine whether both platforms have the power to distinguish between three red wines made using the different cluster types obtained each from three *V. vinifera* cv. Pinot noir putative-clones grown under identical field conditions. In addition, if this is feasible, we are also interested in identifying molecular features which contribute most to the distinction between the different sensory properties of the three types of red wines assessed by an expert panel. Therefore, the selected molecular features can be used as biomarkers for certification or assessment sensorial analysis.

2. Results

2.1. Viticultural Characteristics of Three Pinot Noir Grape Cluster Types

Different fruit size was observed between different cluster types (Figure 1). Grape cluster type iii exhibited a significantly greater berry size in comparison to grape cluster type i, while grape cluster type ii showed an intermediate phenotype, with large, medium, and small grape sizes. Therefore, different ratios between skin and fresh weight were produced by the different types of berries as expected. Although berry size was different, it is apparent that this resulted in an unaltered soluble solids content (Brix), total acidity, and pH of grape berries (Table 1). In wines made from the three grape cluster types, no differences in the alcohol content, residual sugar content, total acidity, and pH were observed.

(A)

(B)

Diameter (cm)	Frecuency of grape diameter per cluster type (%)		
	Cluster type I	Cluster type II	Cluster type III
1.4			13.1 ± 0.7
1.2	4.1 ± 0.8	18.2 ± 0.6	48.3 ± 1.1
1		22.2 ± 0.9	39.2 ± 1.6
0.8	45.2 ± 1.1	34.4 ± 1.8	
0.6	34.3 ± 0.8	18.1 ± 1.5	
0.4	17.1 ± 0.9	8.3 ± 0.4	

Figure 1. (**A**). A comparison of the three Pinot noir grape cluster types. Scale bar = 2 cm. (**B**). Berry diameter at the ripe stage. Values are mean ± standard deviation (n = 50).

Table 1. Grape and wine analysed in the experiment including some basic compositional parameters. Values are mean ± standard deviation (n = 50 and n = 6 for grapes and wines, respectivately.

Season	Cluster Type	Grapes				Wines			
			Average	±	SD		Average	±	SD
		°Brix	24.8	±	2.92	Alcohol (% vol.)	14.4	±	0.31
						Residual sugar (gr/L)	1.3	±	0.17
		Total Acidity (HSO_4 gr/L)	7.5	±	1.11	Total Acidity (HSO_4 gr/L)	4.3	±	0.27
	Type i	pH	3.0	±	0.11	pH	3.4	±	0.01
		Malic acid (mg/L)	3455.2	±	78.82	Malic acid (mg/L)	1352.1	±	45.89
		Tartaric acid (mg/L)	5471.1	±	61.23	Tartaric acid (mg/L)	2622.2	±	53.21
		Total phenolics (mg GAE)	539.7	±	33.31	Total phenolics (mg GAE)	1825.1	±	23.44
		°Brix	23.7	±	0.38	Alcohol (% vol.)	13.9	±	0.12
						Residual sugar (gr/L)	1.3	±	0.31
		Total Acidity (HSO_4 gr/L)	6.9	±	0.38	Total Acidity (HSO_4 gr/L)	4.1	±	0.08
Season 1	Type ii	pH	3.1	±	0.04	pH	3.4	±	0.04
		Malic acid (mg/L)	3322.4	±	86.71	Malic acid (mg/L)	1423.8	±	54.98
		Tartaric acid (mg/L)	5480.5	±	33.11	Tartaric acid (mg/L)	2679.1	±	57.34
		Total phenolics (mg GAE)	576.1	±	47.14	Total phenolics (mg GAE)	1870.1	±	42.32
		°Brix	20.9	±	1.73	Alcohol (% vol.)	12.6	±	0.26
						Residual sugar (gr/L)	1.4	±	0.15
		Total Acidity (HSO_4 gr/L)	7.5	±	0.99	Total Acidity (HSO_4 gr/L)	3.7	±	0.33
	Type iii	pH	3.1	±	0.09	pH	3.4	±	0.12
		Malic acid (mg/L)	3089.3	±	67.32	Malic acid (mg/L)	1200.6	±	66.67
		Tartaric acid (mg/L)	5190.2	±	45.33	Tartaric acid (mg/L)	2499.2	±	71.10
		Total phenolics (mg GAE)	557.9	±	52.41	Total phenolics (mg GAE)	1735.2	±	33.42
		°Brix	23.0	±	0.67	Alcohol (% vol.)	13.5	±	0.89
						Residual sugar (gr/L)	1.3	±	0.10
		Total Acidity (HSO_4 gr/L)	5.5	±	0.82	Total Acidity (HSO_4 gr/L)	4.4	±	0.27
	Type i	pH	3.2	±	0.11	pH	3.4	±	0.12
		Malic acid (mg/L)	3333.3	±	66.62	Malic acid (mg/L)	1487.1	±	87.92
		Tartaric acid (mg/L)	5379.3	±	58.32	Tartaric acid (mg/L)	2598.5	±	58.72
		Total phenolics (mg GAE)	522.1	±	31.23	Total phenolics (mg GAE)	1799.4	±	35.09
		°Brix	24.0	±	1.07	Alcohol (% vol.)	13.1	±	0.65
						Residual sugar (gr/L)	1.3	±	0.25
		Total Acidity (HSO_4 gr/L)	6.0	±	1.01	Total Acidity (HSO_4 gr/L)	4.2	±	0.08
Season 2	Type ii	pH	3.1	±	0.15	pH	3.4	±	0.04
		Malic acid (mg/L)	3322.3	±	88.10	Malic acid (mg/L)	1376.2	±	88.78
		Tartaric acid (mg/L)	5380.2	±	30.22	Tartaric acid (mg/L)	2683.9	±	66.91
		Total phenolics (mg GAE)	518.9	±	48.80	Total phenolics (mg GAE)	1754.3	±	55.42
		°Brix	23.7	±	0.77	Alcohol (% vol.)	12.3	±	0.35
						Residual sugar (gr/L)	1.3	±	0.15
		Total Acidity (HSO_4 gr/L)	5.5	±	0.30	Total Acidity (HSO_4 gr/L)	3.8	±	0.22
	Type iii	pH	3.2	±	0.12	pH	3.3	±	0.09
		Malic acid (mg/L)	3101.3	±	65.31	Malic acid (mg/L)	1246.8	±	86.34
		Tartaric acid (mg/L)	5060.2	±	82.23	Tartaric acid (mg/L)	2501.2	±	79.98
		Total phenolics (mg GAE)	513.1	±	50.22	Total phenolics (mg GAE)	1765.2	±	67.72

2.2. *Influence of Different Pinot Noir Grape Cluster Types on Sensory Evaluation of Wines*

A paired comparison test was carried out to evaluate whether trained panelists were able to note organoleptic differences in taste and aroma between wines made from the three grape cluster types. The sensory data were the mean values of all panelists who took part in every season under evaluation. Results showed significant differences among the three classes of wines for all the sensory properties assessed (Figure 2). It is important to note that no significant differences were observed when data from the two seasons were compared. As revealed in Figure 2, the aroma of wines made with grapes from cluster type i was mainly characterized by red berries, stone fruits, caramel, and coffee/chocolate attributes. By contrast, the herbal and vegetal attributes were especially high in the wine elaborated with grape cluster type iii. The other attributes that varied significantly between samples were spice, woody, and smoky which presented high scores in both samples of wine type i and type ii compared with wine made of grape cluster type iii. In addition, we observed that the taste attributes, acidity, and astringency notes were higher in the wines from grape cluster type i and cluster type ii, while sweetness and bitterness attributes were lower. By contrast, in wine obtained from grape cluster type iii, sweetness and bitterness attributes were enhanced by reduced acidity and astringency. Also, when the panelists were asked about their general impression of these wines, again those made with grapes from cluster type i were the best evaluated. In contrast, larger grapes seem to generate wines with less relative balance, body, and complex attributes. Panelists were also asked to score the wines from 0 to 100 points in terms of enological potential to be used in wine blend and also about their commercial value if those wines were bottled and sold as monovarietal wines. As before, those wines made with cluster type i received the best scores. Statistically significant differences were observed between the qualifications of the wines made with all different cluster types. Overall, given these sensory attributes for the three wine types, the panelists classified as the most well-rounded and with better enological/commercial potential the wine made from grape cluster type i.

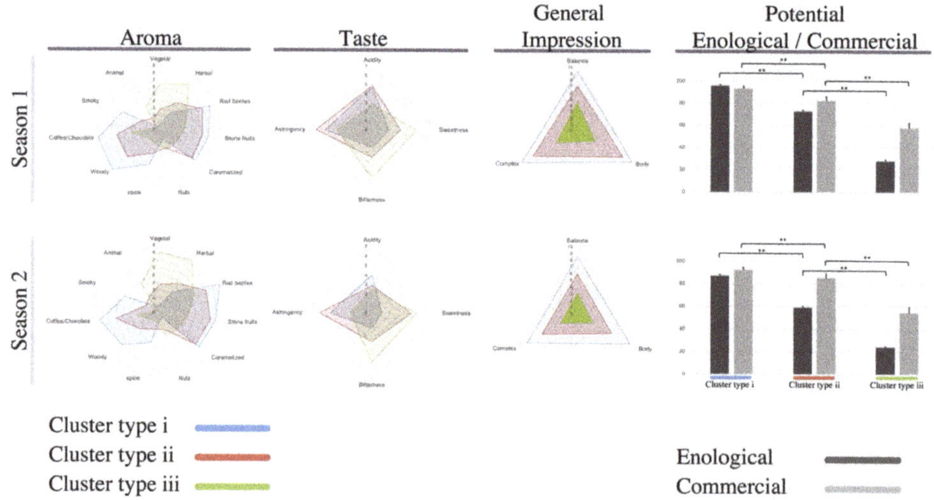

Figure 2. Sensory analysis. Graph of the mean sensory ratings of the three Pinot noir wines studied. Ten experienced Pinot noir wine-makers conducted Duo-Trio test trials to determine the overall difference between wines samples.

2.3. *Untargeted Metabolite Profiles of Three Pinot Noir Wines*

Given the differences in the sensory analysis of the three types of wines, we were next interested in determining whether the unrestricted metabolomics analysis of these types of wines allowed us to significantly associate non-volatile metabolites with specific aromatic and taste attributes. To this end, an exhaustive untargeted analysis of secondary metabolite profiling was conducted on wines produced from the three grape cluster types. Wines from the two seasons were analysed using a UPLC-FT-ICR-MS in positive and negative ionization mode. Based on visual inspection of the raw total ion chromatograms, the wine samples showed clear differences depending on the grape cluster types (data not shown). The data sets were examined by principal component analysis (PCA) (Figure 3A), with two principal components explaining 50.11% and 60.72% of the overall variance of the detected features in positive and negative mode, respectively. The PCA revealed two distinct patterns. The analysis highlighted a sharp season-dependent trend (PC1), while PC2 showed clear differences that were evident between the three types of wines from the same season. This result strongly indicates that the PCA-derived separation was not based only on the presence/absence of grape cluster type specific features but also on quantitative differences within the metabolites present in all wines.

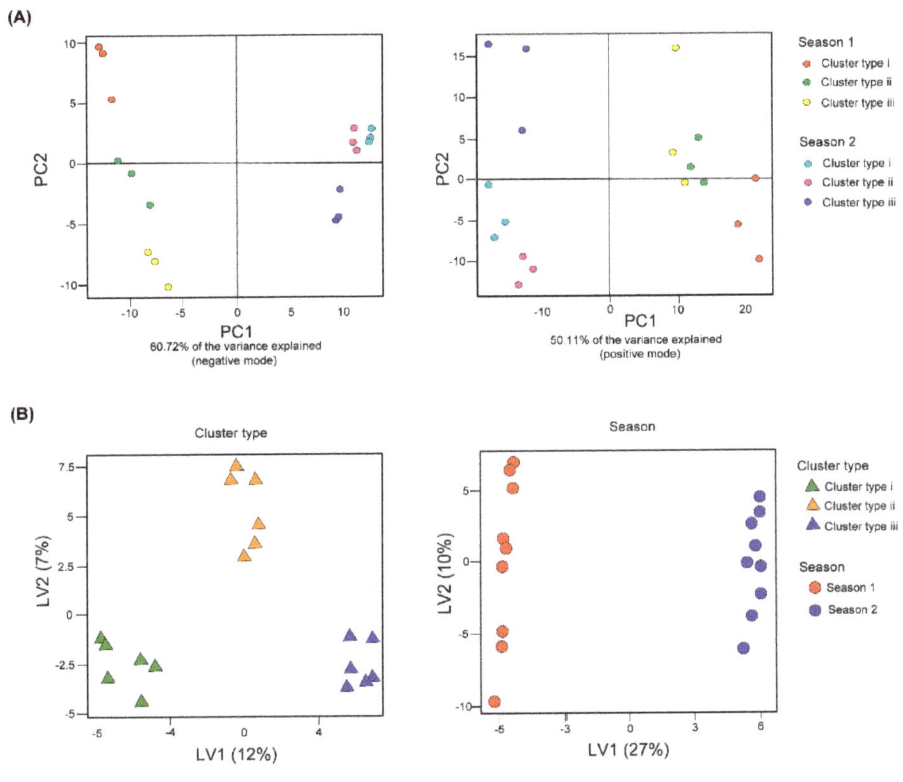

Figure 3. Score plots of all detected *m/z* features of the three Pinot noir wines from two different seasons in positive and negative mode investigated by (**A**) principal component analysis (PCA) and (**B**) the latent variables (LVs) in first discriminant analysis (sPLS-DA).

In light of such results, linear discriminant analysis (sPLS-DA) was used as a supervised classification technique [30] to develop models capable of predicting the classification of wines according to the grape cluster type and harvest year. A plot of the two first discriminant variates (derived from all *m/z* features measured in either positive or negative ionization mode) is presented in Figure 3B. When the discrimination was based on harvest year, a similar pattern was observed compared with the unsupervised PCA (Figure 3A). However, when grape cluster type was considered as a variable, we observed a clear separation into three groups according to the grape cluster type independent of harvest year (Figure 3B).

2.4. Multivariate Discrimination of Three Pinot Noir Wines Based on Metabolic Profiling and Identification of Multiple Metabolic Features as Biomarkers of Wine Quality

Since the quality of the three wine types were scored differently based on their sensory evaluation, we next constructed sparse Partial Least Squares (sPLS) regression models in order to identify *m/z* features (biomarkers) related to sensory attributes and that also distinguished different harvest years and cluster types. An sPLS model was generated for each dependent variable (wine attributes), and the normalized intensity levels of the molecular features were used as predictors. The relative importance of each independent variable (*m/z* feature) is measured in terms of the VIP (variable importance in projection) scores; the higher the VIP scores, the greater the association of the analysed *m/z* feature with respect to the dependent variable (sensory attribute). The *m/z* features with the highest VIP score (> 1.0) are displayed in Table S1. The identified molecular features were then subjected to compound annotation.

Annotation of metabolites based on high resolution mass spectrometry accurate mass provides an initial indication of the potential identification. A match to a potential molecular formula, as defined by the metabolomics standards initiative (MSI), corresponds to a level 2-based identification. Level 3 identification is for *m/z* features that match multiple molecular formulas but within a single class of metabolites, while unknown features are classified as level 4 [31]. In order to classify compounds with higher confidence, we also used the MS2 spectral data (unit mass and peak ratio match) for identification. When MS2 spectral data were not captured for a given metabolite, data acquired with previous grape and wine samples analysed under identical conditions were considered, together with published data. Where a given compound ionized in negative and positive modes and different adduct species were observed for each, the MS2 spectra data for each ion species were manually assessed and compared.

We were able to tentatively identify a total of 26 molecular features with a VIP score higher than 1.0 that exhibited a significant contribution to explain the variation between harvest years. In concordance with the results obtained in the sensorial analysis with respect to the lack of discrimination of wines made from the same cluster type but in different seasons, any of those features were in common with the features found to be associated with all measured sensory attributes.

To better visualize the molecular features that significantly contribute to specific sensory attributes, we performed an association study using networks (Figure 4, Table S1). In each network, sensory attributes and molecular features are represented as nodes of different size (large and small) and are color coded by compound category. An edge represents an association between nodes if the corresponding VIP coefficient is greater than 1.0. The edge thickness and color indicate the strength of the association; darker and thicker lines mean stronger associations. Network analysis emphasized links both between and within the various sensory attributes. From the analysis for sensory evaluation associated with aroma and taste attributes (a total of 15 sensory attributes were addressed; spice, woody, smoky, coffee/chocolate, vegetal, animal, herbal, red berries, stone fruits, nuts, caramel, astringency, acidity, bitterness, and sweetness), we observed a total of 65 polar *m/z* features that displayed good VIP score as listed in Table S2. Following the identification process, 22 secondary metabolites were tentatively assigned to specific compound categories, eight flavonols and flavanols, eight anthocyanins, one hydroxycinnamic acid derivative, two ellagitannins, two cyanogenic glucosides, two classifieds as phenols, and 43 other classes (Figure 4, Table S2).

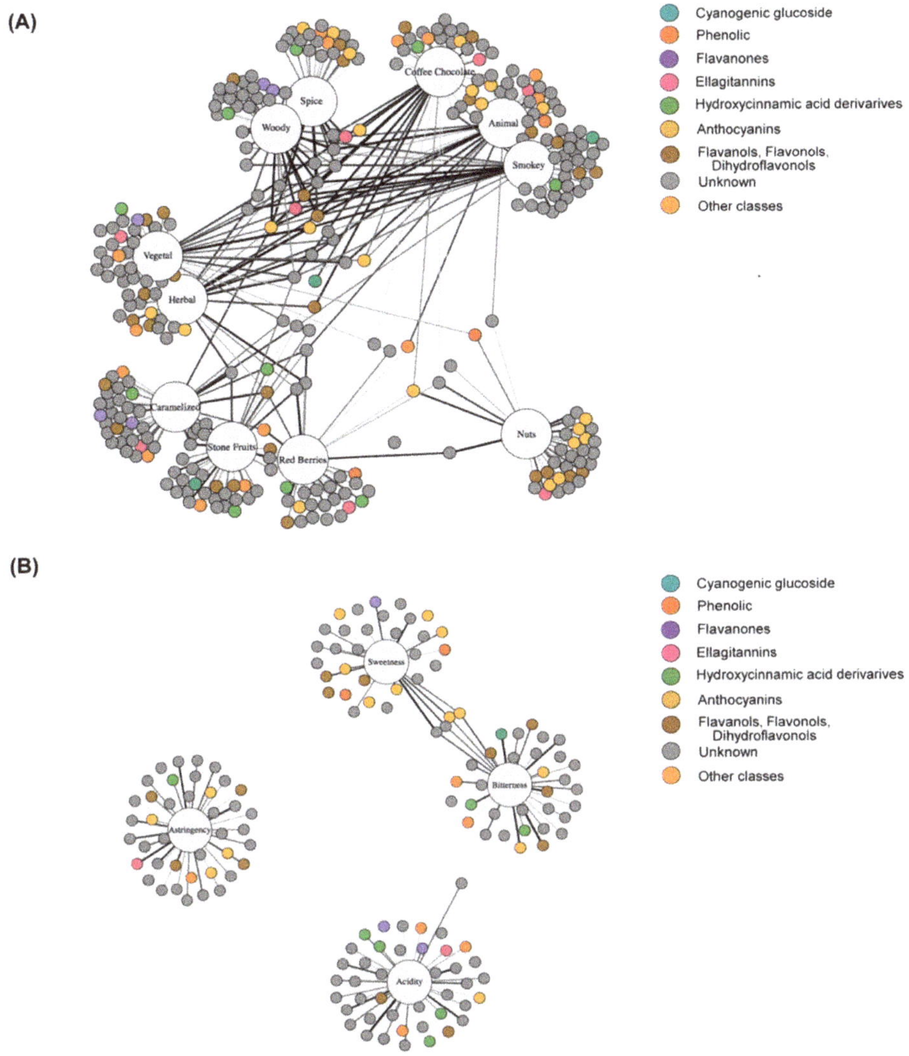

Figure 4. Visualization of sensory attribute-molecular feature correlations in Pinot noir wines. (**A**) Network considering aroma attributes and (**B**) taste attributes. Notes are as follows: sensory attributes are represented as large circles, while molecular features are represented as small circles and are color coded by compound category. The edge thickness and color indicate the strength of the association; darker and thicker lines mean stronger associations (only those that showed a VIP coefficient > 1.0).

Interestingly, the network associated with aroma attributes (Figure 4A) revealed markedly higher connectivity than the network for taste attributes (Figure 4B), suggesting that aroma attributes in wine may represent a higher level of complexity as compared with taste attributes. Comparative network analysis also suggested that molecular features associated with taste attributes were largely modular, while features associated with aroma attributes showed more complex and interconnected associations between specific features. The aroma attributes network analysis revealed five interesting clusters, one of which included red berries, stone fruit, and caramel attributes, another involved spice and woody characteristics, while two other clusters contained herbaceous characters (herbal and vegetal) and coffee/chocolate, smoky, and animal, respectively. Interestingly, we observed a strong link between nut attributes and the two aroma attributes, coffee/chocolate and red berries. For network involving taste attributes, we observed specific *m/z* features strongly associated with the four tested taste attributes. Here, we observed an association between sweetness and bitterness attributes, while acidity and astringency displayed non-clear association with the other analysed attributes. Interestingly, the link between sweetness and bitterness attributes was through four *m/z* features; two of them were annotated as anthocyanins (named catechylpyranopeonidin-3-O-glucose and pyranopeonidin 3-O-coumaroyl-glucose), although in general, Pinot noir wines have been described to have lower anthocyanin concentrations [32].

3. Discussion

The quality of a wine is known to directly correlate with the quality of the grapes used for its production [13]. Therefore, quality and character of a wine should correlate directly with the chemical composition of grape [33].

The starting point of this study was due to an interesting observation made by the grower in an established vineyard, in which a Pinot noir clone grown under the same mesoclimate and treated using the same viticultural practices produced three different grape cluster types (Figure 1).

High-throughput metabolomics facilitates dissecting a phenotype at the metabolite level, potentially allowing a more holistic perspective in monitoring and gaining information on the winemaking process, and thus, this approach can assist in the evaluation of sensorial attributes and can potentially help the improvement of several aspects of wine quality.

The first aim of the present study was to test if a comparative metabolome and sensorial analyses of three Pinot noir wines made by three different grape cluster types were able to discriminate wine types. If this was feasible, we were also interested in better understanding the diversity in the sensory attributes of these wines. For that, we tested a multivariate statistical method to identify metabolites or molecular patterns (biomarkers) related to different sensory attributes, which can be used to assist the sensory analysis postulated by wine expert panelists.

The wine made from grape cluster type iii revealed significant organoleptic differences from the other two wine types. In particular, we observed that wine produced by grape cluster type iii was found to have greater scores for vegetal and herbal aromas. On the other hand, wine types i and ii showed differences by displaying more canned fruity characteristics (red berries, stone fruits, and caramel). Moreover, wine type i was the most intense in woody and coffee/chocolate characters. This strong discrimination between herbaceous and fruity attributes within wines produced from the same grape variety has been previously reported [34–37]. In these studies, the authors also indicated that the descriptive profile of Cabernet Sauvignon wines displayed a dichotomy between herbaceous and fruity characteristics.

With respect to the taste attributes, wine obtained from grape cluster iii was classified as having a reduced mouthfeel in terms of astringency and acidity. It was also noted that wine produced from grape cluster type iii had a noticeable increase in bitterness and sweetness. High levels of organic acids have been associated with an increased perception of astringency. Furthermore, this perception is also pH dependent [38]. Here, although we did not observe significant differences in total acidity and pH, we detected differences in the level of malic acid in wine made from grape cluster type iii compared to the other types. Moreover, astringency of high-molecular-weight polyphenolics (tannins) can be enhanced by the presence of organic acids [38,39]. Therefore, this observation agrees with the hypothesis that different concentrations of tannins in wine type iii can contribute to the observed differences in astringency. Additionally, we observed increased ethanol levels in wine types i and ii, although non-significant, which can have a greater effect on perceived astringency than the difference between acid and high-molecular-weight polyphenol levels [40]. Similarly, a high ethanol level is described to increase the perception of bitterness [41–43]. However, we observed that bitterness was perceived to be higher in wine type iii, although previous studies suggest that the bitterness perception can be suppressed by the sweetness [41]. However, we observed that wine type iii was classified with higher values for sweetness and bitterness attributes, suggesting that these sensory attributes cannot reflect the same chemical properties when comparing different cultivars. Consistent with this hypothesis, we observed no significant changes in residual sugar levels when comparing the three wine types, although it has been described that levels of sugars in wine correlate with the sweetness perception [44].

In view of the fact that wine is a really complex matrix and the levels of individual metabolites are not necessarily as important as their interaction with other compounds, the use of metabolomics allows a holistic perspective in monitoring several aspects of wine quality. Metabolomics have been successfully applied to discriminate and classify wines according to grape variety, geographic origin, age or winemaking practice [15–26]. Therefore, we decided to further use mass-spectrometry metabolomics to investigate if we could establish an association between non-volatile metabolites or features and specific sensory attributes related to wine quality.

For this purpose, wines produced by the three grape cluster types and classified differently according to their quality by wine experts were analysed using UPLC-FT-ICR-MS. This analysis revealed that the different wine types underwent several changes in secondary metabolite concentration, mainly phenolic compounds. We focused on these classes of metabolites since they are important chemical components of wines and can strongly influence the final organoleptic perception [45–47]. The content and presence of these classes of metabolites alone enabled us to discriminate between years as demonstrated by PCA, in which the two analysed seasons can be clearly distinguished, in agreement with previously published studies on other classes of wines [19]. In addition, this analysis was also able to classify wines according to the grape cluster types used for winemaking as shown in the sPLS-DA model (see Figure 3B). In this study, we have described that the putative biomarkers discriminating the wine quality are related to changes in their abundance. Therefore, our results reveal that our metabolite profiling provided sufficient discriminatory power to differentiate wine according to its quality. We were then interested in identifying potential biomarkers associated with sensory wine attributes as scored by a panel of experts. To that end, we applied sparse Partial Least Squares (sPLS) regression models using the experts' scores as dependent variables. From the metabolite-attribute networks constructed from the sPLS models, we observed that the network associated with aroma attributes revealed markedly higher connectivity than the network for taste attributes, suggesting that aroma attributes in wine may represent a higher level of complexity as compared with taste attributes (Figure 4). After that, we focused our attention on the molecular features that were strongly associated with specific sensory attributes (VIP > 10) to assess whether we were able to putatively annotate them.

We found that specific anthocyanins were strongly associated with different aroma attributes (Figure 4 and Table S1). These pigments can be grouped as glycosides (i.e., monoglucosylated anthocyanins), acylated pigments (i.e., acylated anthocyanin monoglucosides, in which the acylated group can be *p*-coumaric acid, caffeic acid or acetic acid), and anthocyanin-derived pigments (i.e., pyranoanthocyanins and tannin-anthocyanin dimers). The Pinot noir anthocyanin profile is unique as it lacks (except in traces) acylated anthocyanins [48,49]. In this study, we found pyranopeonidin-3-O-glc to be associated with caramel, vegetal, herbal, coffee/chocolate, spice, woody, and smoky attributes, while catechylpyranopeonidin-3-O-glc was associated only with the nut character. It has been described that these chemical compounds are produced in wines during fermentation and the aging process [50,51]. Furthermore, two other studies have demonstrated that pyranoanthocyanins can also be found in red grapes postharvest [52,53]. In addition to these anthocyanin compounds, we identified pelargonidin 3-O-glc to be strongly associated with the animal attribute (VIP = 28.603). Interestingly, this anthocyanin has been detected at a very low concentration in comparison to other anthocyanins in red and pink grape berries [54]. This finding can suggest that pelargonidin 3-O-glc is a clear candidate to be a chemical marker associated with the animal attribute in Pinot noir wines. In most f *V. vinifera* varieties, anthocyanins are known to be present only in the skin of red grape cultivars and together with the co-pigments or anthocyanins-related pigments are responsible fof the red colour shown by red wine [55]. Anthocyanins are odorless molecules which can conflict our result. However, anthocyanins are known to modulate the formation of polymeric pigments [56] which are adducts formed during winemaking that provide mouthfeel properties [57]. In addition, anthocyanin-derived pigments have precursors molecules of different chemistries, including vinyl- and ethyl-phenols [58] or tannins [59], and we cannot discard their influence on aroma.

Resveratrol is a non-flavonoid compound in the stilbenoid class. In particular, the biomedical literature focuses attention on *trans*-resveratrol (3,4,5′trihydroxy-*trans*-stilbene) because it shows a strong antioxidant capacity and it has been positively associated with health effects, for instance, in cardiovascular disease and colon and breast cancers [60,61]. In wine, this compound has been found to be present around 10 times higher in red varieties than in white varieties [62]. Piceid, the β-D-glucoside form of resveratrol (resveratrol 3-O-β-D-glucoside), is also found in wines [63] and has been physiologically described as important as *trans*-resveratrol [64,65]. Interestingly, our data revealed that the piceid level in wine was associated with coffee/chocolate. We are unaware of any previous research that showed how the piceid level influenced wine quality; however, it has been described that a high level of *trans*-resveratrol did not alter the flavor or aroma profile of Cabernet Sauvignon [66].

Following the same strategy to identify chemical compounds associated with mouthfeel attributes, we focused our interest on those features with a VIP score > 10. We observed that the sPLS model associated with taste was less dense that the model from aroma characters. The model clearly showed that the compounds responsible for the discrimination between the four tested characters (sweetness, bitterness, acidity, and astringency), at least from the putative annotated features, are mainly represented by anthocyanins, flavanols, and flavanols species (Figure 4). It is well understood that monomeric flavan-3-ols and oligomeric tannins, particularly catechin and epicathechin species, are associated with astringency perception of red wines [41,67,68]. Interestingly, we identified catechin (VIP = 17.19) that correlated with the astringency attribute. Therefore, we can suggest that the level of catechin can act as a chemical marker for the astringency attribute in Pinot noir wines. However, the level of variation of this metabolite that can be associated with this sensory character will need further investigation.

From this study, we can demonstrate that wine quality in term of sensory analysis is a complex process in which the individual level of volatile and non-volatile compounds but also their interaction play an important role. Here, we presented a metabolic pipeline methodology allowing

(i) efficient discrimination of different wine quality
(ii) identification of *m/z* features associated with specific sensory attributes.

However, further research on the validation of the identified *m/z* features (putative annotated and non-annotated) related to aroma and taste characters is needed to release metabolic biomarkers to assist panelists with wine quality discrimination.

4. Materials and Methods

4.1. Chemicals

All solvents were of High Performance Liquid Chromatography (HPLC) grade and were obtained from VWR International. Other chemicals were of the highest purity grade available and were obtained from Sigma–Aldrich. Water was of Milli-Q quality (Millipore Corp., Bedford, MA, USA).

4.2. Experimental Design

An established commercial vineyard located in Casablanca valley, Chile (33°18′59′′ S 71°27′30′′ W) was used in this study. The grapevines were clonally multiplied, and the field was established seven years before this study. The experiments were performed on *V. vinifera* cv. Pinot noir during the 2008/2009 and 2009/2010 seasons. Vine spacing was 1.2 m between vines and 2.2 m between rows.

Vine management was performed by the grower according to the regional appropriate viticultural practices for this *V. vinifera* cultivar. The vines were cane pruned for a vertical shoot-positioning trellis system. No rain occurred after fruit set. Irrigation was applied to vines in order to maintain midday leaf water potential between −0.6 and −0.8 MPa throughout the season. Considering the phenotypic cluster difference among plants in the field, each experimental unit consisted of a few interior vineyard panels in a randomized complete block design. Individual grapevine plants and cluster samples were selected at random from throughout a vineyard block at the ripe stage.

4.3. Wine Samples

Clusters from the different selected blocks were used to make microvinifications. Wines were made from each selection group according to standard winemaking practices (Centro Tecnológico de la Vid y el Vino, Universidad de Talca) with four biological and three technical replicates. Each replicate consisted of 100 ± 1 kg of hand-harvested fruit collected from the field-selected blocks. Grapes were destemmed, and the berries were inoculated with yeast (*Saccharomyces bayanus*, EC-1118, Lallemand, 1620 Rue Préfontaine, Montréal, QC H1W 2N8, Canada) at a rate of 20 g hL^{-1} in stainless steel fermenters. Cap management was performed twice per day by manual punch-downs. The fermentation temperature was maintained between 22 °C and 23 °C at a density of 990 g/L. After alcoholic fermentation, wines were pressed and racked into carboys and inoculated with Enoferm Alpha (Lallemand, 1620 Rue Préfontaine, Montréal, QC H1W 2N8, Canada) to start malolactic fermentation. At the end of malolactic fermentation, $K_2S_2O_5$ was added to 30 ppm of SO_2 in the wine. In addition, a cold stabilization at 2 °C was performed before bottling. Bottling and capping were performed manually.

4.4. Wine Sensory Analysis

Duo-Trio test trials were conducted to determine the overall difference between wines samples. Ten experienced Pinot noir wine-makers, highly trained and experienced as Pinot noir enologists, formed the wine panelists (four females and six males between the ages of 30 and 58). To minimize carry-over of flavors, the panelists were asked to rinse their mouth with water between samples, and the test was repeated on different days. One month after bottling, the wines were evaluated.

4.5. Metabolite Measurements

Chromatographic separation, mass spectrometric measurements, and data analysis were performed using a Waters Acquity Ultra High Resolution Liquid Chromatography (UPLC) system (Waters, Milford, MA, USA) using an HSS T3 C_{18} reverse phase column (100 × 2.1 mm i.d., 1.8 µm particle size; Waters) which operated at a temperature of 40 °C. The mobile phases consisted of 0.1% formic acid in water (Solvent A) and 0.1% formic acid in acetonitrile (Solvent B). The flow rate of the mobile phase was 400 µL/min, and 2 µL of sample was loaded per injection. The gradient was: 0–1 min isocratic flow at 99% A, 1–12 min linear gradient from 95% to 65% A, 12–13.5 min linear gradient from 65% to 20% A, 13.5–14.5 min linear gradient from 20% to 1% A, 14.5–16 min isocratic flow at 1% A, 16–17 min linear gradient from 1% to 95% A, and 17 to 20 min isocratic flow at 95% A, to re- equilibrate the column before to the next sample could be injected. The UPLC was connected to an Ultra High Resolution Mass Spectrometry (FT-ICR) via a TriVersa NanoMate (Advion, Ithaca, NY, USA). The spectra were recorded alternating between full-scan and all ion-fragmentation-scan modes, covering a mass range from 100 to 1500 *m/z* using the LTQ FT-ICR-Ultra Mass Spectrometer (Thermo-Fisher, Bremen, Germany). The resolution was set to 35,000, and the maximum scan time was set to 250 ms. The sheath gas was set to a value of 50, while the auxiliary gas was set to 20. The transfer capillary temperature was set to 250 °C, while the heater temperature was adjusted to 350 °C. MS spectra were recorded from minute 0 to 19 of the UPLC gradient. All samples were randomized prior to mass spectrometric analyses to avoid any experimental drifts. A number of internal standards (chloramphenicol, corticosterone, and ampicillin), added to each sample just prior to injection, were used to control experimental variability. One month after bottling, the wines were chemically evaluated.

4.6. Metabolite Annotation and Statistical Analysis

Pre-processing of raw chromatograms was performed using Expressionist Refiner MS 10.0 (GeneData; http://www.genedata.com) with an established workflow. This workflow included baseline correction, removal of chemical noise, and chromatogram alignment. As output, a list of molecular features, that is, a retention time and a *m/z* ratio pair, and a data matrix containing relative intensities for each feature and for each chromatogram were obtained. All further computations, data manipulations, and plot generation were carried out using the R programming language (http://www.r-project.org).

The data matrix was further filtered for low intense features whose mean intensity across all samples was less than 0.1% of the highest average. Moreover, features with more than 30% missing values were also removed as well as features that eluted before 0.45 min. After filtering, the *m/z* feature data were normalized by sample volume and sample median intensity. Due to machine sensitivity variation across different measurement runs, each feature was normalized by dividing its intensity in a sample by the median intensity across all measurements within a batch in order to compensate for said effect.

Principal component analysis (PCA) was performed in R using the package pcaMethods [69] using five components and unit variance scaling. Sparse partial least squares (sPLS) and Sparse partial least squares discriminant analysis (sPLS-DA) were performed using the package mixOmics [70]. The sPLS-DA was used for categorical variables (season and cluster type) while the sPLS was used for quantitative data (sensorial attributes). In both cases, the metabolite levels were considered predictors. For each response variable (categorical or quantitative), a single sPLS/sPLS-DA model was established. To determine the optimal number of components and variables of a given model, we searched the parameter space spanned by 1 to 12 (3 for sPLS-DA) components and 50, 100, 200, 300, 500, 1000 selected variables. For each such component/variable combination, 100 iterations of 5-fold cross-validation rounds were tested and the pair that resulted in the lowest (sPLS-DA) or highest R2 (sPLS) classification error was taken as the optimal parameters. Once an optimal number of components and variables was determined for each response variable, we computed the respective sPLS/sPLS-DA model and using this, we obtained the variable importance in projection (VIP) coefficient for each metabolite. Finally, we used the VIP coefficients to generate networks in which a node represents either a molecular feature or a sensorial parameter, and an edge was drawn if the respective VIP coefficient was greater than zero. All networks manipulations were performed using the igraph package [71].

Molecular features were putatively annotated by searching the *m/z* value against the KEGG compound database. A maximum tolerance of 5 ppm was allowed, considering the following potential known adducts: M+H, M+Na, M+NH_4 for positive mode and M-H, M+AcOH-H for negative mode. A custom Python script was written for this task. The MS/MS fragmentation of the metabolites was compared with candidate molecules found in databases and verified with earlier literature on similar compounds, especially when the presence of the metabolite was reported in grapes and wines.

Supplementary Materials: The following are available online at http://www.mdpi.com/2218-1989/10/6/220/s1, Table S1. Table of all features associated with specific attributes (VIP score > 1.0). Table S2. Identification of all molecular features that significantly contribute to specific sensory attributes (VIP score > 1.0).

Author Contributions: J.G.V., Y.M.-S. and L.W. conceptualized the study. J.G.V. and C.V.-A. carried out the field trials and sensory analysis. J.G.V. and Á.C.-I. processed samples and performed the data analysis. J.G.V. wrote the manuscript which was proofread, corrected, and approved by all authors. All authors have read and agreed to the published version of the manuscript.

Funding: This research was funded by an INNOVA-CORFO-Chile grant and by the Max Planck Society, Germany.

Acknowledgments: We would like to thank Agricola Santa Sara, Casablanca Valley, Chile, for providing the grape samples for this study. Also, we would like to thank Änne Michaelis, Max Planck Institute of Molecular Plant Physiology, for her help with wine chromatography.

Conflicts of Interest: The authors declare no conflict of interest.

References

1. Maul, E.; Topfer, R. Vitis International Variety Catalogue (VIVC): A cultivar database referenced by genetic pro les and morphology. *BIO Web Conf.* **2015**, *5*, 01009. [CrossRef]
2. Lacombe, T. Contribution à L'étude de L'histoire Évolutive de la Vigne Cultivée (*Vitis vinifera* L.) par L'analyse de la Diversité Génétique Neutre et de Gènes D'intérêt. Ph.D. Thesis, Institut National d'Etudes Supérieures Agronomiques de Montpellier, Montpellier, France, 2012.
3. OIV-2. Distribution of the World´s Grapevine Varieties, in Focus OIV 2017. Available online: http://www.oiv.int/public/medias/5888/en-distribution-of-the-worlds-grapevine-varieties.pdf (accessed on 31 January 2020).
4. Armanino, C.; Casolino, M.C.; Casale, M.; Forina, M. Modelling aroma of three Italian red wines by headspace-mass spectrometry and potential functions. *Anal. Chim. Acta* **2008**, *614*, 134–142. [CrossRef] [PubMed]
5. Bavaresco, L.; Gatti, M.; Pezzutto, S.; Fregoni, M.; Mattivi, F. Effect of leaf removal on grape yield, berry composition and stilbene concentration. *Am. J. Enol. Viticult.* **2008**, *59*, 292–298.
6. Vallarino, J.G.; Gainza-Cortés, F.; Verdugo-Alegría, C.; González, E.; Moreno, Y.M. Abiotic stresses differentially affect the expression of O-methyltransferase genes related to methoxypyrazine biosynthesis in seeded and parthenocarpic fruit of Vitis vinífera (L.). *Food Chem.* **2014**, *154*, 117–126. [CrossRef]
7. Torrens, J.; Riu-Aumatell, M.; Lopez-Tamames, E.; Buxaderas, S. Volatile compounds of red and white wines by headspace-solid-phase micro-extraction using different fibers. *J. Chromatogr. Sci.* **2004**, *42*, 310–316. [CrossRef]
8. Sabon, I.; de Revel, G.; Kotseridis, Y.; Bertrand, A. Determination of volatile compounds in Grenache wines in relation with different terroirs in the Rhone Valley. *J. Agric. Food Chem.* **2002**, *50*, 6341–6345. [CrossRef]
9. Koundouras, S.; Marinos, V.; Gkoulioti, A.; Kotseridis, Y.; van Leeuwen, C.J. Influence of vineyard location and vine water status on fruit maturation of nonirrigated cv. Agiorgitiko (*Vitis vinifera* L.). Effects on wine phenolic andaroma components. *J. Agric. Food Chem.* **2006**, *54*, 5077–5086. [CrossRef]
10. Alessandrini, M.; Gaiotti, F.; Belfiore, N.; Matarese, F.; D'Onofrio, C.; Tomasi, D. Influence of vineyard altitude on glera grape ripening (*Vitis vinifera* L.): Effects on aroma evolution and wine sensory profile. *J. Sci. Food Agric.* **2017**, *97*, 2695–2705. [CrossRef]
11. Esti, M.; Tamborra, P. Influence of winemaking techniques on aroma precursors. *Anal. Chim. Acta* **2006**, *563*, 173–179. [CrossRef]
12. Gambetta, J.M.; Cozzolino, D.; Bastian, S.E.P.; Jeffery, D.W. Towards the creation of a wine quality prediction index: Correlation of chardonnay juice and wine compositions from different regions and quality levels. *Food Anal. Methods* **2016**, *9*, 2842–2855. [CrossRef]
13. Smith, P. Assessment of relationships between grape chemical composition and grape allocation grade for cabernet sauvignon, shiraz and chardonnay. *Aust. N. Z. Grapegrow. Winemak.* **2015**, *620*, 30–32.
14. Clarke, R.J.; Bakker, J. *Wine Flavour Chemistry*; Blackwell Publishing: Oxford, UK, 2004.
15. Son, H.S.; Hwang, G.S.; Kim, K.M.; Kim, E.Y.; van den Berg, F.; Park, W.M.; Lee, C.H.; Hong, Y.S. ^{1}H NMR-based metabolomic approach for understanding the fermentation behaviors of wine yeast strains. *Anal. Chem.* **2009**, *81*, 1137–1145. [CrossRef] [PubMed]
16. Son, H.S.; Kim, K.M.; van den Berg, F.; Hwang, G.S.; Park, W.M.; Lee, C.H.; Hong, Y.S. ^{1}H nuclear magnetic resonance-based metabolomic characterization of wines by grape varieties and production areas. *J. Agric. Food Chem.* **2008**, *56*, 8007–8016. [CrossRef] [PubMed]
17. Zhang, J.; Li, L.; Gao, N.; Wang, D.; Gao, Q.; Jiang, S. Feature extraction and selection from volatile compounds for analytical classification of Chinese red wines from different varieties. *Anal. Chim. Acta* **2010**, *662*, 137–142. [CrossRef]
18. Green, J.A.; Parr, W.V.; Breitmeyer, J.; Valentin, D.; Sherlock, R. Sensory and chemical characterization of Sauvignon Blanc wine: Influence of source of origin. *Food Res. Int.* **2011**, *44*, 2788–2797. [CrossRef]
19. Cuadros-Inostroza, A.; Giavalisco, P.; Hummel, J.; Eckardt, A.; Willmitzer, L.; Pena-Cortés, H. Discrimination of wine attributes by metabolome analysis. *Anal. Chem.* **2010**, *82*, 3573–3580. [CrossRef]
20. Rochfort, S.; Ezernieks, V.; Bastian, S.E.P.; Downey, M.O. Sensory attributes of wine influenced by variety and berry shading discriminated by NMR metabolomics. *Food Chem.* **2010**, *121*, 1296–1304. [CrossRef]

21. Robinson, A.L.; Boss, P.K.; Heymann, H.; Solomon, P.S.; Trengove, R.D. Development of a sensitive non-targeted method for characterizing the wine volatile profile using headspace solid-phase microextraction comprehensive two-dimensional gas chromatography time-of-flight mass spectrometry. *J. Chromatogr. A* **2011**, *1218*, 504–517. [CrossRef]
22. Arapitsas, P.; Scholz, M.; Vrhovsek, U.; Di Blasi, S.; Biondi Bartolini, A.; Masuero, D.; Perenzoni, D.; Rigo, A.; Mattivi, F. A Metabolomic approach to the study of wine micro-oxygenation. *PLoS ONE* **2012**, *7*, e37783. [CrossRef]
23. Corso, M.; Ziliottoa, F.; Rizzinia, F.M.; Teob, G.M.; Cargnello, G.; Bonghi, C. Sensorial, biochemical and molecular changes in Raboso Piave grape berries applying "Double Maturation Raisonnée" and late harvest techniques. *Plant Sci.* **2013**, *208*, 50–57. [CrossRef]
24. Castro, C.C.; Martins, R.C.; Teixeira, J.A.; Silva Ferreira, A.C. Application of a high-throughput process analytical technology metabolomics pipeline to Port wine forced ageing process. *Food Chem.* **2014**, *143*, 384–391. [CrossRef] [PubMed]
25. Roullier-Gall, C.; Witting, M.; Gougeon, R.D.; Schmitt-Kopplin, P. High precision mass measurements for wine metabolomics. *Front. Chem.* **2014**, *2*, 102. [CrossRef] [PubMed]
26. Rocchetti, G.; Gatti, M.; Bavaresco, L.; Lucini, L. Untargeted metabolomics to investigate the phenolic composition of Chardonnay wines from different origins. *J. Food Compost. Anal.* **2018**, *71*, 87–93. [CrossRef]
27. Welke, J.E.; Manfroia, V.; Zanus, M.; Lazzarotto, M.; Alcaraz Zini, C. Differentiation of wines according to grape variety using multivariate analysis of comprehensive two-dimensional gas chromatography with time-of-flight mass spectrometric detection data. *Food Chem.* **2013**, *141*, 3897–3905. [CrossRef]
28. Gougeon, R.D.; Lucio, M.; Frommberger, M.; Peyron, D.; Chassagne, D.; Alexandre, H.; Feuillat, F.; Voilley, A.; Cayot, P.; Gebefügi, I.; et al. The Chemodiversity of wines can reveal a metabologeography expression of cooperage oak wood. *Proc. Natl. Acad. Sci. USA* **2009**, *106*, 9174–9179. [CrossRef]
29. Vaclavik, L.; Lacina, O.; Hajslova, J.; Zweigenbaum, J. The use of high performance liquid chromatography-quadrupole tome-of-flight mass spectrometry coupled to advanced data mining and chemometric tools for discrimination and classification of red wines according to their variety. *Anal. Chim. Acta* **2011**, *685*, 45–51. [CrossRef]
30. Chun, H.; Keles, S. Sparse partial least squares regression for simultaneous dimension reduction and variable selection. *J. R. Stat. Soc. Ser. B Stat. Methodol.* **2010**, *72*, 3–25. [CrossRef]
31. Sumner, L.W.; Amberg, A.; Barrett, D.; Beale, M.H.; Beger, R.; Daykin, C.A.; Fan, T.W.M.; Fiehn, O.; Goodacre, R.; Griffin, J.L.; et al. Proposed minimum reporting standards for chemical analysis. *Metabolomics* **2007**, *3*, 211–221. [CrossRef]
32. Piccardo, D.; Favre, G.; Pascual, O.; Canals, J.M.; Zamora, F.; González-Neves, G. Influence of the use of unripe grapes to reduce ethanol content and pH on the color, polyphenol and polysaccharide composition of conventional and hot macerated Pinot Noir and Tannat wines. *Eur. Food Res. Technol.* **2019**, *245*, 1321–1335. [CrossRef]
33. Longbottom, M.; Simos, C.; Krstic, M.; Johnson, D. Grape quality assessments: A survey of current practice. *Wine Vitic. J.* **2013**, *28*, 33–37.
34. Chapman, D.M.; Matthews, M.A.; Guinard, J.X. Sensory attributes of Cabernet Sauvignon wines made from vines with different crop yields. *Am. J. Enol. Viticult.* **2004**, *55*, 325–334.
35. Chapman, D.M.; Roby, G.; Ebeler, S.E.; Guinard, J.X.; Matthews, M.A. Sensory attributes of Cabernet Sauvignon wines made from vines with different water status. *Aust. J. Grape Wine Res.* **2005**, *11*, 339–347. [CrossRef]
36. Falcão, L.D.; Revel, G.D.; Perello, M.C.; Moutsiou, A.; Zanus, M.C.; Bordignon-Luiz, M.T. A survey of seasonal temperatures and vineyard altitude influences on 2-methoxy-3-isobutylpyrazine, C13- norisoprenoids, and the sensory profile of Brazilian Cabernet Sauvignon wines. *J. Agric. Food Chem.* **2007**, *55*, 3605–3612. [CrossRef]
37. Preston, L.D.; Block, D.E.; Heymann, H.; Soleas, G.; Noble, A.C.; Ebeler, S.E. Defining vegetal aromas in Cabernet Sauvignon using sensory and chemical evaluations. *Am. J. Enol. Viticult.* **2008**, *59*, 137–145.
38. Hufnagel, J.C.; Hofmann, T. Quantitative reconstruction of the nonvolatile senso-metabolome of a red wine. *J. Agric. Food Chem.* **2008**, *56*, 9190–9199. [CrossRef]

39. Joscelyne, V.L.; Downey, M.O.; Mazza, M.; Bastian, S.E.P. Partial shading of Cabernet Sauvignon and Shiraz Vines altered wine color and mouthfeel attributes, but increased exposure had little impact. *J. Agric. Food Chem.* **2007**, *55*, 10889–10896. [CrossRef] [PubMed]
40. Fontoin, H.; Saucier, C.; Teissedre, P.L.; Glories, Y. Effect of pH, ethanol and acidity on astringency and bitterness of grape seed tannin oligomers in model wine solution. *Food Qual. Prefer.* **2008**, *19*, 286–291. [CrossRef]
41. Noble, A.C. Bitterness in wine. *Physiol. Behav.* **1994**, *56*, 1251–1255. [CrossRef]
42. Vidal, S.; Courcoux, P.; Francis, L.; Kwaitkowski, M.; Gawel, R.; Williams, P.; Waters, E.; Cheynier, V. Use of an experimental design approach for evaluation of key wine components of mouth-feel perception. *Food Qual. Prefer.* **2004**, *15*, 209–217. [CrossRef]
43. Jones, P.R.; Gawel, R.; Francis, I.L.; Waters, E.J. The influence of interactions between major white wine components on the aroma, flavour and texture of model white wine. *Food Qual. Prefer.* **2008**, *19*, 596–607. [CrossRef]
44. Nurgel, C.; Pickering, G. Contribution of glycerol, ethanol and sugar to the perception of viscosity and density elicited by model white wines. *J. Texture Stud.* **2005**, *36*, 303–323. [CrossRef]
45. Blanco-Vega, D.; Gomez-Alonso, S.; Hermosin-Gutierrez, I. Identification, content and distribution of anthocyanins and low molecular weight anthocyanin- derived pigments in Spanish commercial red wines. *Food Chem.* **2014**, *158*, 449–458. [CrossRef] [PubMed]
46. Aleixandre-Tudo, J.L.; Du Toit, W. Cold maceration application in red wine production and its effects on phenolic compounds: A review. *LWT-Food Sci. Technol.* **2018**, *95*, 200–208. [CrossRef]
47. Setford, P.C.; Jeffery, D.W.; Grbin, P.R.; Muhlack, R.A. Factors affecting extraction and evolution of phenolic compounds during red wine maceration and the role of process modelling. *Trends Food Sci. Technol.* **2017**, *69*, 106–117. [CrossRef]
48. Dimitrovska, M.; Bocevska, M.; Dimitrovski, D.; Murkovic, M. Anthocyanin composition of Vranec, Cabernet Sauvignon, Merlot and Pinot noir grapes as indicator of their varietal differentiation. *Eur. Food Res. Technol.* **2011**, *232*, 591–600. [CrossRef]
49. Casassa, L.F.; Sar, S.E.; Bolcato, E.A.; Diaz-Sambueza, M.A.; Catania, A.A.; Fanzone, M.L.; Raco, F.; Barda, N. Chemical and sensory effects of cold soak, whole cluster fermentation, and stem additions in Pinot noir wines. *Am. J. Enol. Viticult.* **2019**, *70*, 19–33. [CrossRef]
50. Arapitsas, P.; Perenzoni, D.; Nicolini, G.; Mattivi, F. Study of Sangiovese wines pigment profile by UHPLC-MS/MS. *J. Agric. Food Chem.* **2012**, *60*, 10461–10471. [CrossRef]
51. Arapitsas, P.; Speri, G.; Angeli, A.; Perenzoni, D.; Mattivi, F. The influence of storage on the "chemical age" of red wines. *Metabolomics* **2014**, *10*, 816–832. [CrossRef]
52. Marquez, A.; Dueñas, M.; Serratosa, M.P.; Merida, J. Formation of vitisins and anthocyanin-flavanol adducts during red grape drying. *J. Agric. Food Chem.* **2012**, *60*, 6866–6874. [CrossRef]
53. Marquez, A.; Dueñas, M.; Serratosa, M.P.; Merida, J. Identification by HPLC-MS of anthocyanin derivatives in raisins. *J. Chem.* **2012**, *2013*, 274893. [CrossRef]
54. Castillo-Muñoz, N.; Fernández-González, M.; Gómez-Alonso, S.; García-Romero, E.; Hermosín-Gutiérrez, I. Red-color related phenolic composition of Garnacha Tintorera (*Vitis vinifera* L.) grapes and red wines. *J. Agric. Food Chem.* **2009**, *57*, 7883–7891. [CrossRef] [PubMed]
55. Castellarin, S.D.; Di Gaspero, G. Transcriptional control of anthocyanin biosynthetic genes in extreme phenotypes for berry pigmentation of naturally occurring grapevines. *BMC Plant Biol.* **2007**, *7*, 46. [CrossRef] [PubMed]
56. Kilmister, R.L.; Mazza, M.; Baker, N.K.; Faulkner, P.; Downey, M.O. A role for anthocyanin in determining wine tannin concentration in Shiraz. *Food Chem.* **2014**, *152*, 475–482. [CrossRef] [PubMed]
57. Weber, F.; Greve, K.; Durner, D.; Fischer, U.; Winterhalter, P. Sensory and chemical characterization of phenolic polymers from red wine obtained by gel permeation chromatography. *Am. J. Enol. Vitic.* **2013**, *64*, 15–25. [CrossRef]
58. Schwarz, M.; Wabnitz, T.C.; Winterhalter, P. Pathway leading to the formation of anthocyanin–vinylphenol adducts and related pigments in red wines. *J. Agric. Food Chem.* **2003**, *51*, 3682–3687. [CrossRef]
59. Somers, T.C. The polymeric nature of wine pigments. *Phytochemistry* **1971**, *10*, 2175–2186. [CrossRef]

60. Schneider, Y.; Vincent, F.; Duranton, B.; Badolo, L.; Gosse, F.; Bergmann, C.; Seiler, N.; Raul, F. Anti-proliferative effect of resveratrol, a natural component of grapes and wine, on human colonic cancer cells. *Cancer Lett.* **2000**, *158*, 85–91. [CrossRef]
61. Wang, Z.; Huang, Y.; Zou, J.; Cao, K.; Xu, Y.; Wu, J.M. Effects of red wine and wine polyphenolic resveratrol on platelet aggregation in vivo and in vitro. *Int. J. Mol. Med.* **2002**, *9*, 77–79.
62. Waterhouse, A.L. Wine phenolics. *Ann. N. Y. Acad. Sci.* **2002**, *957*, 21–36. [CrossRef]
63. Mattivi, F.; Reniero, F.; Korhammer, S. Isolation, characterization, and evolution in red wine vinification of resveratrol monomers. *J. Agric. Food Chem.* **1995**, *43*, 1820–1823. [CrossRef]
64. Chung, M.I.; Teng, C.M.; Cheng, K.L.; Ko, F.N.; Lin, C.N. An antiplatelet principle of *Veratrum formosanum*. *Planta Med.* **1992**, *58*, 274–276. [CrossRef] [PubMed]
65. Orsini, F.; Pelizzoni, F.; Verotta, L.; Aburjai, T. Isolation, synthesis, and antiplatelet aggregation activity of resveratrol 3-*O*-d-glucopyranoside and related compounds. *J. Nat. Prod.* **1997**, *60*, 1082–1087. [CrossRef] [PubMed]
66. Gaudette, N.J.; Pickering, G.J. Sensory and chemical characteristics of trans-resveratrol-fortified wine. *Aust. J. Grape Wine Res.* **2011**, *17*, 249–257. [CrossRef]
67. Peleg, H.; Gacon, K.; Schlin, P.; Noble, A.C. Bitterness and astringency of flavan-3-ol monomers, dimers and trimers. *J. Sci. Food Agric.* **1999**, *79*, 1123–1128. [CrossRef]
68. Casassa, L.F.; Larsen, R.C.; Beaver, C.W.; Mireles, M.S.; Keller, M.; Riley, W.R.; Smithyman, R.; Harbertson, J.F. Impact of extended maceration and regulated deficit irrigation (RDI) in Cabernet Sauvignon wines: Characterization of proanthocyanidin distribution, anthocyanin extraction, and chromatic properties. *J. Agric. Food Chem.* **2013**, *61*, 6446–6457. [CrossRef]
69. Stacklies, W.; Redestig, H.; Scholz, M.; Walther, D.; Sellarge, J. pcaMethods—A Bioconductor package providing PCA methods for incomplete data. *Bioinformatics* **2007**, *23*, 1164–1167. [CrossRef]
70. Rohart, F.; Gautier, B.; Singh, A.; Le Cao, K.A. mixOmics: An R package for 'omics feature selection and multiple data integration. *PLoS Comput. Biol.* **2017**, *13*, e1005752. [CrossRef]
71. Csardi, G.; Nepusz, T. The igraph software package for complex network research. *Inter. J. Complex Syst.* **2006**, *1695*, 1–9. Available online: http://igraph.org (accessed on 30 November 2019).

MDPI
St. Alban-Anlage 66
4052 Basel
Switzerland
Tel. +41 61 683 77 34
Fax +41 61 302 89 18
www.mdpi.com

Metabolites Editorial Office
E-mail: metabolites@mdpi.com
www.mdpi.com/journal/metabolites

www.ingramcontent.com/pod-product-compliance
Lightning Source LLC
Chambersburg PA
CBHW061841210926
51850CB00006B/945
* 9 7 8 3 0 3 9 4 3 0 1 4 7 *